TRUTH
BEAUTY
WONDER

Toward A New Conception Of Mathematics
A theory of the mind-world interplay

First edition 2018

Published by Truth Beauty Wonder Press.

https://truthbeautywonder.com

Copyright © 2018 Davide Lo Vetere
All rights reserved.

https://davidelovetere.com

This work is subject to copyright. No part of this publication may be reproduced, distributed, or transmitted in any form or by any means, including photocopying, recording, or other electronic or mechanical methods, without the prior written permission of the publisher.

ISBN-13 978-1-9165020-1-7

Also available in Paperback, Hardcover and eBook.

Toward A New Conception Of Mathematics

A theory of the mind-world interplay

Davide Lo Vetere

To you who are holding this book

*We can't solve problems
by using the same kind of thinking we used
when we created them.*

Albert Einstein

*There is no art which does not
conceal a still greater art.*

Percival Wilde

*Discovery is the privilege of the child:
the child who has no fear of being once again wrong,
of looking like an idiot,
of not being serious,
of not doing things like everyone else.*

Alexander Grothendieck

*Familiar things happen,
and mankind does not bother about them.
It requires a very unusual mind
to undertake the analysis of the obvious.*

Alfred North Whitehead

TRUTH BEAUTY WONDER PRESS

Truth Beauty Wonder Press is an independent press dedicated to publishing writings that push the boundaries of modern thought and catalyze the evolution of culture and society.

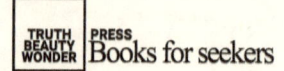

https://truthbeautywonder.com

About The Author

Davide Lo Vetere is an Italian author, futurist and culture hacker focusing on the structure of human progress. He writes about the evolution of intelligence, the rediscovery of human nature, and the frontiers of culture and society.

His fascination with the origin of mathematics led him to study in the mid-90's the uncharted realm of the mind where we form our mathematical intuition of reality. Since then, he has expanded his quest to other aspects of the human experience, and is now on a mission to define new vision of human purpose.

To learn more visit
https://davidelovetere.com

Books by Davide Lo Vetere

Toward a New Conception of Mathematics (2018)

The Journey of the Self (2019)

Archetypes of the Modern World (2019)

The Mystery of Mathematics (2019)

Contents

Preface	**xiii**
Introduction	**xvii**
Notes To The Reader	xix
Structure Of The Book	xxii
Conventions And Terminology	xxvii
The Nousaurus	xxix
Part I. Phenomenology Of Mathematization	**1**
Introduction	**5**
A Model Of Abstract Mind	7
1 Mind, Mathematics And Reality	**13**
1.1 Mathematics As Cognitive Technology	15
1.2 Why We Think That Reality Might Be Mathematical	21
1.3 The Strong Math-Reality Connection	24
1.4 The Weak Math-Reality Connection	24
1.5 The Fundamental Question To Ask	26
2 What Mathematics Tells Us About Reality	**31**
2.1 Worlds And Theories	33
2.2 What We See And What We Comprehend	36
2.3 The Structure Of What We Comprehend	37
2.4 Theory-induced Blindness	39
2.5 Comprehension-preserving Changes	41
2.6 Testing What We Comprehend	42
2.7 Theories That Describe What We Comprehend	44
2.8 Contrasting Interpretations Of The World	46
3 The Observation Of The Abstract Mind	**49**
3.1 Worldviews And Thought	50
3.2 Two Ways To Grasp Concepts	52

	3.3 Observational Structure Of Intentionality	60
	3.4 What Does It Mean To Observe The Abstract Mind?	64

4 The 4 Lenses And The Architecture Of Mathematics   **71**
    4.1 The Generative Lens 75
    4.2 The Performative Lens 79
    4.3 The Motivic Lens 85
    4.4 The Methodic Lens 98

Part II. Inside The Abstract Mind   105

Introduction   **109**
    The Structure Of Appearance 111
    Metaillusions 113
    A Technique To Probe The Abstract Mind 116

5 A Metaphysical Structure Of Concepts   **119**
    5.1 Stipulative And Observational Modes Of Existence 121
    5.2 Observational Identities 124
    5.3 Relational And Pattern Identities 126
    5.4 The Bus Bunching Experiment 128
    5.5 Definitional Interference 132

6 The Lazy Boy Experiment   **137**
    6.1 The Thought Experiment 138
    6.2 Object Classifiers 139
    6.3 A First Concept Of Space 140
    6.4 The Dynamic Structure Of Space 143

7 The Yarn-Ball Experiment   **145**
    7.1 The Thought Experiment 146
    7.2 How To Wind A Ball Of Yarn 147
    7.3 Precision 148
    7.4 Taming Infinity 149
    7.5 What Have We Discovered? 150

8 A Synthetic Structure Of Intentionality   **155**
    8.1 An Analytic Structure Of Intentionality 156
    8.2 Validation Systems 161
    8.3 Interaction Models 163
    8.4 The Limits Of Formal Systems 165
    8.5 Gödel's Argument 168

	8.6	Lawvere's Fixed Point Theorem	169
	8.7	Local Identities	170

Part III. An Introduction To Interaction Theory — 173

Introduction — 177
  Postulates . . . 179
  The Fundamental Correspondence . . . 185
  Structure Of Interaction Theory . . . 186

9 The Common Mathematical Intuition Of Reality — 191
  9.1 A Strategy To Visit The Mathematical Zoo 192
  9.2 The Architecture Of Mathematical Definitions . . 193
  9.3 The Objects Pattern . . . 196
  9.4 The Structure Pattern . . . 198
  9.5 The Properties Pattern . . . 200
  9.6 Why We Define Mathematical Gadgets In This Way 202
  9.7 The Identity-based Mode Of Thought . . . 204
  9.8 The Deep Structure Of The Concept Of Set 206

10 $\mathcal{M}$-signatures — 211
  10.1 Thinking In And Thinking About . . . 215
  10.2 How $\mathcal{M}$-signatures Define Intelligible Thought . . 218
  10.3 **concrete** And **abstract** Concepts . . . 220
  10.4 Classification Of $\mathcal{M}$-signatures . . . 224
  10.5 How **concrete** Concepts Define Intelligibility . . . 226
  10.6 How Intelligibility Defines Objectivity . . . 231
  10.7 The Structure Of $\mathcal{M}$-signatures . . . 234

11 Cognitive Architectures — 239
  11.1 Conceptions And Definitions Of Knowledge 240
  11.2 The Basic Intuition Of Cognitive Architectures . . 242
  11.3 How Cognitive Architectures Define Knowledge . . 245
  11.4 Classification Of Interaction Models . . . 247
  11.5 Minimal Interaction Models . . . 250

12 Epistemology Of The Abstract Mind I — 259
  12.1 Generative Knowledge . . . 263
  12.2 The Structure Of Knowledge . . . 267
  12.3 Knowledge In Physics And Mathematics . . . 270
  12.4 Why Knowledge Is Locally Superadditive 272
  12.5 The Structure Of Comprehension . . . 275

12.6 Meaning And Semantogenesis 280

13 Epistemology Of The Abstract Mind II **295**
13.1 The Principle Of Epistemic Relativity 297
13.2 Explanations And Explanatory Systems 298
13.3 Epistemic Continuity, Truth And Reality 301
13.4 A Canon Of Good Explanations 304
13.5 The Explanatory Process 308
13.6 Explanations, Meaning And Ignorance 311
13.7 The Metastructure Of Explanations 314
13.8 The Structure Of The Knowledge-Reality Interface 318
13.9 Knowledge, Computation And Reality 320

14 Epistemology Of The Abstract Mind III **325**
14.1 Modes Of Consciousness 326
14.2 The Structure Of Awareness 328
14.3 Consciousness And Spatialization 330
14.4 A Basic Description Of Consciousness 334
14.5 The Archetypical Structure Of Communication . . 336
14.6 How Data Becomes Message 339
14.7 Metaphysics Of Language And Computation . . . 346
14.8 Conceptions And Foundations Of Mathematics . . 352

Part IV. Three Symbolic Intuitions Of Reality **357**

Introduction **361**

15 Identity-based $\mathcal{M}$-*signatures* **363**
15.1 What Is Identity? 364
15.2 Identity-based Intelligibility 367
15.3 The `identiton` 371
15.4 The Minimal Interaction Model 378
15.5 The Structure Of Knowledge By Division 380
15.6 Identity-based Comprehension 385
15.7 Meaning And Semantogenesis 386
15.8 Explanatory Systems 388
15.9 Process And Causal Efficacy 391
15.10 Identity And The Language Of Change 397
15.11 Conception And Foundations Of Mathematics . . . 403
15.12 Features Of Consciousness 405

16 Finiteness-based $\mathcal{M}$-*signatures* **409**

Contents

16.1 What Is Finiteness? 411
16.2 Finiteness-based Intelligibility 413
16.3 The **finiton** 418
16.4 The Minimal Interaction Model 425
16.5 The Structure Of Knowledge By Inclusion 427
16.6 Finiteness-based Comprehension 432
16.7 Meaning And Semantogenesis 433
16.8 Explanatory Systems 435
16.9 Encompassment And Nearby Efficacy 439
16.10 Finiteness And The Language Of Proximity 444
16.11 Conception And Foundations Of Mathematics . . . 450
16.12 Features Of Consciousness 453

17 Cohesiveness-based $\mathcal{M}$-signatures   **459**
17.1 What Is Cohesion? 460
17.2 Cohesiveness-based Intelligibility 463
17.3 The **cohesivon** 470
17.4 The Minimal Interaction Model 476
17.5 The Structure Of Knowledge By Similarity 479
17.6 Cohesiveness-based Comprehension 483
17.7 Meaning And Semantogenesis 485
17.8 Explanatory Systems 487
17.9 Morphogenesis And Formal Efficacy 490
17.10 Cohesion And The Language Of Form 496
17.11 Conception And Foundations Of Mathematics . . . 502
17.12 Features Of Consciousness 505

Part V. Toward A New Conception Of Mathematics 509

Introduction   **513**

18 Higher Epistemology   **515**
18.1 Submodalities 518
18.2 Higher Knowledge, Higher Explanations 521

19 Interlude: Definitions Of Mathematics   **529**
19.1 Classic And Contemporary 532
19.2 Interaction Theories 538
19.3 Mathematical Intuitions Of Reality 543

20 Information As Measure Of Relative Knowledge   **547**
20.1 Common Information Bearers 548

Contents

    20.2 Information In Interaction Theory 558

21 The Rediscovery Of Intellect   **561**
    21.1 Mathematical Physicalism 565
    21.2 Principles Of Metaphysical Realism 568
    21.3 The Unity Of Knowledge 573

How Big Is An Elephant?   **577**

Bibliography   **579**

Index   **596**

Preface

In mathematics you don't understand things.
You just get used to them.

John Von Neumann (Zukav 1979)

There is nothing which comes closer to true humility than the intelligence. It is impossible to feel pride in one's intelligence at the moment when one really and truly exercises it.

S. Weil and Miles 1986

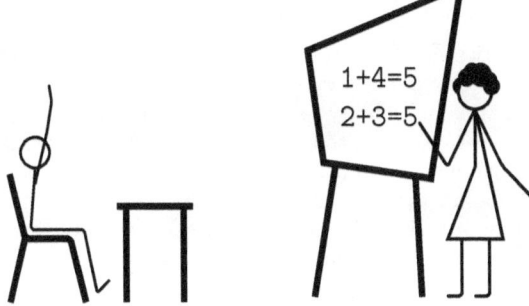

It's 1979. In an elementary school of Northern Italy, a teacher is explaining addition, and has just written two equations on the blackboard. One of the students, a 7-year-old boy, is staring at those equations in awe. He is transfixed. He knows those equations are correct, but his intuition tells him that something doesn't add up. He keeps thinking about the mystery that's catching his attention so vividly, but he cannot explain what it is. Finally, after enormous soul searching, and with all the trepidation and anticipation of discovering what's puzzling him, he raises his hand and asks: "Why did you write 2+3=5? You've already used 5 to write 1+4=5!" The teacher gives him a funny look, and goes on with the lesson.

That kid was me, and the question I was asking some 40 years ago, contained an intuition about the nature of Mathematics that I was too young to articulate because its premises were too abstract. What I wanted to ask was: Why does it make sense to use the same symbol to denote the sum of the elements of two distinct pairs of numbers, and to destroy, by doing so, the information that those pairs of numbers are different? Mathematics is, essentially, a vast collection of techniques to reconstruct the information that we destroy like in the 1+4=5 example. Why can we describe the things in the real world by combining symbols in this way? Why does it make sense to construct our mathematical knowledge of the world in this way? What are we really doing with those symbols? and What is it in the human mind that gives us the ability to create elegant mathematical constructions that seem to connect us to the deep structure of physical reality? All those symbols on the blackboard seemed like magic to me, and I couldn't find a satisfactory explanation of our mathematical intuition of reality. That's when I promised myself that one day I would fill this gap.

Mathematics has always been explained to me, first at school and later at university, as a mental skill that anyone can develop thanks to the human mind's ability to manipulate and represent symbols, and I have always found this explanation utterly unsatisfactory because it doesn't relate Mathematics to the workings of the human mind: it doesn't explain our ability to connect with the world symbolically, it takes it for granted. By the time my studies had completely convinced me that there are hidden cognitive structures from which our mathematical intuition of reality originates, and that we know almost nothing about them, I had also realized with glaring clarity six things that would guide me in my research:

1. We can direct the mind outward, and interact with the physical world, and we can direct the mind inward, and interact with the ideas in our head, and these two faculties constitute the phenomenology of *consciousness and intentionality*. Consciousness is the basic feature of any subjective

reality, and intentionality is the most sophisticated mechanism to interact with consciousness. We don't know what role consciousness and intentionality play in our ability to mathematize.

2. We have an innate symbolic intuition of the things we become conscious of, and this symbolic intuition transcends perception and is proto-mathematical in ways we do not comprehend. It is a direct metaphysical intuition about the conceptual nature of the objects of thought or perception, that is not mystic or revelatory: it is simply the *most authentic experience* of reality.

3. Mathematics is much more than a cognitive technology to study physical phenomena and to create elegant abstractions of the world, it is a window to the human mind, a sailing vessel to explore the deep *structure of human thought*, and that's how we should study it.

4. We don't need to know the nature of Mathematics to do Mathematics, and we seem to accept passively that our ability to form mathematical abstractions of our symbolic understanding of reality happens to us like *perception*.

5. The importance of the metaphysical origin of mathematical thought and intuition isn't sufficiently recognized in Western culture, and there is the almost complete absence of a serious definition of mind in the modern conception of Mathematics, despite the fact that Mathematics is universally regarded as a product of the mind. Consequently, Mathematics is not conceived and developed in the broader context of the study of the *mind-world interactions*.

6. To comprehend the nature of Mathematics, I had to focus on the medium of knowing: the *mind*. "Know the knower" had to become my mantra, my own personal version of an "inveniam viam" that would lead me to explore the invariants of human experience from which our mathematical intuition of reality originates.

So, in the mid-90's, I embarked on a solitary quest to explain the nature of the symbolic intuition of reality from which Mathematics originates.

I began writing this book in 2013, after about 18 years of

intermittent attempts to define the symbolic intuition of reality, and the structure of the mind-world transactions. Despite the abstractness of this research, this is not a metaphysical fugue, it is a journey to the foundations of human thought, head-on, and like any real journey, it has all kinds of discoveries. The main challenges I encountered are of a cultural nature, because the best part of this work pertains to the deconstruction and reformulation of the metaphysical structure of human thought, and of the universal principles of human knowledge, and is, therefore, distinctively interdisciplinary. And the interdisciplinary nature of this research is, at its core, extreme and necessary and unyielding and intransigent and uncompromising in liberating and revelatory ways, that put the ordinary mode of thought under enormous stress, before revealing the majestic horizon of a new intellectuality based on a model of mind-world interplay.

The new framework that emerges from this research is called Interaction Theory, and is based on a body of speculations of my own, and on my interpretation of the work of distinguished philosophers and mathematicians. Interaction Theory aims to be to Mathematics what science is to technology. It is a way to map and present in an organic and rigorous form, the emergence of our symbolic intuition of reality from the deeply unintuitive structures of the mind-world interplay. It is a self-contained system of thought, a worldview, a way of thinking and a style of reasoning, where Mathematics as we know it today can be described like digestion, photosynthesis or any other natural process, and where knowledge, objectivity, comprehension, meaning, explanation, communication, information, computation and consciousness have deeper and unexpected meanings and interconnections. This account is free from serious technicalities, but certain metaphysical and mathematical concepts have an intrinsic, ineliminable complexity. To convey counterintuitive ideas I use mathematical analogies, and have designed metaphors and thought experiments.

I cannot think of a better time to present this research to *you*, dear reader, to your metaphysical and mathematical sensitivity, and to invite you to explore with me new foundations of human thought based on a theory of the mind-world interplay.

Davide Lo Vetere
Dublin, July 10, 2018

Introduction

> A scientific world-view which does not profoundly
> come to terms with the problem of conscious minds
> can have no serious pretensions of completeness.
> Consciousness is part of our universe, so any
> physical theory which makes no proper place for it
> falls fundamentally short of providing a genuine
> description of the world.
>
> Penrose 1995

One of the deepest mysteries of our Universe is the human mind's ability to connect us to the World.

I call this ability the *symbolic intuition of reality*, because it connects the subjective reality of our thoughts to the objective reality of the World. An example of symbolic intuition of reality is language, because through words and symbols we are able to relate our desires and intentions to the objects and states of affairs in the World. An example of language is Mathematics. Mathematics is a language that allows us to make precise statements about ideas and eliminate ambiguity; it allows us to synthesize concepts that describe simultaneously the features common to a range of different notions, transcend the limitations of ordinary intuition, and reason about ideas of arbitrary complexity.

The main aim of this book is to present a model of symbolic intuition of reality in which we can define the foundations of human thought. Roughly, we can think of a *foundation of human thought*

Introduction

as a self-contained object where the terms knowledge, objectivity, comprehension, meaning, explanation, communication, language, computation, mathematics, information and consciousness are defined based on the particular structure of a mind-world transaction. When the structure of the mind-world transaction changes, the foundation of human thought changes, and, with it, the way we connect to the World.

The Indian parable of "The blind men and an elephant" describes eloquently what we will do in this book. This is a beautiful parable present in numerous texts of the Hindu tradition. The version I quote here is in Peter T Johnstone 2002a.

> *"Four men, who had been blind from birth, wanted to know what an elephant was like; so they asked an elephant-driver for information. He led them to an elephant, and invited them to examine it; so one man felt the elephant's leg, another its trunk, another its tail and the fourth its ear. Then they attempted to describe the elephant to one another. The first man said, 'The elephant is like a tree'. 'No,' said the second, 'the elephant is like a snake'. 'Nonsense!' said the third, 'the elephant is like a broom'. 'You are all wrong,' said the fourth, 'the elephant is like a fan'. And so they went on arguing amongst themselves, while the elephant stood watching them quietly."*

I think that the human intellect is very much like the group of blind men in this parable, and the World is like the elephant. Each blind man represents a particular symbolic intuition of reality, and through that intuition we connect to reality and think that what leaves an impression in our mind's eye is a faithful representation of the World.

Notes To The Reader

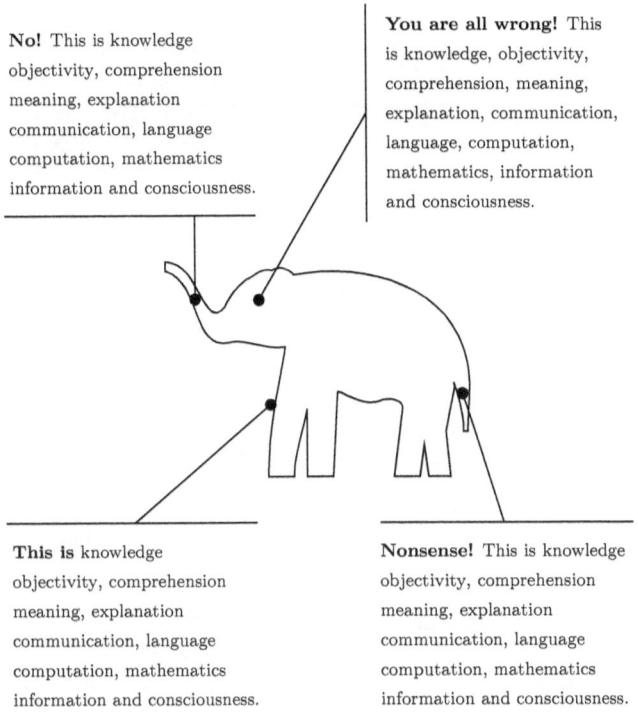

Figure 1: The aim of this book depicted as the blind men in *"The blind men and an elephant"* parable that stop arguing among themselves and find a way to describe the entire elephant.

In this book we present a method to describe the entire elephant, and in the process, we redefine the foundations of human thought and the modern conception of Mathematics.

Notes To The Reader

In the first half of the book (Parts I and II), we are going to visit a wild and secluded and unforgiving territory of the mind at the intersection of Metaphysics and Mathematics, where concepts are in their purest form because language and imagination can't touch them. Without language and imagination, the human mind perceives concepts as metaphysical intuitions, as hunches that feel both familiar and vividly incomprehensible. Thoughts just happen. The mind is intensely aware of them, but can't grasp them or decode their features or modes of existence. That place of the mind, utterly impervious to language and imagination, and yet so intuitively familiar, is, perhaps, the closest thing to Plato's heaven, but in a completely non idealized form, because it is, as you will discover, as real as mountains and oceans.

In the second part of the book (Parts III and IV), you will have a new pair of eyes to contemplate a majestic landscape unfold before your mind, and a language to relish what you'll see.

There are a couple of things I want you to know before we begin this journey to the center of the mind, things that I feel the need to clarify, because my desire as an author is to take you gently to the second half of the book, not to parachute you there. I want you to know what happens when you think abstractly, and the techniques I use to work with concepts in this book.

What Happens When We Think Abstractly

It is useful to know that when the human mind tries to grasp highly rarefied concepts, there is a very special class of thought processes called *spatialization*, that becomes more prominent than anything else. Spatialization is the process of projecting an object of thought onto an arena of quantitative and qualitative becoming. By spatializing an object of thought, the human mind ascribes to a concept some of the properties of space, and in this way it develops quickly the capacity to reason *about* that concept. How the mind is able to spatialize is a mystery. Spatialization is for an untrained intellect the only experience of abstract thought, and is such an ingrained aspect of any ordinary cognitive process, that it can be easily confused with a reflex or an instinct. For example, in Physics, time—which is dimensionless—is spatialized as an arrow, the so called arrow of time—which is a one-dimensional object—to study the evolution of physical systems. In Mathematics we have invented numbers as a spatialization of quantity, and Geometry as a spatialization of shape, structure and relation. The list goes on.

Notes To The Reader

Three Techniques To Work With Concepts

What is needed to ease the first steps of this journey, is a technology to navigate through the spaces that will spring from the mind's attempts to grasp rarefied concepts. Throughout the text, we will handle concepts in three ways that can be seen as three distinct and interrelated cognitive technologies: knowing how and why we use these technologies should help you decode what we do.

We will *fold* concepts like paper to create origamis that are more than the paper they are made of. These are the parts of the text where I use philosophical and mathematical arguments to reveal certain hidden structures of thought and to deconstruct concepts. Very often, I use mathematical reasoning to create analogies that describe philosophical arguments. The human mind appears to grasp concepts with great clarity when they are given shape, either literally or metaphorically. This is what makes analogy a powerful tool: because it creates spatial relations between parts of the same concept, thus revealing its hidden nature.

We will *carve* concepts like a stonemason, and arrange them to form structures of thought. To do this, we use a tool I call *The Sieve* that I created to cut through concepts to reveal their internal structure. The Sieve rests on a distinct worldview on the nature of concepts that I use extensively in my research, and that consists in regarding concepts as technological gadgets. This worldview is my own interpretation of one of the tenets of the philosophical movement called Pragmatism. The version of The Sieve I use in this book consists of a set of three questions that can be applied to any idea to explain its nature and purpose.

Any concept can be described by the following three questions:

- What problem does it solve?
- How does it solve the problem it was invented to solve?
- What knowledge is needed to construct it?

Inherent in every concept created by the human mind is an answer to each of these three questions. Anything that does not contain answers to these tree questions is not a product of the human mind.

We will *immerse* hard-to-grasp concepts into a large reservoir of thoughts and arguments and questions until their skin softens and eventually opens up revealing what's inside. This metaphor, called "The Rising Sea", is due to the great mathematician Alexander Grothendieck, and is described in his memoir Récoltes et Semailles (Grothendieck 1985). There, he compares a theorem to be proven to a nut to be opened. The translation from French that follows is in McLarty 2003:

> *"I can illustrate the second approach with the same image of a nut to be opened. The first analogy that came to my mind is of immersing the nut in some softening liquid, and why not simply water? From time to time you rub so the liquid penetrates better, and otherwise you let time pass. The shell becomes more flexible through weeks and months—when the time is ripe, hand pressure is enough, the shell opens like a perfectly ripened avocado! [...] A different image came to me a few weeks ago. The unknown thing to be known appeared to me as same stretch of earth or hard marl, resisting penetration [...] the sea advances insensibly in silence, nothing seems to happen, nothing moves, the water is so far off you hardly hear it [...] yet it finally surrounds the resistant substance"* Grothendieck 1985, pp. 552–553

Structure Of The Book

This book consists of five parts, and the organization of its content reflects, to some extent, the structure of the physical investigation outlined below.

- There is a *phenomenon we observe*: our symbolic intuition of reality, our ability to connect with reality through our thoughts.

- To *study* the symbolic intuition of reality, we analyze the genesis of our mathematical intuition of reality, because there the cognitive processes underpinning the symbolic intuition of reality are more apparent.

- We formulate a *theory* to explain the origin of our symbolic intuition of reality.

- We *apply* our theory to three types of symbolic intuition of reality, to comprehend how it works, and what we can uncover with it about the deep structure of the human mind.

- We review what we have discovered of the nature of our symbolic intuitions of reality, and *put forward new principles* of human knowledge.

Part I. Phenomenology Of Mathematization

Part I is *observational*. We present the project of this book, and situate the mathematical and philosophical discussion about the study of the symbolic intuition of reality by introducing a model of mind-world interplay called Abstract Mind. The aim of this part is to lay the conceptual foundations to study the genesis of our symbolic intuition of reality like we study digestion, photosynthesis, nuclear reactions or any other natural process.

Chapter 1 sets the scene for the study of the symbolic intuition of reality with a fundamental question: What do we do when we do mathematics? This question leads to the analysis of Chapter 2, where we use elementary mathematics in a somewhat advanced way, to explore the delicate relation between mathematical concepts and the realities that they describe. The syntax of our mathematical representations of reality leads to the framework we introduce in Chapter 3, where we ask how we can observe the genesis of our symbolic intuition of reality *as a whole*, in what is, perhaps, best described as a very abstract bracketing exercise (*bracketing* à la Husserl) to mark out the boundary between mathematics and mathematization, and to capture the creative tension between symbol and meaning.

The first observation of a symbolic intuition of reality based on the framework of Chapter 3 is purely phenomenological, and is the observation of mathematization we present in Chapter 4 with a tool called The 4 Lenses, which gives a cognitive and evolutionary view of the edifice of mathematics as we know it today.

Part II. Inside The Abstract Mind

Part II is *analytical*. Its goal is the study of the structure of the mind-world interplay that gives rise to our symbolic intuitions of reality.

Chapter 5 is one of the chapters with the highest density of philosophical content. It's aim is to give a metaphysical characterization of concepts, and to introduce a way to think about them called *definitional interference*. In The Lazy Boy Experiment (Chapter 6), and in The Yarn Ball Experiment (Chapter 7), we examine a special type of mind-world transactions. The goal is to reveal the hidden cognitive structures that transform perception into the fabric of experience. These two chapters pave the way for the ideas we present in Chapter 8, where we work out the general structure of those transactions, and put forward a theory of the mind-world interplay called Synthetic Structure of Intentionality.

Part III. An Introduction To Interaction Theory

Part III is *theoretical*. We introduce a model of mind-world interplay based on the Synthetic Structure of Intentionality called Interaction Theory. The central theme of Interaction Theory is the description of the symbolic intuition of reality through a reformulation of the foundations of human thought based on models of *intelligibility*.

Chapter 9 introduces the basic form of mathematical intuition of reality that gives rise to the modern conception of mathematics. The objective is to begin to construct an answer to the question "What do we do when we do mathematics?" we asked in Chapter 1.

Chapters 10 and 11 lay down the foundations of Interaction Theory by describing the archetypical structure of intelligibility and knowledge.

The construction of Interaction Theory continues in Chapters 12, 13 and 14, where we define the foundations of human thought by describing the meaning that the terms knowledge, comprehension, objectivity, meaning, explanation, information, communication, computation and consciousness acquire based on a given conception of what is intelligible.

At the end of Part III, we have a conceptual toolkit to look at the World through an arbitrary conception of what a mind can grasp intuitively, and we have a language and a style of reasoning to articulate question and answers about what we see.

Part IV. Three Symbolic Intuitions Of Reality

Part IV is *experimental*. We apply the principles and methods of Interaction Theory to describe three symbolic intuitions of reality based on three distinct models of intelligibility defined by the concepts of: *identity*, *finiteness* and *cohesiveness*.

Each symbolic intuition of reality is a self-contained *foundation of human thought* where we characterize the basic cognitive structures that define

- knowledge, comprehension, meaning, explanation and consciousness
- more advanced cognitive functions such as communication, language and computation
- and a foundation of mathematics.

Note: Chapters 15, 16 and 17 have the same structure and, in certain parts, they are identical, word-by-word. This is a stylistic choice to emphasize the parts of a concept that do not change across multiple Symbolic Intuitions of Reality, and the moving parts that change. For this reason, some readers may find it useful to read these chapters horizontally rather than in the order in which they are presented—e.g. Sections 15.3, 16.3, 17.3, then Sections 15.4, 16.4, 17.4, and so on. To the reader who wishes to do so, we recommend to read the first two sections of each chapter, and then switch to a horizontal way of reading.

Part V. Toward A New Conception Of Mathematics

Part V is *visionary*. We use the tenets of Interaction Theory to revisit a list of select topics. The foundations of human thought defined by Interaction Theory have several ramifications in the Philosophy of Mathematics and Physics, in Mathematical Ontology and Epistemology, in the Philosophy of Mind, and in the modern conceptions of information and computation, just to mention a few fields of research touched by the reformulation of the fundamental structure of human thought we present here.

In Chapter 18 we examine briefly the general structure of the Interaction-theoretic foundations of mathematics, and interpret the dynamic behavior of the Abstract Mind.

Introduction

Chapter 19 is an interlude in which we review the evolution of some definitions of mathematics, from Aristotle and Plato to the modern interpretations of mathematics, to the reformulation of mathematics presented in this text. These views of mathematics from the altitude of definitions are highly suggestive and inspiring, because from up there it is possible to see broader trajectories of thought and avenues of research.

Chapter 20 revisits some of the main theories of information from the point of view of Interaction Theory, and presents a unifying notion of information.

Chapter 21 draws the book to a close by offering some reflections on the nature of the human intellect, and on the purpose and meaning of human knowledge.

Conventions And Terminology

Typefaces and capitalization

Italic is used for *emphasis*.

American Typewriter is used to denote special concepts, such as for example `success condition`, `binary object` and `cohesivon`. These concepts are not capitalized when they are used at the beginning of a paragraph.

Capitalization is used to denote concepts, such as for example in $\mathcal{M}$-*signature*, Minimal Interaction Model and Actuation Mechanism, and to distinguish them from ordinary words, such as Organizing Principle and organizing principle.

Italic typeface may be combined with capitalization for emphasis, or when a term is introduced for the first time.

Lists

Lists are used to emphasize particularly complex ideas, when it is deemed useful to suspend the rhythm of prose to let concepts sediment into the reader's mind.

Diagrams

In this book I use diagrams to illustrate relations between concepts in two ways:

- as a visual tool to convey philosophical or mathematical concepts
- in a higher category-theoretic sense, where the nodes denote mathematical objects and the arrows denote morphisms between them, and where there can be arrows between arrows, and arrows between arrows between arrows and so on.

Diagrams are graphs such as the one below: they are collections of vertices, some of which may be connected by arrows

Introduction

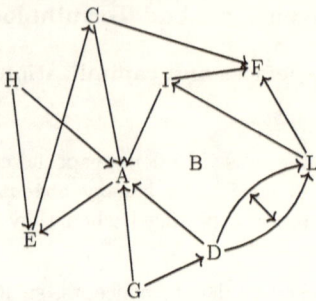

I use freely the terms bullet and arrow as synonyms of respectively: vertex and edge, object and arrow, node and arrow, node and arc, vertex and arc.

Special words

Nature is the stage where existence takes place in the broadest possible sense, and is not tied to a particular conception of physical reality or philosophical or intellectual tradition.

Universe denotes physical reality, the reality described by our physical theories and, to a certain extent, revealed by our measurements of physical quantities.

World denotes the reality involved in the transactions with the human mind. The World is part of Nature and the Universe.

End-of-chapter Notes

Each chapter ends with a short bibliographical note where I cite some of the authors that influenced my work, or whose work I mention in the chapter. Some notes may contain or reference technical material. The bibliographical notes are, in general, incomplete, because the multidisciplinary nature of this research demanded me to gather a vast amount of information from more sources than I can remember. My apologies for factual errors or inaccuracies in the resources I mention.

The Nousaurus

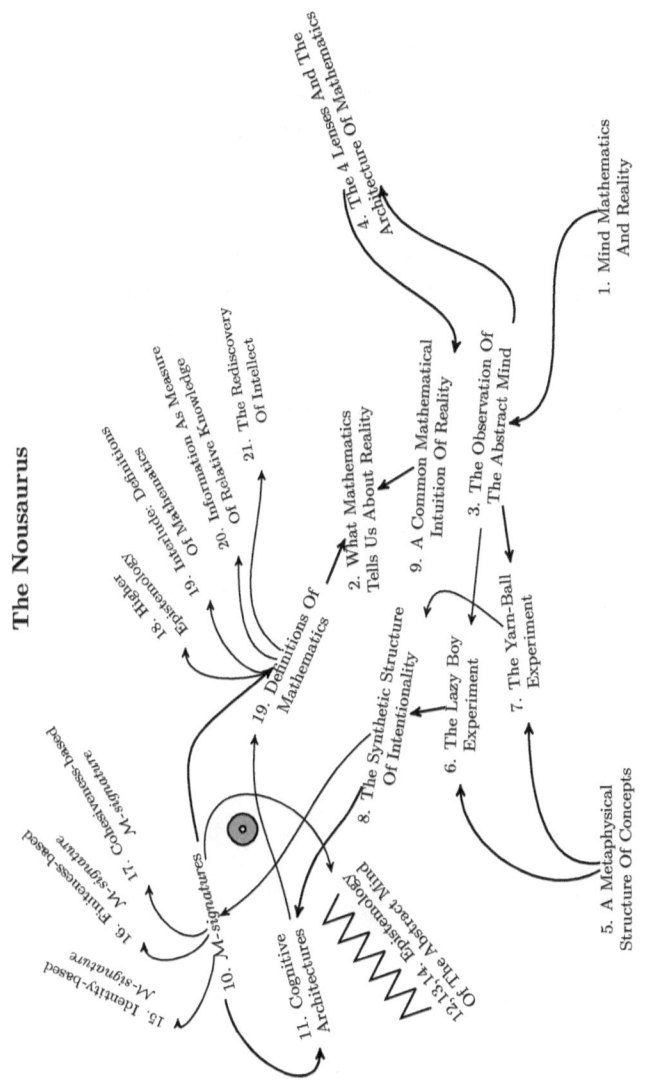

The *Nousaurus*, from Greek νοῦς (*nous*, mind, intellect) and σαῦρος (*sauros*, lizard) is a transcendental creature of Plato's heaven, a metaphysical muse that embodies the cosmic features of human thought.

This is what I know about this creature.

- Its footprints are the concepts that connect us with the World.
- The safest way to approach it is to think about its tail.
- Its guts contain the principles of human knowledge.
- Its head and teeth contain the principles of human thought.
- It uses the spikes on its head and on its back to communicate with us, and with other Nousauruses.
- Its excrements are all we comprehend of this World.

PART I

Phenomenology Of Mathematization

Can we observe the interplay between the mind and the world?

Introduction

> Nature uses only the longest threads to weave her patterns, so each small piece of her fabric reveals the organization of the entire tapestry.
>
> Feynman 1965

To begin this journey we need an explorer's mindset. And to think like an explorer, we need to know how to wander guided only by observation. This is why the central theme of this first part is *observation*. We want to learn to observe the manifestations of the human mind's ability to turn objects of thought or perception into symbolic intuitions about the nature and structure of reality.

But where do intuitions of reality come from?

To answer this question, we need to observe the human mind. We think of the mind as filled with thoughts, memories and emotions. Like your last birthday party. Your plans for tomorrow night. The last movie you saw, or how you felt when you met that long-time friend. We as a species have the ability to use the symbolic content of the mind to connect us with each other and with the World around us in meaningful ways. And we have invented awesome technologies to do this, like art, music, literature and dance. Art connects us with the World through the language of colors and shapes, music uses the language of sounds and rhythm, literature uses the language of emotions and metaphors, and dance uses the movements of the human body.

There is one technology we have invented to connect us with the World that stands out from any other known human invention: it's Mathematics. Unlike art, music, literature and dance, mathematics does not connect us with the world through the language of colors, sounds, emotions or metaphors, but through the mysterious language of *structure*. The language of structure is a system to arrange symbols that represent concepts or objects in the physical world, in such a way that certain sentences of the language express something about those concepts and objects, by virtue of a correspondence between the syntactic structure of those sentences, and certain regularities in the evolution of the concepts or phenomena they refer to. The language of structure is mysterious because it gives us symbolic access to the things in the physical world without us having to make experiments. With the language of structure we can recreate a version of reality in our mind, and simulate what we want to explore in the physical world.

We know very little about the origin of our ability to connect with the World symbolically through the language of structure. The region of the mind where our symbolic intuition of reality originates doesn't even have a name, because it is the backdrop against which we create the foundations of human thought by asking fundamental questions like: What is knowledge? What does it mean to understand? What is meaning? What is an explanation?

To get a feel for what I'm talking about, imagine watching all the thoughts, memories and emotions in your mind evaporate. And not just your plans for tomorrow night or the last movie you saw. I mean everything. All the things you've been stacking in your mind since the day you were born. All the things you've learned. The books you've read. The people you've met. Even the image of yourself. Imagine watching them all disappear. What would be left is not nothing, it's something. It's *the stage* where your mind creates all your experiences, where all your thoughts continuously bubble into existence and vanish. That stage is where our symbolic intuition of reality originates, and it has a structure and properties we will describe in this book. That stage is where our mathematical intuitions of reality come from. It is where the mind-world interplay takes place.

To observe the manifestations of the mind-world interplay that produce symbolic and mathematical abstractions of reality, we need a method to observe not the genesis of mathematics but the genesis of *mathematization*. The distinction between mathematics and mathematization is equivalent to the distinction between product and process of production, but with an important difference: the distinction between mathematics and mathematization isn't always obvious, because mathematics itself can be studied with mathematical methods. Mathematics is an example of a process where the product (mathematics) is also the raw material of the process of production (mathematization). When that's the case, and the object of study of a mathematical theory is another mathematical theory, we speak of *metamathematics* and *metatheories*. Consequently, some of the features of mathematical reasoning can also be the features of mathematics itself, or generalizations of the same. Yet mathematics and mathematization are still, from a cognitive viewpoint, products of the mind-world transactions, and therefore, a method to describe the genesis of mathematization must be intrinsically different from mathematics and from mathematization.

To begin to orient ourselves in the phenomenology of the mind-world interplay from which the symbolic and mathematical intuitions of reality originate, we need to know the fundamental structure of mathematics, and we need to comprehend the sophisticated mechanism through which we relate to the World mathematically. We will do this in Chapter 1. We also need to know the structure of observation, so that we can articulate what it means to *observe* the mind-world interplay. We will do this in Chapter 3. With these tools, we can describe the phenomenology of mathematical thought. We will do this in Chapter 4 with a theoretical device called The 4 Lenses, which I created to describe the phenomenology of mathematics from a cognitive and evolutionary viewpoint.

But first, we need a *language* to articulate what a mind-world transaction is. The elephant we met in the Introduction is cute but cumbersome. We need something simple that we can can carry around in our heads like a favorite tune: we need a *model* of Abstract Mind.

A Model Of Abstract Mind

The Abstract Mind is a theoretical model of human mind that uses the language of process to describe the genesis of the fundamental structure of human thought. The idea behind the Abstract Mind model is that the fundamental structure of human thought is the manifestation of a *natural process* that can be studied like we study digestion, photosynthesis, nuclear reactions or any other natural process. The natural processes modeled by the Abstract Mind are the mind-world transactions that give rise to the fundamental structure of human thought, that in this research we call Symbolic Intuition of Reality, and that from now on will be capitalized.

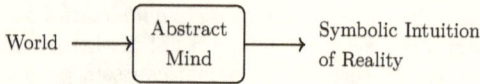

Figure 2: The Abstract Mind model.

In this research we assume that we have direct access only to the Symbolic Intuitions of Reality, and that the World can be comprehended through the study of the properties of certain types of Symbolic Intuitions of Reality.

The World

The *World* (Figure 2) is a theoretical device that represents the neutral, objective, unfiltered, metaphysical, raw experience of objects of thought or perception. It is the experience of a mode of objective existence. It is the objective reality of things, the contemplative experience of objects of thought or perception as they are, before the human mind superimposes concepts on them. The World is an archetypical notion represented in all major philosophical traditions. It is what Greek philosophers called *Chaos*, from Greek χάος (*kaos*, emptiness, disorder). It is the Dharma in Buddhist thought, the Brahman in the Vedic tradition, and the quiddity, the inherent essence of things in the metaphysics of Ibn Sīnā. It is a vivid image of reality that symbolizes what we know without having to comprehend it, and denotes the theoretical limit of what a human mind can grasp in a given state of consciousness. It is an experience of reality where we do not try or need to find or create patterns, order, structure, regularities or meanings to develop knowledge, and where the significance of knowledge is unrelated from any requirement to be predictive or quantitative.

A Model Of Abstract Mind

What I would see if I could observe these three dots without any conceptual lens would be this page in its entirety. The three dots would still be there, and I'd see them, but they'd just *be*, without being dots, and they'd be there with the same intensity as any other element on this page. My eyes could linger on them, but my unbiased perception of those three dots would not change: they'd still be, without being dots.

The Symbolic Intuition Of Reality

The *Symbolic Intuition of Reality* (Figure 2) is a theoretical device that represents what we see when we observe objects of thought or perception through the lens of concepts. Our intuition and experience of those objects of thought or perception is defined by the meaning of the terms knowledge, comprehension, meaning, explanation and consciousness, and has various qualities and intensities that depend on how we acquire knowledge of those objects. It is a symphony of structures and a cacophony of unexpressed futures. The Symbolic Intuition of Reality is what Greek philosophers call *Cosmos*, from Greek κόσμος (*kosmos*, order). It is a map of reality that resonates with our knowledge, and with the purpose of our observations. In particular, when the concepts through which we observe reality are mathematical concepts, what we see is often predictable, and full of structure and order, and seems designed according to spectacularly sophisticated aesthetic canons.

I see three dots when I observe these three dots, and I know intuitively that they are not the same dots I would see if I could observe them without any conceptual lens, because the structure of my awareness of the three dots would be radically different. I can observe the musings in my mind as it explores infinite trajectories through these dots, and as it tries to draw imaginary lines to connect them. I don't just see three dots: I *recognize* them. I know that they are dots because they have a certain correspondence with something in my mind that I call dot, and that feels like a dot when I see one. My experience of those dots is direct, intuitive, intentional and abstract, and its clarity and intensity are completely defined by the fluctuations in my awareness of those dots.

The Structure Of The Symbolic Intuition Of Reality

To comprehend the difference between the World and the Symbolic Intuition of Reality, we need to introduce the basic structure of the Symbolic Intuition of Reality. The Symbolic Intuition of Reality is made of two components called *intuitive*, and *epistemic*.

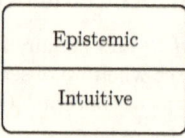

Figure 3: The structure of the Symbolic Intuition of Reality.

The *intuitive* part symbolizes a specific *type of mind-world transaction* by defining what is *intelligible*, and denotes the fundamental way in which we become aware of an object of thought or perception as a result of the interaction between the human mind and the World. In this sense, the intuitive component of the Symbolic Intuition of Reality is like a lens through which we can observe the World: what we cannot see depends entirely on the physical and optical features of the lens. However, the intuitive part alone does not produce any *experience*: it represents mere potential. In the Abstract Mind model, to create any experience of an object of thought or perception, the human mind needs a cognitive apparatus, which is the content of the epistemic part of the Symbolic Intuition of Reality.

The *epistemic* part contains a model of the cognitive structures at the foundation of human thought, which we identify with the meaning of the following 5 terms: *knowledge, comprehension, meaning, explanation* and *consciousness*. The cognitive structures that define the meaning of those 5 terms, depend on the content of the *intuitive* part of the Symbolic Intuition of Reality, and are responsible for the genesis of experience.

The Abstract Mind

The project of describing the Symbolic Intuition of Reality, is the project of explaining the nature of the intuitive and the epistemic framework through which we create our abstract knowledge of the World. Our mathematical intuition of reality has a privileged place in this research because is produces a very special type of knowledge that seems to penetrate the deep structure of the physical world.

The box in Figure 2 sitting between *World* and *Symbolic Intuition of Reality* indicates that the Abstract Mind model describes the origin of our Symbolic Intuition of Reality as a *process*: we perceive an object of thought or perception, and something transforms that perception into a symbolic representation through which we construct our abstract knowledge of reality. The Abstract Mind, in this sense, denotes an archetypical intuition about the World, and symbolizes the project of coming at the intrinsic cognitive structures that yield the ability to create Symbolic Intuitions of Reality, like we come at the project of baking a pizza, or getting a fire started, or building a skyscraper, or weaving a rug.

The description of the Abstract Mind is the description of the structure and the limits of the cognitive processes through which we are able to transform our experience of the *World* into an intelligible, predictable instance of human thought, and ultimately into knowledge.

Key Concepts Of Part I

1. The Abstract Mind is a theoretical model of human mind that uses the language of process to describe the genesis of the Symbolic Intuition of Reality.

2. The genesis of the mind's ability to mathematize can be described like we describe digestion, photosynthesis, nuclear reactions or any other natural process.

3. What we see when we look at the World through the lens of Mathematics, are models of the metaphysical structure of our own fundamental thought processes.

4. To comprehend the origin of the mind's ability to mathematize, we need a framework to answer the question: What

do we do when we do mathematics?

5. It is possible to recognize the syntax of mathematical thought by observing the phenomenology of mathematization through a gadget called The 4 Lenses, which gives a cognitive and evolutionary description of how we use mathematical technologies to acquire knowledge.

Chapter 1

Mind, Mathematics And Reality

> I recall a restaurant in New York in which customers, cooks, waiters, and the cashier may speak different languages, yet rapid operation is achieved without any written orders nor bills by simply stacking used dishes according to shape.
>
> Lawvere 1992

> The world of ideas is not revealed to us in one stroke; we must both permanently and unceasingly recreate it in our consciousness.
>
> Thom 1971

In the history of human thought, there is an idea that is so radical, that it forms a kind of horizon in which a large number of questions can be asked. This idea is the idea that the fabric of reality is accessible to the human mind. It is the idea that the Universe and everything in it are designed in such a way that the creatures of a little blue dot at the periphery of the Milky Way can understand how it's built. I call this idea the *intelligible universe hypothesis* because it has the character of a metaphysical intuition about how the faculties of the human mind connect us to reality at a fundamental level. The intelligible universe hypothesis is a profound intuition about the nature of the human mind, that has turned the discovery of reality into an exciting game between Man and Nature, and has had two crucially important consequences for the intellectual, spiritual, scientific and technological development of our species.

1 Mind, Mathematics And Reality

First, it has redefined the mind from mere survival tool, to metaphysical gadget that allows us to transcend the structure of appearance. This intuition about the transcendental nature of the human mind, has elevated the pursuit of knowledge to an act that reunites Man with Nature through the structure of human thought. In this sense, the structure and content of human thought have become a way to reach for reality rather than the mere manifestation of the activity of human consciousness and intentionality.

Second, the intelligible universe hypothesis has determined, enabled, influenced and undermined our ability to conceive, seek, and acquire knowledge as a result of the prevailing conception of mind, and of its role in defining the purpose of human existence. This second consequence has provided, and continues to provide, the conceptual backdrop against which we develop our cognitive technologies. Among the many cognitive technologies invented by the human species, one stands out for having the remarkable property of producing knowledge purely by deduction from basic abstract principles: mathematics. Mathematics is an extremely successful cognitive technology because it allows us to make very accurate predictions about objects and states of affairs in the world, and because it allows us to reason about things that we are not able to grasp with our imagination, such as a space with 26 dimensions or the quirks of Quantum Mechanics.

However, the fact that mathematics is effective in decoding physical reality is problematic because it clashes with the predominant materialistic view of the world. The problem is: How do we get an account of our ability to make accurate mathematical predictions about the physical world, that we can make consistent with the notion that mathematics is a product of the mind? How can mere subjective processes of thought such as the hierarchies of mathematical concepts that we create when we mathematize, result in any knowledge about independently existing objects and states of affairs in the world?

I argue that in order to comprehend the deep nature of mathematics, we need to humbly turn our attention to the mind's ability to mathematize, and reframe the study of mathematics within the larger context of the thought processes that produce mathematical concepts as we know them today. To begin to understand mathematics as a product of the mind, we must first familiarize ourselves with the phenomenology of mathematization. To this end, in this chapter we touch on the key themes that define the structure and manifestation of mathematical thought, such as the structure of mathematics as a cognitive technology, and the ramifications of the intuition that mathematics and reality are related.

1.1 Mathematics As Cognitive Technology

In the introduction to this chapter, we said that mathematics is a cognitive technology, but what does it mean, exactly, that mathematics is a cognitive technology?

Let's ignore for the moment *why* we are talking about mathematics as a cognitive technology, because that is a topic on its own right, that rests on philosophical positions that we will explain in a separate discussion, and let's focus on the statement "Mathematics is a cognitive technology".

I think there is a very effective way to think about just about anything as if it were a technology. Just ask these three questions:

1. What problem does this it solve?

2. How does it solve the problem it was invented to solve?

3. What knowledge do we need to develop it?

Think about the telephone. The telephone was invented to allow people to have a conversation when they are too far apart to hear each other. A telephone contains a speaker and a microphone, because we hear as a result of the pressure waves that hit our eardrums, and converts sound into electric signals, because this is how we are able to transmit and receive data over wire or radio. So to solve the problem of letting people have a chat, we employed our knowledge of electromagnetism and our knowledge of the human body. The telephone, like any other technology, is the result of the application of our knowledge of the physical world to solve a problem. In most cases, technology allows us to overcome the limitations of our body, and the limitations that the environment we live in imposes on our lives.

Mathematics is a technology that allows us to transcend the limitations of our own imagination, and very often, of our own intelligence and intuition. It allows us to reason about things that we are not able to grasp with our mind's eye, such as a space

1 Mind, Mathematics And Reality

with 26 dimensions, and tackle problems that defy our wildest physical and symbolic intuition, such as Quantum Mechanics, the physics of black holes or Fermat's last theorem. But mathematics wasn't invented to amplify our intellect or to overcome the limitations of our three-dimensional imagination, those are secondary effects of the use of mathematics on the human intellect. Mathematics was invented to count, classify and measure, but it became hugely more powerful than that: it has become a way to overcome the intransigent appearance of the world with the power of imagination.

Very much like phones, which are a technological manifestation of our current idea of mobile communication based on our scientific knowledge of some physical laws, mathematics is a cognitive technology that embodies our current conception of rational knowledge based on the predominant conception of mind, and on our understanding of what the human mind does when we think. However, I think it's fair to say, our knowledge of the workings of the human mind is hard to compare with our scientific knowledge of the physical world. The reason why we know more about the physical world than about mental realities—at least in the Western culture—is that our current conception of scientific knowledge is purely quantitative, whereas the phenomenology of the human mind is predominantly of a qualitative nature. Nonetheless, and this is a truly remarkable fact, the way we have been able to interpret the workings of the human mind has produced mathematical technologies that have been spectacularly successful in describing the physical world.

The current conception of mathematics stems from a quantitative description of the cognitive processes that are believed to be involved in abstract thought, and gives us a picture of the world as a linguistic structure. The *language* of mathematics is the technology currently used to *construct* mathematics. The modern technology used to build mathematics is called *formal system*, and is designed to give a quantitative description of the cognitive processes of abstract thought. The purpose of formal systems is to allow us to translate ideas into computational problems. The building blocks of a formal system (Table 1.1) are called *alphabet*, *syntax*, *inference rules* and *axioms*, and represent the philosophical and cognitive themes of *access*, *identity*, *change* and *belief system* respectively. A convenient way to regard formal systems is to think of them as a technology to describe the purpose of an observation.

1.1 Mathematics As Cognitive Technology

In this discussion we want to emphasize the structure of formal systems from a cognitive viewpoint rather than their computational properties, therefore, in the explanation that follows, we will give an *observational* description of the components of the technology of formal systems, and leave to the notes at the end of this chapter a computational example.

Technology	Function	Purpose
Alphabet	*Represent an object in a way that is consistent with the purpose of the observation encoded by the formal system.*	*Provide a description of an object of state of affairs consistent with the purpose of the observation.*
Syntax	*Define a concept of identity for the objects described by the formal system.*	*Provide a concept of identity.*
Inference rules	*Define the changes that preserve the identity defined by the syntax.*	*Provide a concept of change consistent with the purpose of the observation.*
Axioms	*Define existence and truth values in the context of the formal system.*	*Underlying belief system.*

Table 1.1: The building blocks of mathematical technologies.

An *alphabet* is a collection of symbols that represent the distinct elements of an object or state of affairs that we want to observe. For example, if we want to observe the faces of a coin, we need to find a representation of the coin that is consistent with the purpose of our observation. Since a coin has two faces—if we exclude its side—then anything with two distinct states can be a representation of a coin consistent with the purpose of the observation of the two faces of a coin. So the observation is the problem and the alphabet is part of the technology used to build a solution to solve that problem. For, we can describe the faces of a coin as a result of the application of an "alphabet technology" to represent its faces. The *syntax* provides a concept of identity for the particular object of state of affairs one wants to observe, that is: it defines the meaning of "is" and "isn't" in the context of the observation. A syntax defines a concept of identity by

1 Mind, Mathematics And Reality

dictating what sequences of symbols of the alphabet belong to the formal system and what don't. It is useful to think of the concept of identity and of the notion of syntactic correctness, as interchangeable concepts used in formal systems to represent a basic form of coherence condition. For, coherence and identity share the feature of defining a notion of consistency on the object to which they are applied. For example, in a formal system that describes the evolution of a process, we might use the alphabet $\{Start, B, C, D, End\}$ to indicate the phases of the process, and define a syntax to dictate that only the sequences of symbols of the alphabet that begin with *Start* and end with *End* are consistent with the process. Thus for instance the sequence of symbols $B \to End \to C \to Start$ does not describe a process, according to this syntax, because it does not begin with *Start* and end with *End*. This is how syntax provides a rudimentary concept of identity: by defining the universe of all possible descriptions of an object or state of affairs that are consistent with the purpose of a specific observation. The *inference rule* define the transformations between sequences of symbols of the alphabet of a formal system that are consistent with the purpose of the observation for which the formal system is used. As we saw earlier in the explanation of the syntax with the example of a process, only certain sequences of symbols of the alphabet are consistent with the purpose of a given observation. The inference rules expand on the notion of identity introduced by a syntax by *dictating what changes preserve that identity*. In this sense, the inference rules provide higher notion of identity by defining a *concept of identity through change*. For example, with reference to the example made earlier, an inference rule may define that when a valid sequence of symbols ends with $x \to D \to B \to End$, where x is any valid sequence of symbols, then the sequence of symbols $C \to B \to x \to End$ is also valid. The *axiom* define what we assume to be consistent with the purpose of our observation, and encode the underlying worldview that motivates and justifies the observation itself. They are the irreducible units of thought from which we build a conceptual framework to reason about the portion of reality encoded by the formal system. An example of axiom could be that which states the existence of certain objects or the truth of certain facts or statements: this is how axioms express the belief system of the observer. We use formal systems because we have reason to believe that they model the human thought. This belief is supported by the following empirical observations:

1.1 Mathematics As Cognitive Technology

1. When we think we seem to arrange clusters of data as sequences of symbols. Symbols can be anything from individual thoughts to clusters of ideas: this observation gives rise to the definition and use of axioms and alphabets of symbols in the construction of formal systems.

2. When we think we seem to follow patterns and rules to form clusters of ideas: this observation gives rise to the definition and use of axioms, syntax and inference rules in the construction of formal systems.

3. Taken individually, symbols, axioms, syntax and inference rules are meaningless conceptual gadgets, but when we consider the strings of symbols produced by a formal system collectively, we notice the emergence of higher types of order that are almost completely invisible otherwise.

The phenomenon we observe as the emergence of order produced by a finite set of symbols and rules from a large number of entities is the archetypal form of the notion of *meaning*. The function of meaning is to mark that something is consistent with what we know. For, whenever we discover a new type of coherence in our knowledge of the world, we begin to assign new meanings to our experience. Meaning contains the idea of an implicit order that is encoded in the simple rules and symbols of a formal system, and that manifests itself only when the formal system produces a sufficiently large amount of strings of symbols. The idea behind formal systems is a powerful intuition that echoes the behavior of biological system and of dynamical systems in general. So how do we use the technology of formal systems to acquire knowledge? Figure 1.1, called ∞-diagram, illustrates how we use mathematics as a cognitive technology.

A. We translate a problem or an idea from the natural language into the language of formal systems, and the result of this translation is what we call a mathematical theory.

B. We construct a model of the mathematical theory $\mathcal{T}$ we create in A, and study its features and properties.

C. We interpret the features and properties of the model we construct in B as the features and properties of the world $W_{\mathcal{T}}$ we want to describe with our mathematical theory.

D. We believe that the world **W** has a nature that we can de-

1 Mind, Mathematics And Reality

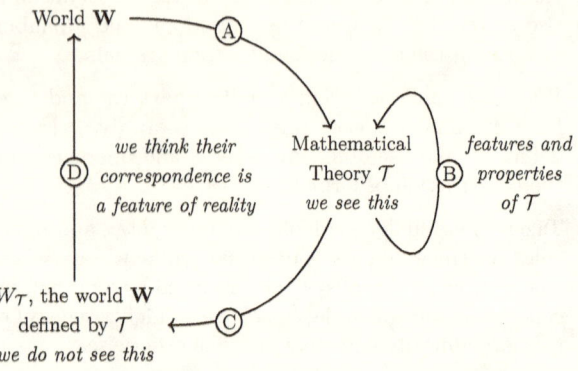

Figure 1.1: Illustration of the use of mathematics as a cognitive technology.

code with our mathematics—or is the same world we recreate with our mathematics altogether—whenever our mathematical predictions about the world **W** are corroborated by facts or experiments that we can explain in $W_\mathcal{T}$.

We don't know why we observe correspondences between the features and properties of a mathematical theory, and the objects and states of affairs in the world, but since in most cases we do observe these correspondences, we are motivated and justified in using mathematics as a cognitive technology. We will revisit formal systems in the third part of the book in a much broader conceptual framework called Interaction Theory. There we will dig a lot deeper into the nature of formal systems and the genesis of mathematical gadgets. For now, it is sufficient to take away from this discussion the key ideas about the components of the technology of formal systems and how we use this technology to acquire our mathematical knowledge of the world. Let us now go back to the telephone analogy: How do we know that a telephone works? Well, you call a friend on the phone, and if her phone rings, and if she picks up the phone, and if you can hear her when she speaks, and if she can hear you when you speak, then your phone works. To answer this question, we do not need to know how to build a phone and a telecommunication system that makes phone calls possible, all we need is the knowledge of how to *use* a phone. This is because the knowledge needed to use a technology is very different from the knowledge needed to create that technology. This observation about the different types of knowledge needed to use or create a technology applies also to mathematical technologies. There are two types of knowledge involved in this introductory discussion about the phenomenology of mathematics: the knowledge needed to determine when mathematics works as a cognitive technology, which we'll discuss briefly in the next Section, and the knowledge needed to comprehend what mathematics *is* at the most fundamental level, which we will develop in greater detail in this book. The latter is not mathematics or metamathematics. It is something radically different that demands us to understand the nature of the problem of studying the genesis of our ability to construct mathematical technologies.

1.2 Why We Think That Reality Might Be Mathematical

The practical and conceptual nature of mathematics have been at the center of intense philosophical and mathematical speculations for more than two thousand years, but what mathematics really is, remains a mystery. We don't know why we can write equations that describe the motion of a body, the behavior of a smart grid, the evolution of a biological network, the geometry of spacetime, or what happens on the surface of a black hole or at the center of a star. What we do know, is that in most cases mathematics allows us to create extremely accurate descriptions of the physical world, and that without its extraordinary descriptive power, any scientific and technological progress would be very hard to imagine. It is precisely because mathematics seems to decode physical reality so well that philosophers, physicists and mathematicians have often come to the conclusion, or have assumed or implied in their work, that there must be some sort of connection between mathematics and the physical world.

We think that mathematics and reality are related for four interconnected reasons:

- Because mathematical knowledge has a predictive character.

1 Mind, Mathematics And Reality

- Because we believe that exact knowledge must be predictive.
- Because the current conception of knowledge is based on and entertains only quantitative arguments.
- Because we believe that the fundamental structure of reality is either not involved in the evolutionary process of the Universe, or is an invariant of that process, and can, therefore, be explained, at least in part, by an eternal, non-evolutionary symbol system such as Mathematics.

There are numerous examples in the history of human thought where the idea about the existence of a connection between mathematics and reality surfaces—or where it seems reasonable to interpret certain statements made by a philosopher or a mathematician about the deep nature of reality, as evidence that the author must have been thinking about some sort of math-reality connection.

Pythagoras (572-ca.500 BCE), believed that everything that emerges and happens in the world could be measured by means of numbers. Plato (427-347 BCE), influenced by the Pythagorean tradition, associated ideas with numbers, as reported by Aristotle on various occasions. When Aristotle introduced the notion of "formal cause" to explain what characterizes the shape or appearance of the moving or changing things in the world, he used ratios (e.g. 2:1) and numbers to illustrate how knowing the form or structure of an object is a necessary condition for understanding that object. In "Il Saggiatore" (english: The Assayer) (1623), Galileo Galilei (1564-1642) captured the idea that mathematics and reality share a common nature in this famous passage:

> *"Philosophy [i.e. physics] is written in this grand book - I mean the universe - which stands continually open to our gaze, but it cannot be understood unless one first learns to comprehend the language and interpret the characters in which it is written. It is written in the language of mathematics, and its characters are triangles, circles, and other geometrical figures, without which it is humanly impossible to understand a single word of it; without these, one is wandering around in a dark labyrinth."*

Galileo goes even further than just maintaining that the universe is written in the language of mathematics, he also warns us that it

1.2 Why We Think That Reality Might Be Mathematical

is humanly impossible to have epistemic access to the world unless we choose to look at the world through the lens of mathematics. And again, 1933 Physics Nobel laureate Paul A. M. Dirac, one of the fathers of Quantum Mechanics, writes that God *"used beautiful mathematics in creating the world"* (Kursunoglu and E. P. Wigner 1990), and 1965 Physics Nobel laureate Richard P. Feynman writes, as quoted in Feynman 1965,

> *"To those who do not know mathematics it is difficult to get across a real feeling as to the beauty, the deepest beauty, of nature. [...] If you want to learn about nature, to appreciate nature, it is necessary to understand the language that she speaks in."*

and again German physicist Heinrich Hertz writes:

> *"One cannot escape the feeling that these mathematical formulae have an independent existence and an intelligence of their own, that they are wiser than we are, wiser even than their discoverers, that we get more out of them than was originally put into them."*

and Albert Einstein in Einstein 1921

> *"How can it be that mathematics, being after all a product of human thought independent of experience, is so admirably adapted to the objects of reality?"*

Eugene Wigner (1902-1995), 1963 Physics Nobel laureate, in his famous 1960 paper "The Unreasonable Effectiveness of Mathematics in the Natural Sciences" (E. Wigner 1960) gives a modern account of the lack of any rational explanation for why mathematics works. In another passage taken from the same paper, Wigner describes what he calls the "miracle of helium", and provides a great example of the awe that strikes every scientist when it seems that Nature speaks mathematics. To quote E. Wigner, mathematics surprises us every time *"we get something out of the equations that we did not put in"*. This is a recurring theme in the study of the practical and conceptual foundations of mathematics, and is without doubt *the* most puzzling feature of mathematics.

I think that there are two ways to interpret the belief that mathematics and reality are connected, as described in Table 1.2.

1.3 The Strong Math-Reality Connection

The first interpretation, which I call the *strong math-reality connection*, is a philosophical position that claims that the universe, at a fundamental level, is designed and organized in a mathematical language—though not necessarily in the mathematical language as we know it today. This conviction seems to be crystallized in A. N. Whitehead's words *"The ultimate goal of mathematics is to eliminate any need for intelligent thought."* This position is strongly anthropic—in a sense which I clarify in Section 3.2. According to this mathematical picture of the universe, which interprets the success of mathematics in describing natural phenomena as an indication that physical reality itself has a mathematical nature, we can essentially describe everything that exists with a few mathematical formulas written on a napkin. The strong math-reality connection, combined with our current conception of mathematics, finds its most advanced expression in the idea that computation is a universal feature of Nature. This is a profound philosophical and physical principle known as the *universality of computation*, that when fully embraced, radically modifies how we see the world in ways that I find rather difficult to fully comprehend. The universality of computation is a reductionist picture of reality that tells us that everything that exists can be distilled into a few rules that allow us to systematically describe every objective reality. It is the idea that physical systems are essentially systems that compute. It is the idea that Nature itself has a computational origin, and that computation is the very fabric of everything that exists. If we knew all the rules needed to decode the computational matrix of the Universe, and if we had a computer or any other physical system capable of carrying out the necessary computations, we could explain anything from particles physics, to cosmology, to intelligence, feelings, elections, gossip, dreams, consciousness and spirituality.

1.4 The Weak Math-Reality Connection

The second interpretation, which I call the *weak math-reality connection*, is the view that the image of reality that mathematics gives us is intrinsically incomplete because mathematics reflects the limited nature of human thought. This idea appears, for example, in J. J. Sylvester as *"[t]he object of pure Mathematic [is] that of unfolding the laws of human intelligence."* According to the weak

1.4 The Weak Math-Reality Connection

	Strong math-reality connection	**Weak math-reality connection**
Leads to	*Universality of computation.*	*Mathematics as a dialect of Nature.*
Nature of things	*Everything that exists has a computational nature. Mathematical ontology corresponds to objective ontology.*	*Mathematical ontology is a reference to objective ontology.*
Epistemic access to reality	*Self-referential access to reality.*	*Local access to subjective reality.*
Worldview	*Anthropic.*	*Participatory.*

Table 1.2: Comparison of two interpretations of the math-reality connection.

math-reality connection, the fact that mathematics is effective in producing a predictive description of reality should be taken as an indication that, to the extent to which the human intelligence can be trusted, a portion of physical reality is indeed intelligible. That is to say that the intensional properties of a mathematical image of reality should not be construed as the ontological features of reality itself, instead, they should be interpreted as a math-reality correspondence that suggests that there is a certain degree of mutual intelligibility between the language in which Nature is written and the mathematical language. Thus mathematics, as a cognitive technology, is effective in doing only what it is designed to do, that is, give us a picture of reality that is necessarily compatible with the intelligence that invented mathematics, but nothing more than that. I am inclined to believe that the weak interpretation of the math-reality connection gives a more realistic, down-to-earth view of what mathematics is and can do. There is a sense in which the mathematical dialogue between Man and Nature is a like a conversation between a Portuguese and a Spanish: the two languages are related, and have a certain degree of mutual intelligibility, but they are not the same.

1.5 The Fundamental Question To Ask

In this introductory chapter we have discussed how mathematics as a cognitive technology seems to have privileged epistemic access to the world, and have described the practical and philosophical tension created by our current conception of mathematics and by the mind-world dichotomy present in the Western view of the World and its scientific culture. Let's summarize these ideas here for convenience.

1. Mathematics is a cognitive technology in which knowledge is acquired purely by deduction from basic principles that express the fundamental cognitive structures of human thought and a worldview on the human experience of reality.

2. We accept that mathematics is a product of the mind-world interplay, but in our current conception of mathematics there isn't an actual concept of mind.

3. We explain the fact that mathematical theories give us accurate descriptions of the physical world by postulating that, at a fundamental level, physical reality has a mathematical structure, or a structure that is epistemically accessible to the human mathematics.

The very act of constructing mathematical concepts the way we do seems to have the features that make it effective in describing the world, even though when we look at the world through the lens of mathematics all we really see is the structure of our mathematical arguments. What is it that is actually occurring in the mind when we create mathematical concepts of the world? What are the processes, the dynamics, the laws and the limitations of the mind's ability to recreate structures of thought that seem to mimic and decode physical reality so well?

I think that in order to understand the nature of the mind-world interplay that gives rise to mathematics as we know it today, we need to understand what makes mathematics effective in describing the physical world. I think that to understand the effectiveness of mathematics in describing the world, and the mystery of what seems to be the mathematical nature of reality, we need to understand how we mathematize. Thus, the questions we should be asking are not so much What is mathematics? Why is mathematics effective in describing the world? and so on and

1.5 The Fundamental Question To Ask

so forth, the question we should be asking is:

"What do we do when we do Mathematics?"

What is it about the very act of mathematizing that connects us to reality? We will explain with the analogy we introduced in Section 1.1 why answering this question is probably more important than giving an explanation of what is perceived as a connection between mathematics and the deep structure of physical reality.

Smartphones are a remarkable piece of technology, and more and more people are starting to use them. Yet, not all smartphone users know how to design and manufacture a smartphone, and how to design and deploy a mobile communication network. Not all smartphone users know the mathematical and physical theories that make smartphone technology possible. Not smartphones users know how to write the software that runs on the smartphone and that makes it the powerful and versatile gadget that it is. And the reason why not all smartphones users know how to build a smartphone is simple: *they don't have to*. Smartphone users don't need to know how to build a smartphone and a mobile communication network, because the knowledge to *design and build* an object, and the knowledge to *use* that object are in general two very different and distinct things that belong to two very different epistemic domains.

There is a sense in which the knowledge we have of mathematics is the same knowledge that the laymen has of smartphones, cars, airplanes and computers. We *use* our conception of mathematics much like the laymen uses a smartphone: without having to know how the human mind is able to mathematize. The ability to mathematize just happens to be there, in our mind, and it's ready to use if we are willing to develop it: all we have to do is learn how mathematics is built, and develop the necessary degree of mathematical sensitivity and sophistication to be able to conceive and navigate very broad hierarchies of abstract concepts. Those who decide to learn how to use mathematics go to a school of mathematics, become professional mathematicians and learn very sophisticated methods to use this extraordinary ability of the human mind. But they do not have to learn *how* their mind gives them the ability to mathematize, because that is, so to speak, a given. Yet, I think it's fair to say, most mathematician would have no problem in acknowledging that mathematics is a product of the mind and that, in our current conception of mathematics, there is no concept of mind, with the exception of the structure of formal systems, that are an attempt to reproduce the *effects* of rational thought rather than its structure.

We have never had to design or build this extraordinary ability of the human mind to create abstractions of our own ideas, in a language so elegant and rarefied that makes us think that, perhaps, it is the language used to create everything that exists. This is why I think there is a compelling philosophical and mathematical need to revisit our conception of mathematics within the larger context of the mind-world interplay, because it is there that we can learn how the human mind is able to mathematize, and is it there that we can learn what mathematics really is, and what it is there to tell us about reality and about our place in the World.

Bibliographical Notes

The theme of the intelligible universe hypothesis is mildly inspired by a general structuralist orientation in the Philosophy of Science, and by the themes of Structural Realism (Worrall 1989, Russell 1927, Carnap 1928, McCabe 2007). It is difficult to trace all the authors that have influenced and shaped my views on the practical and conceptual nature of Mathematics. Without doubt, the works of MacLane 1986, Lawvere and Rosebrugh 2003, N. U. Salmon 2005, have been particularly important in the preparation of the structure of this Chapter. Also Hintikka 1969, Hintikka 1997, Feferman 1966, Feferman et al. 1984, Tarski 1941, Shields 2012, Hopcroft and Ullman 1979, Bohm 2002, E. Wigner 1960, Galilei 1988, Hofstadter 1980,Lawvere 1966 have been central in the early stages of my research.

The theme of formal systems and of the model-theoretical views of the philosophical foundations of mathematics were primarily influenced by Hodges 1997, Lawvere 1996b, Lawvere 1969a, Lawvere 2007.

The theme of the universality of computation is in Minsky 1988 and in various works in Quantum Information Theory. Probably the most advanced example of the mathematical and philosophical use of the principle of the *universality of computation* is Deutsch and Marletto 2014, in which the entire structure of reality is reconstructed in terms of an extended definition of computation. Deutsch's vision has helped me reframe my ideas on the computa-

1.5 The Fundamental Question To Ask

tional nature of thoughts and cognition in the larger context of physical phenomena, but I think it doesn't give an account of why, for examples, we use formal systems to conceptualize the world, and why the genesis of mathematical thinking is the way it is.

The theme of the math-reality connection as is presented here is in Lo Vetere 2011, Lo Vetere 2013b.

Example Of Formal System

The KAT-System is defined in Hofstadter 1980

Alphabet: $\{K, A, T\}$

Syntax: All strings over the given alphabet are in the language

Axioms: KA is a string of **The "KAT" System**

Inference rules: Let x and y be arbitrary strings of zero or more symbols, then

1. xAT follows from xA
2. Kxx follows from Kx
3. xTy follows from $xAAAy$
4. xy follows from $xTTy$

Thus, for example, by applying the rules above we obtain the following strings of symbols over the alphabet $\{K, A, T\}$:

1. KA follows from the axiom
2. KAA follows from 1) and inference rule 2
3. $KAAAA$ follows from 2) and inference rule 2
4. $KAAAAT$ follows from 3) and inference rule 1
5. $KTAT$ follows from 4) and inference rule 3
6. $KTATTAT$ follows from 5) and inference rule 2

The strings $KA, KAA, KAAAA, KAAAAT, KTAT$ and $KTATTAT$ are theorems of the KAT-system.

Chapter 2

What Mathematics Tells Us About Reality

> The fairest thing we can experience is the mysterious. It is the fundamental emotion which stands at the cradle of true art and true science. He who knows it not and can no longer wonder, no longer feel amazement, is as good as dead, a snuffed-out candle.
>
> Einstein 1934

In the previous chapter we described briefly the technology of formal systems, and how they allow us to create mathematical pictures of reality by translating questions into computational problems. In this chapter, we argue that to comprehend the human mind's ability to mathematize, and to comprehend how mathematics describes the reality that we think it describes the way it does, we need to reframe our conception of mathematics within the larger context of the mind-world interplay. Thus, we want to explain what we see when we look at the World through the lens of a mathematical theory, and show the practical and theoretical necessity to encode a concept of mind in our conception of mathematics.

The argument we present here is intended as a *mathematical allegory*, in that our use of a mathematical style of reasoning is, to a certain extent, metaphorical. It is allegoric because it is an attempt to use the rarefied symbolism of mathematics to

2 What Mathematics Tells Us About Reality

describe the metaphysical tension between symbol and meaning where mathematization originates. The cognitive processes at the basis of mathematization exist only at the intersection between Mathematics and Metaphysics, where the ordinary structure of human thought is inadequate to articulate concepts and to formulate questions. The place where Mathematics and Metaphysics coexist is the realm of various forms of symbolic and semantic intuitions of reality. It is a secluded and brutal and majestic corner of the mind where the language of mathematics is the language of metaphysics, where knowledge is only intuitive, where mathematical and metaphysical perceptions spring into existence without syntactical inhibitions or prejudices, and are braided in patterns that the mind vividly recognizes without the need to comprehend them through imagination and language.

The ideas we present in this chapter are allegories when they hint at the metaphysical nature of the mathematical facts that they describe, and are literal, direct, unmediated statements about the nature of mathematics and human thought, when they describe the structure of those mathematical facts.

It is, perhaps, useful to outline the general structure of the argument we present. First, we challenge the conceptual machinery set in motion when we use mathematics as a cognitive technology, by producing two contrasting descriptions of the ontological structure of a world. Second, we maintain that what seems to be an ontological paradox is a property of our conception of mathematics that cannot be studied within the context of the conception of mathematics in which this property manifests itself as a paradox. The contents of the this chapter reflect this rationale.

- To describe the relation between a mathematical theory and what we think that theory describes, we characterize how a mathematical theory acquires its ontological status (existence) and its definitional structure (identity).

- We challenge the conceptual framework in which we develop the first points, by producing two theories that give us conflicting descriptions of the metaphysical structure of the same world.

This chapter contains a somewhat advanced use of elementary mathematical concepts that some readers may find not entirely straightforward, and can be skipped without impacting the understanding of the text.

2.1 Worlds And Theories

Worlds and theories symbolize the metaphysical tension between symbol and meaning where our mathematical intuition of reality originates.

We need two definitions to work with this mathematical allegory. Let **W** be a World and $\mathcal{T}$ a mathematical theory that describes **W** or part of it.

- The World we *see through* $\mathcal{T}$, which we indicate with $W^\mathcal{T}$, is the object of $\mathcal{T}$. $W^\mathcal{T}$ is what $\mathcal{T}$ describes of **W**, and we have epistemic access to it.

- The World *defined by* $\mathcal{T}$, which we indicate with $W_\mathcal{T}$, is the result of an interpretation of $\mathcal{T}$. $W_\mathcal{T}$ denotes what we *comprehend* of **W**. It is how $\mathcal{T}$ helps us think about **W**, and we do *not* have direct epistemic access to it. $W_\mathcal{T}$ represents what we learn, or what we have reason to believe about **W**, by means of the cognitive technology $\mathcal{T}$ applied to **W**.

For example, let us consider the following theory $\mathcal{T}$ about perfect squares:

$$\mathcal{T} := \text{``The sum of the first } n \text{ consecutive odd numbers is equal to } n^2\text{''}$$

In this example, the World **W** is the set of all integers $1, 2, 3, \ldots$ and $\mathcal{T}$ is a theory about perfect squares $1, 4, 9, \ldots$. When we look at **W** *through* $\mathcal{T}$ we *see* a World $W^\mathcal{T}$ made of all perfect squares. To characterize $W_\mathcal{T}$, the World defined by $\mathcal{T}$, we ask: What is the insight into the world **W** codified by $\mathcal{T}$? $\mathcal{T}$ tells us that every time we *see* a perfect square n^2 in **W**, we can *think* of it as the sum of the first n consecutive odd numbers in **W**: that's $W_\mathcal{T}$.

The general structure of the theories we are interested in in this discussion is *"You can think of $W^\mathcal{T}$ as $W_\mathcal{T}$"* where the $W_\mathcal{T}$ part contains the new bit of information about the description of **W** given by $W^\mathcal{T}$. Historically important examples of this pattern are, among others, Einstein's General Relativity, "You can think of gravity ($W^\mathcal{T}$) as a geometric property of spacetime ($W_\mathcal{T}$)", or

2 What Mathematics Tells Us About Reality

the Copenhagen interpretation of Quantum Mechanics, "You can think of matter particles ($W^\mathcal{T}$) as probability waves ($W_\mathcal{T}$)".

Let's get now to a description of how we *use* mathematical theories.

In Figure 2.1:

- The **Worlds** region contains the objects described by what is in the **Theories** region. Thus, the **Worlds** region consists of entities of type $W_\mathcal{T}$, and the **Theories** region consists of entities of type $W^\mathcal{T}$, where $\mathcal{T}$ is any valid theory.

- We *do not* see what's in the **Worlds** region, we only see what's in the **Theories** region.

- The transitions from **Worlds** to **Theories** are called *translations* (Figure 2.1-A).

- The transitions from **Theories** to **Worlds** are called *interpretations* (Figure 2.1-C).

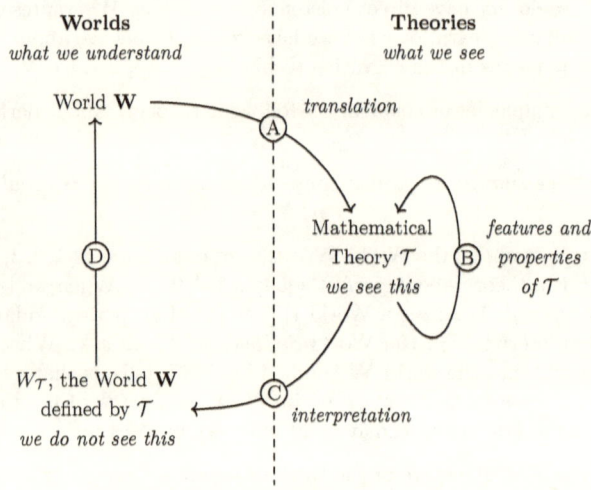

Figure 2.1: Detailed illustration of the use of mathematics as a cognitive technology.

An example of translation is the description of the physical world

2.1 Worlds And Theories

as a place made of geometrical shapes such as lines, squares, circles and triangles, in the language of a formal system, such as Euclidean geometry for example. In contrast, an example of interpretation is the ascription of the geometrical properties of lines, squares, circles and triangles to the nature of reality.

One might rightly ask: Isn't it perhaps the case that you can observe the world as a place made of geometrical shapes because you have already those geometrical ideas in your mind? So what comes first, the world or our mathematical intuition of it? This question points to the apparent circularity of the translation-interpretation description we have given. However, the circularity is really only apparent. We never observe reality in its totality. Our experience of reality is the result of a conscious, intentional act that we perform by observing the world for a specific purpose. Any direct observation of reality, exists by virtue of being consistent with the structure of human thought. In this sense, pure, unmediated observations are axiomatic in that they rest entirely on the fundamental structure of the human thought. When we translate an observation into a mathematical theory, we build on the axiomatic substrate that constitutes the structure of our observation: and this is an abstraction process which produces a theory based on that axiomatic substrate. In contrast, when we ascribe to reality the features and properties of a mathematical theory, we do the opposite: we regard reality as a model of our theory. So there is a sense in which translations are to interpretations what axioms are to formal theories.

When we use mathematics as a cognitive technology[1], we follow the steps described in Figure 2.1:

A. We translate what we want to observe of a World in the language of a formal system, and the result of that translation process is a mathematical theory $\mathcal{T}$.

B. We construct a model of $\mathcal{T}$, and study its features and properties.

C. We interpret the features and properties of $\mathcal{T}$ as the features and properties of the World we want to describe with $\mathcal{T}$.

---

[1] The exact same pattern applies to the study of non-physical entities. As a matter of fact, mathematics itself is studied according to the same scheme: we create mathematical theories to study other mathematical theories.

D. We think that W_T and W_T have a certain degree of correspondence, although the nature of this correspondence is unknown: this is the math-reality connection we saw in Section 1.2.

2.2 What We See And What We Comprehend

To characterize the ontic and epistemic content of a theory T, we study its definitional structure, and describe what we see and what we comprehend through T.

A general description of how the definitional structure of an entity is related to the entity's identity and ontological status is given in Chapter 5. For the moment, let us think that a theory is defined into existence by the very process by means of which we articulate its definition. We characterize T's ontological structure in three steps.

K.1 In this section we describe what we see (W^T) and what we understand of what we see (W_T).

K.2 In Sections 2.3 and 2.4 we study the definitional structure of what we understand (W_T).

K.3 In Section 2.5 we use the data in **K.2** to describe the changes in the structure of what we understand (W_T) that are consistent with the description of the World **W** given by T.

Let us see how these three steps work on a real theory. We have already introduced the theory T *"The sum of the first n consecutive odd numbers is equal to n^2"*. T describes a relation between perfect squares and certain sets of odd numbers. Thus, the World we see *through* T is made of all the perfect squares $1, 4, 9, 16, 25, \ldots$ and T tells us that we can *think* of any square n^2 as the sum of the first n consecutive odd numbers: this is the insight that T gives us about squares and odd numbers.

2.3 The Structure Of What We Comprehend

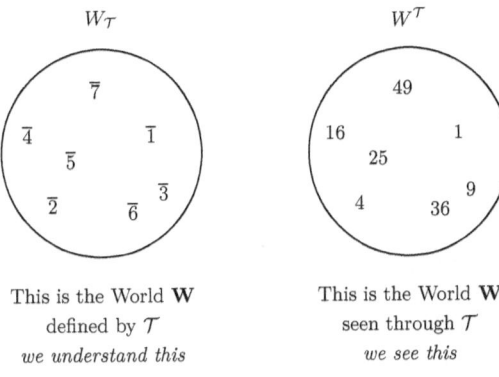

Figure 2.2: The World **W** seen through $\mathcal{T}$ (right), and the world $W_\mathcal{T}$ (left), which is how $\mathcal{T}$ describes **W**.

In Figure 2.2 $W^\mathcal{T}$ represents what we see of **W**. It is our reference to what in this allegory represents the "real" World **W**, and consists of the set of all squares. $W_\mathcal{T}$, which we do not see directly, but only as an interpretation of $\mathcal{T}$, is the set of all the sets of the first n consecutive odd numbers, where $n = 1, 2, 3, \ldots$, and in this allegory represents what we *learn* about **W** thanks to the mathematical technology $\mathcal{T}$. For convenience, we represent the elements of $W_\mathcal{T}$ as $\overline{1}, \overline{2}, \overline{3}, \overline{4}, \ldots$, which represent respectively the sets of the first $1, 2, 3, 4 \ldots$ consecutive odd numbers. Thus $\overline{1} = \{1\}, \overline{2} = \{1, 3\}, \overline{3} = \{1, 3, 5\}, \overline{4} = \{1, 3, 5, 7\}$ and so on.

To define how $W_\mathcal{T}$ acquires its ontological status, we want to give an internal characterization of $W_\mathcal{T}$, and of the changes that preserve its definitional structure. An *internal* characterization of an entity does not rely on the features and properties of a theory that defines that entity, but only on the features and properties of the entity itself. In the notes at the end of the chapter there is an example that illustrates the difference between *internal* and *external* characterization of $W_\mathcal{T}$.

2.3 The Structure Of What We Comprehend

*Here we determine what holds together the conceptualization of **W** constructed via $\mathcal{T}$.*

2 What Mathematics Tells Us About Reality

To find out what holds together our comprehension of **W** through $\mathcal{T}$, we identify each individual structural component of the world $W_{\mathcal{T}}$, and ask: How can we characterize this component? $W_{\mathcal{T}}$ consists of two parts: there are its elements $\overline{1}, \overline{2}, \overline{3}\ldots$, which are collections of integers with certain characteristics, and there is a rule σ that assigns each element $\overline{n}$ of $W_{\mathcal{T}}$ to an integer that equals the sum the elements of $\overline{n}$. For example $\sigma(\overline{3}) = \sigma(\{1, 3, 5\}) = 1 + 3 + 5$. Thus we use the notation $W_{\mathcal{T}} = (W, \sigma)$ to indicate that $W_{\mathcal{T}}$ consists of two parts, the set W of elements of $W_{\mathcal{T}}$—the sets $\overline{1}, \overline{2}, \overline{3}\ldots$—and a certain rule σ to assign integers to the elements of W.

Let's start with σ: How can we characterize it? Characterizing σ means describing what must be defined about σ to determine its identity, and strictly within the context in which σ is applied, in this case, the elements of W. What this question is asking is: How can we say that another integer-valued rule η defined on the elements of W is identical to σ? The answer is simple, in order for η and σ to be the same rule on W, they must assign the same integer to each element of W. This means that the only way in which σ can change without breaking the definitional integrity of $W_{\mathcal{T}}$ is by remaining identical to itself.

The other component of $W_{\mathcal{T}}$ is the set of numbers W. To characterize W, we must describe the properties that a set of integers must satisfy to be an element of W. Since the elements of W are all the sets of the first n consecutive odd numbers, if we break down the structure of *a* set of n consecutive odd numbers into its logical components, we find that the following three conditions must be satisfied simultaneously to construct an element of W: the n integers are *odd, consecutive*, and 1 is a member of the collection.

In summary, the definitional structure of $W_{\mathcal{T}}$ is:

W.1 All the numbers in W are odd

W.2 All the numbers in W are consecutive

W.3 1 is in W

W.4 the rule σ that assigns to $\overline{n}$ the sum of its elements

2.4 Theory-induced Blindness

To show the structure of our comprehension of a theory, we break that theory's definitional structure.

To complement the reflections of the previous section, let us see how a change that *breaks* the definitional structure or integrity of $\mathcal{T}$ affects what we understand about **W**. The change we introduce to break the definitional structure of $\mathcal{T}$ consists in violating condition **W.3**. The new definitional structure resulting from dropping condition **W.3**, gives rise to a new theory $\mathcal{R}$ about the sum of consecutive odd numbers.

To discover what $\mathcal{R}$ has to say about the sum of consecutive odd numbers, let us introduce the notation $S_{k,n}$ to denote the sum of n consecutive odd numbers starting from the kth. For example, given the sequence of the first 8 consecutive odd numbers $1, 3, 5, 7, 9, 11, 13, 15$, we denote with $S_{3,4}$ the sum $5 + 7 + 9 + 11$, that is, the sum of the 4 consecutive odd numbers of the sequence starting from the 3rd.

Our original theory $\mathcal{T}$, expressed in the $S_{k,n}$ notation becomes

$$S_{1,n} = n^2$$

that is, *"The sum of n consecutive odd numbers starting from the 1st equals n^2"*, and from this observation, we discover that we know already how to compute $S_{k,n}$ when $k = 1$.

If we observe how $S_{k,n}$ is defined, we notice that the sum of two consecutive sequences of consecutive odd numbers is still a sequence of consecutive odd numbers, which means that we can still use $S_{k,n}$ to express the sum of two consecutive sequences of consecutive odd numbers. If we translate this observation in the $S_{k,n}$ notation, we discover that we can express $S_{k,n}$ via $S_{1,n}$ as

$$S_{k,n} = S_{1,n+k-1} - S_{1,k-1}$$

From this identity, and from the knowledge that $S_{k,n} = n^2$ when $k = 1$, we discover that $\mathcal{R}$ tells us that the sum of n consecutive odd numbers is the *difference of two squares*, namely, $(n+k-1)^2 - (k-1)^2$. But then, given the difference of any two squares $p^2 - q^2$, with $p > q \geq 0$, the parameters n and k such that $S_{k,n} = p^2 - q^2$

2 What Mathematics Tells Us About Reality

are $n = p - q$ and $k = q + 1$, as one can easily verify by substitution. In other words
$$S_{q+1, p-q} = p^2 - q^2$$

Thus, the world we *see* through $\mathcal{R}$ is made of all the differences between perfect squares $p^2 - q^2$, where $p > q \geq 0$, and $\mathcal{R}$ tells us that we can *think* of any difference of squares $p^2 - q^2$ as the sum of the $p - q$ consecutive odd numbers starting from the $(q+1)^{th}$: this is the insight that $\mathcal{R}$ gives us about differences of squares and odd numbers.

With these ideas in mind about $\mathcal{R}$, and with what we already know about $\mathcal{T}$, we want to compare the images of the World **W** provided by these two theories. On the one hand, we have $\mathcal{T}$ that tells us that the sum of the first n consecutive odd numbers is a perfect square, and on the other hand we have $\mathcal{R}$ that tells us that the sum of n consecutive odd numbers is the difference of two squares.

The differences between $\mathcal{T}$ and $\mathcal{R}$ are, superficially, both in how we *see* the World **W** through these theories, and in what we *comprehend* about it. However, on closer inspection, the difference between $\mathcal{T}$ and $\mathcal{R}$ is more subtle. For there are infinitely many cases where $p^2 - q^2$ is itself a perfect square: when p, q and r form the following Pythagorean triple $r^2 + q^2 = p^2$. In these cases, the images of the World **W** we form through $\mathcal{T}$ and $\mathcal{R}$ *coincide* because both theories make us *see* r^2, as illustrated in Figure 2.3, yet what we *comprehend* about **W** is different. When we think of r^2 with $\mathcal{T}$, we understand *"sum of the first r consecutive odd numbers"*, when we think of r^2 with $\mathcal{R}$, we understand *"sum of $p - q$ consecutive odd numbers starting from the $(q+1)^{th}$, where $r^2 = p^2 - q^2$"*.

2.5 Comprehension-preserving Changes

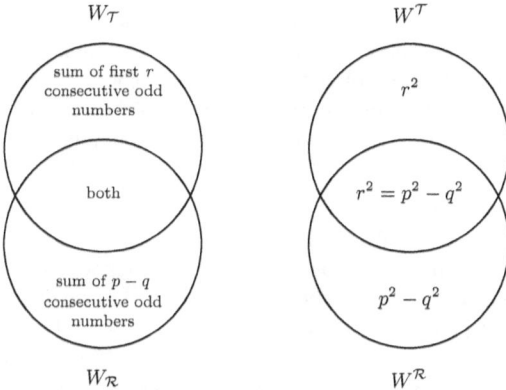

Figure 2.3: Example of theories through which we ascribe different meanings to the same partial image of a World.

This allegory shows how the comprehension of the World we derive from our mathematical theories, not only is an expression of the definitional structure of those theories, but also of the robustness of those definitional structures to change.

But then, how well do we comprehend the World through our mathematical theories? To answer this question, we must first establish what changes our theory-laden comprehension of the World can withstand.

2.5 Comprehension-preserving Changes

We express the robustness of the definitional structure of $\mathcal{T}$ by characterizing the changes that preserve our comprehension of $\boldsymbol{W}$ through $\mathcal{T}$.

Given $W_{\mathcal{T}} = (W, \sigma)$, let us start by examining the changes that preserve the definitional structure of W described by the conditions **W.1**, **W.2** and **W.3**. We notice that there are two ways in which we can change a collection of numbers: we can alter the number of elements of the collection, or we can replace some of them. Let's translate this observation into the types of changes compatible with the definitional structure of $\mathcal{T}$.

To *increase* the number of elements of the set of the first n consecutive odd numbers and preserve the definitional integrity of $W_\mathcal{T}$, we don't need to know that 1 is in $W_\mathcal{T}$ (condition W.3), since we can only add elements greater than 1 to the collection. Let's indicate with $\uparrow$ this transformation. So for example, the action of a $\uparrow$-type transformation could be $\uparrow \{1,3,5\} = \{1,3,5,7,9\}$.

To *decrease* the number of elements of the set of the first n consecutive odd numbers and preserve the definitional integrity of $W_\mathcal{T}$, we don't need to know that all the numbers in $W_\mathcal{T}$ are odd (condition W.1) because the only way in which we can decrease the number of elements of $W_\mathcal{T}$ and preserve $W_\mathcal{T}$ at the same time, is by removing the greatest element of the collection, and to do so, we don't need to know if it's odd or even. Let's indicate with $\downarrow$ this transformation. Again, as an example, a $\downarrow$-type transformation could be $\downarrow \{1,3,5,7\} = \{1,3\}$

To *replace* some of the elements of the set of the first n consecutive odd numbers, and preserve the definitional integrity of $W_\mathcal{T}$, we can do one thing and one thing only: a permutation of the elements. Let's indicate with $-$ this transformation. As an example, the action of a $-$-type transformation is $-\{1,3,5\} = \{3,5,1\}$.

For what concerns the condition **W.4**, a rule defined on W either is or is not σ, therefore σ cannot change without breaking the definitional integrity of $W_\mathcal{T}$.

2.6 Testing What We Comprehend

To test the metaphysical content of a theory, we use a simple form of introspection expressed in the language of that theory.

We want to challenge the definitional structure of **W** as we know it from $\mathcal{T}$. Let's recall the tenets of the mathematical allegory we are developing here are: we don't know **W**, and we do not have direct epistemic access to $W_\mathcal{T}$. So, how can we study the correspondence of Figure 2.1-D?

2.6 Testing What We Comprehend

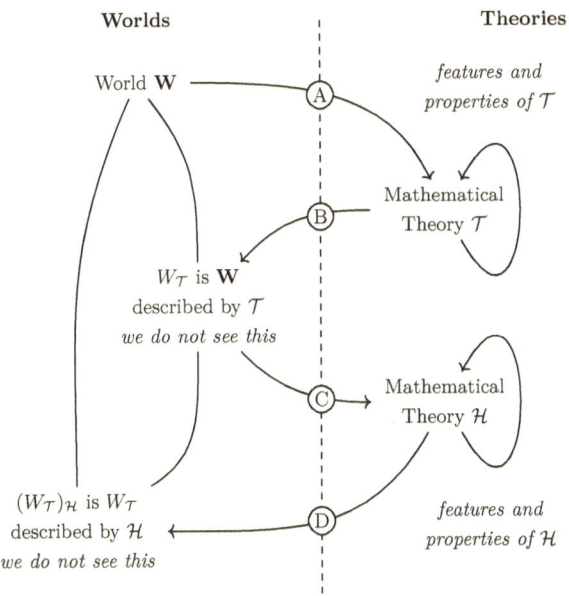

Figure 2.4: Interpretation of a mathematical theory via the properties of the world that we think the theory describes.

A way to begin to comprehend the nature of the correspondence of Figure 2.1-D rests on the following observation: there are two worlds, there is $W^{\mathcal{T}}$, and there is the part of **W** that we do not see through $\mathcal{T}$. To characterize $W^{\mathcal{T}}$, we can construct a theory $\mathcal{H}$ that describes how $W_{\mathcal{T}}$ acquires its definitional structure. Intuitively, $\mathcal{H}$ characterizes the distinction between what we describe with $\mathcal{T}$, that is, $W_{\mathcal{T}}$, and what we do not describe with $\mathcal{T}$. For, to comprehend the ontological structure of $W_{\mathcal{T}}$, and how it relates to the part of **W** that we don't see with $\mathcal{T}$, we need to understand how $W_{\mathcal{T}}$ acquires its own ontological status in a way that is consistent with the way in which $W_{\mathcal{T}}$ is defined as an interpretation of $\mathcal{T}$, but *independently* of the features and properties of $\mathcal{T}$—once again, it is crucial that the description of how $W_{\mathcal{T}}$ acquires its ontological status happens inside $W_{\mathcal{T}}$, and not as a result of the interpretation of the features and properties of $\mathcal{T}$.

Here are the practical steps that we'll follow, which are essen-

2 What Mathematics Tells Us About Reality

$$- \uparrow\downarrow\downarrow - \uparrow\uparrow -- \uparrow\downarrow - \uparrow -- \uparrow\downarrow\uparrow - \uparrow - \downarrow -- \downarrow\uparrow - \uparrow - \downarrow - ...$$

Figure 2.6: An element of $W_{\mathcal{T}}$ seen through $\mathcal{H}$

To fully comprehend why what we see through $\mathcal{H}$ challenges the ontological structure of **W** that $\mathcal{T}$ seems to suggest, we need to take one final step: we need to construct a correspondence between $W^{\mathcal{T}}$ and $(W_{\mathcal{T}})^{\mathcal{H}}$.

2.8 Contrasting Interpretations Of The World

We are not able to reconcile the idea of a mathematically designed World, with the existence of mathematical theories that describe that World accurately, and that, at the same time, offer contradicting metaphysical descriptions of it.

We have established that $(W_{\mathcal{T}})^{\mathcal{H}}$ is made of all the sequences of three types of symbols $\uparrow, \downarrow, -$. Let $0 \leq i \leq j$ be integers and ω be any sequence in $(W_{\mathcal{T}})^{\mathcal{H}}$. We introduce the notation $(i,j)_\omega$ to indicate the subsequence of symbols of ω from the ith position to the jth position. So for example, if $\omega = \uparrow - \downarrow\uparrow -$, then $(1,3)_\omega = - \uparrow\downarrow$. We have constructed $\mathcal{H}$ so that $(i,j)_\omega$ is itself an element of $\mathcal{H}$.

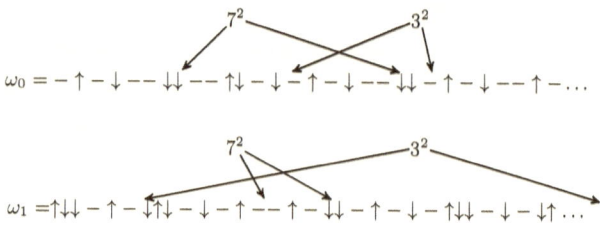

Figure 2.7: The elements of $W_{\mathcal{T}}$ as we see them from $\mathcal{H}$.

Since any symbol of $(W_{\mathcal{T}})^{\mathcal{H}}$ represents a transformation between elements of $W_{\mathcal{T}}$, and since $(W_{\mathcal{T}})^{\mathcal{H}}$ contains infinitely many infinite sequences, any element of $W^{\mathcal{T}}$ is represented by infinitely many positions in infinitely many infinite sequences of $(W_{\mathcal{T}})^{\mathcal{H}}$. This means that for any given element n^2 of $W^{\mathcal{T}}$, there are infinitely many subsequences of a given infinite sequence ω of $(W_{\mathcal{T}})^{\mathcal{H}}$, such

2.8 Contrasting Interpretations Of The World

that $codomain(i,j)_\omega = n^2$ for some indices i,j. There are infinitely many infinite sequences ω, and for each of them it is possible to find infinitely many pairs (i,j) such that $codomain(i,j)_\omega = n^2$. Figure 2.7 illustrates two of these sequences.

When we look at **W** through $\mathcal{T}$ we see squares, and when we look at $W_\mathcal{T}$ through $\mathcal{H}$ we see infinitely many sequences of arbitrary length made of three types of symbols. But the squares that we see through $\mathcal{T}$ lose their definite identity when we look at them from $\mathcal{H}$, because there are infinitely many sequences in $(W_\mathcal{T})^\mathcal{H}$ with the same codomain. This conundrum is summarized in Figure 2.8.

$$\begin{array}{ccc} \mathbf{W} & \xrightarrow{see} & W^\mathcal{T} \\ \| & & \downarrow{?} \\ W_\mathcal{T} & \xrightarrow{see} & (W_\mathcal{T})^\mathcal{H} \end{array}$$

Figure 2.8: Summary of the construction of two contrasting pictures of **W**.

A way to read the diagram of Figure 2.8 is to think about the analogy between Classical Mechanics and Quantum Mechanics. In Newtonian Mechanics, it makes perfectly sense to ask the question "Where is the particle n^2?" whereas in Quantum Mechanics, the same question does not make any sense because particles are interpreted as probability waves. Similarly, in the picture of the world **W** offered by $\mathcal{T}$, **W** is made of squares $1^2, 2^2, 3^2, \ldots$ whereas in the picture of the world $W_\mathcal{T}$ offered by $\mathcal{H}$, each square is literally infinitely many sequences of $\uparrow, \downarrow, -$, each of which *appears to us* as $1^2, 2^2, 3^2, \ldots$ each time we pick an instance of those sequences.

Believing that **W** and $W_\mathcal{T}$ share the same metaphysical structure poses an ontological problem because the elements of **W** that we see from $\mathcal{T}$, and the elements of $W_\mathcal{T}$ that we see from $\mathcal{H}$, appear to exist in two radically different ways that we are not able to reconcile, even though the theories $\mathcal{T}$ and $\mathcal{H}$, taken individually, present us with a consistent, coherent description of **W**.

The metaphysical clash between *symbol* and *reference* we have attempted to explain in this chapter with the mathematical allegory of **Worlds** and **Theories**, suggests that the problem of explaining the nature of the relation between mathematics and

3 The Observation Of The Abstract Mind

3.1 Worldviews And Thought

The ability to observe conceptual or material realities depends on an array of worldviews that determine, influence or undermine the capacity to observe and discern facts and ideas, the ability to formulate questions, and the strategies devised to solve problems and interpret their solutions. Worldviews have a potent presence in the human thought processes involved in the *definition* of the object of observations, and determine the type of attainable *knowledge* of the facts and ideas we intend to observe, even before the observation takes place. Thus, to comprehend how worldviews determine, influence or undermine our ability to acquire knowledge, we need to comprehend in what ways they exert their influence on thought processes, and why this influence may not be completely manifest. For example, in the history of Physics, we find statements such as "everything is a process", or "events occur and are caused locally", or "physical space has more than three dimensions", or "space has no properties" and so on. These are examples of worldviews that profoundly shape and determine our ability to ask questions and to reason in general about Nature. Thinking that everything is a process relies on a notion of causality, which makes it difficult to reconcile with this worldview the facts that we are unable to describe causally. This worldview, like most worldviews, is embedded in the broader worldview that Nature is governed by laws. The view that events occur and are caused locally is one of the beliefs that was shattered by the quirks of Quantum Mechanics, where it has been observed that the cause-effect propagation is not limited by the speed of light—contrary to what stated in a principle of classical mechanics known as *local realism*, which combines the finite speed at which effects can propagate with the assumption that values of physical quantities exist before measurement. The view that space has no properties is part of the Newtonian cosmology, where space is seen as a completely inactive stage where physical phenomena occur and evolve. Other examples of worldviews are even more subtle and elusive, such as the fact, elegantly explained in "The Character of Physical Law" (Feynman 1965), that our current description of the physical world is based on what we have assumed to be the basic constituents of physical reality, namely, energy, mass, forces and spacetime, but that there is no necessity for that to be so.

The observation of the Abstract Mind poses precisely a problem

3.1 Worldviews And Thought

about what worldviews are suitable to define the Symbolic Intuition of Reality from a theoretical and operational perspective, to formulate questions about its origin and structure, and to reason about it. To get a feel for how a worldview can mislead thought, let's consider the following analogy that shows how causation can fail to explain a physical fact.

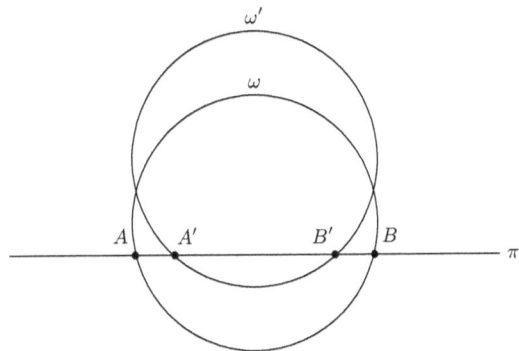

Figure 3.1: Example of false causation.

In Figure 3.1, ω and π denote two one-dimensional planets that can move freely on this page according to certain physical laws, and that may intersect each other in A and B as a result of their continuous movement. The inhabitants of ω and π are point-like people. A strange feature of one-dimensional worlds is that people don't move: when they want to know what happens, they ask their neighbors. Each citizen of a one-dimensional world has two neighbors: left and right. When ω and π intersect, the people in A and B feel a shock wave, and they continue to think that they have just two neighbors in their own world, left and right, not knowing that left, or right, or both may belong to the world that has just flown through their own planet. Unbeknownst to them, when they ask their neighbors what caused the shock wave, they always get the same answer: "I don't know, but there is another fellow on the other side of this planet that felt a shock wave too." To make this situation even more confusing, sometimes the communication between A and B occurs along a straight line in π, sometimes along an arc in ω, sometimes it's mixed, a question asked by A is transmitted over π, and the answer is received over ω, and vice versa. So depending on the paths taken by the communication between A and B, and on the paths taken by questions and answers, sometimes A gets a question from B before he asks its neighbors what happened, sometimes he thinks that the shock wave he felt caused the shock wave felt by B, sometimes B thinks that it was the shock wave he felt to cause the shock wave felt by A, and so on. To add insult to injury, ω and π are always moving. When the world ω is in position ω', shock waves are felt by A' and B', which come up with their own explanation as to what caused the shock wave. In an attempt to explain this strange phenomenon, the scientists in ω and π invented a method to measure time very accurately, and discovered that A and B feel the shockwave at the same time, but they haven't been able to determine what mysterious law of Nature connects the shock waves measured by A to those measured by B, and how these seem to propagate instantly from A to B.

3.2 Two Ways To Grasp Concepts

Worldviews can be difficult to detect because they bias the assumptions underpinning the explanations through which we connect with the fabric of reality. Their effect on our ability to discern between our method of questioning, and the objects of intellectual inquiry, can be subtle and unintuitive. This realization, briefly exemplified in Section 3.1 by the worlds ω and π, and the elusive character of worldviews, raise a pressing question: Is there an overarching worldview that we should consider for its potential to influence the genesis and adoption of other worldviews?

A first step to answering this question, consists in analyzing two worldviews that are cosmically universal in the history of human thought, and that symbolize two ways to interpret the relation between Man and Nature. These two universal worldviews are called: *anthropic* and *participatory*. We will not indulge in a detailed description of these two worldviews because this discussion is purely instrumental to the introduction of the observation of the Abstract Mind. Cosmic worldviews, or, more accurately, types of intelligibility, will be examined in greater detail in Parts III and IV. My use of the terms anthropic and participatory is not to be confused with the various forms of anthropic and participatory principles used in Cosmology—but it is indeed possible to spot analogies between my use of these terms and the use of these terms in Cosmology. The anthropic and participatory worldviews are ways to interpret the place of human consciousness in the Universe. As such, they should not be misconstrued as innate cognitive skills.

The Anthropic Worldview

The anthropic worldview uses a naïve and simplistic cosmology we have invented to describe our own mental states at the alert, problem-solving state of consciousness, to explain everything else in the Universe. In the anthropic worldview the human intellect is identified with rational thought and logic, and plays a *referential* role: its processes are direct references to objects and states of affairs in the world. The anthropic worldview is based on the dichotomy "Man vs. Reality", regards Man and its intellect as the "measure of all things"—but not quite in the way Greek philosopher Protagoras (490-420 BC) intended—and ascribes to all the things in Universe the features and properties of rational human thought. But what are the essential properties of human thought at the alert, problem-solving state of consciousness? The anthropic mode of thought appears to be the default way in which we learn to think. We seem to have developed most of the basic cognitive skills that we could ascribe to the anthropic worldview, as a result of an adaptation process to what we consider, rightfully or wrongfully, to be the fundamental themes of the material aspects of human experience, namely: space, time, and beginning and end. These themes, or categories—which we use here as conceptual placeholders rather than as references to their physical counterparts—are summarized in Table 3.1, and determine and influence how we use our core cognitive skills to articulate and comprehend our own thoughts.

3 The Observation Of The Abstract Mind

Space	Time	Beginning-and-End
It is the archetype of the notion of identity, whose function is to help us reason about the intensive and extensive properties of other concepts.	*It is the archetype of the notion of order, and is used to create hierarchies of concepts and descriptions of reality such as causality and universality.*	*It is the archetype of the notion of continuity, and is used to construct the coherence conditions with which we validate our experience.*

Table 3.1: The components of the anthropic worldview.

The Theme Of Space

Space encodes the mental states that relate us to the physical space and to mental realities via an archetypical form of *identity*. The function of this theme is to assign an independent ontological status to the mental and psychological concept of "I" by marking out the perimeter that defines man and world as two ontologically and epistemically distinct entities. At the most fundamental level, the theme of space provides the practical and conceptual demarcation between the "I" and the objects and states of affairs in the world. The keywords of space are: *identity* and *identification*.

The Theme Of Time

Time encodes the impermanent features of the human experience, such as the irreversibility of many phenomena and states of affairs in the world. Time means that we choose to interpret our observation that "things don't go back in time" as an indication that the features of our interpretation of the human experience must apply to the larger context of everything that exists. Thus, time represents that particular way of arranging thoughts according to our experience of the physical time, and is the archetypical form of the concept of order. The keywords of time are: *order* and *direction*.

The Theme Of Beginning-And-End

Beginning-and-End is the mental framework at the basis of the primary features of human existence as we perceive it. The purpose of the concepts of Beginning-and-End is that of providing the tools

3.2 Two Ways To Grasp Concepts

to reason about continuity, especially ontological continuity. The keywords of beginning-and-end are: *continuity* and *evolution*.

Phenomenology Of The Anthropic Worldview

In the anthropic mode of thought, Space, Time and Beginning-and-End form a cognitive pattern around which the human intellect constructs any experience. A way to visualize this pattern is a tree structure, as in Figure 3.2. Three essential features of a tree, the tree structure itself, the nodes, and the uniqueness of a path from one branch to another, provide an effective visual metaphor to interpret the anthropic mode of thought. According to this visual interpretation, the tree structure symbolizes Time and the archetype of order, the nodes symbolize Space and the archetype of identity, and the path between two nodes symbolizes the Beginning-and-End theme, which captures a primitive form of the notion of continuity.

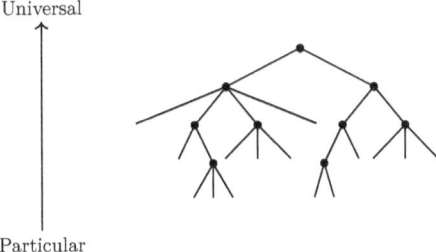

Figure 3.2: Diagrammatic illustration of the anthropic worldview.

Thinking in an anthropic fashion means reasoning in terms of things that must have a principle, an origin. Things that must have a definite identity. Things that either are or aren't. And things that cohere with reality by virtue of being continuous with respect to a representation of the categories of space and time. Another consequence of thinking in an anthropic way is the tendency to assign, or look for, an origin of things. An anthropic mode of thought is often characterized by the search for universal laws and principles, and by the conceptual need of coherence conditions, which are a form of ontological continuity. Examples of anthropic mode of thought are the notions of physical law and law of Nature,

3 The Observation Of The Abstract Mind

which combines coherence with universality. Causality, which combines order and continuity. The structure of natural language, where we find again an implicit coherence condition in the subject-object dichotomy.

The Participatory Worldview

The participatory worldview is a holistic mode of thought built on the notions of unity and wholeness, and on an archetypical conception of the interplay between the part and the whole. In the participatory mode of thought, the human intellect plays a purely *representational* role, and does not employ its features and properties to translate or interpret its own experience of reality. What we mean when we say that in the participatory mode of thought the intellect plays a purely representational role, is that the intellect functions in a way that does not require it to project on or ascribe to reality any of the features and properties of mental states. When the intellect functions in a purely representational fashion, its role is to passively "walk us through our experience of reality" without superimposing on that experience a conceptual frame of reference. How is this possible? How can we dispense with the mental categories of Space, Time, Beginning-and-End, and with the language we have invented to reason about our own mental states and about the world, and still be able to construct mental realities and conceptualize and rationalize our experience? And how does the intellect function passively?

The point here is that what fails when we try to reason exclusively by means of the notion of unity and wholeness, is a characterization of unity and wholeness in the language of Space, Time, Beginning-and-End that we can make consistent with an experience of reality that does not rely on notions of identity, change and order. There isn't a descriptive and explanatory way of reasoning about unity and wholeness in the natural language, because natural language is an anthropic language whose syntax is built to generate infinitely many anthropic thoughts. However, this failure should not be taken as an indication that the human intellect is limited or inadequate to study and comprehend reality in a participatory manner: not in this sense and not in this case. This failure is a property of the system of ideas that we use to conceptualize and describe what our intuition tells us about unity and wholeness. So there is a sharp distinction to be made between

3.2 Two Ways To Grasp Concepts

what we might call the "pure human intellect" and the anthropic or participatory *use* of it: confusing the human intellect with the way we use it to conceptualize our experience is an error. The very explanation that Space, Time, Beginning-and-End are mental categories upon which we formulate any other system of ideas is itself an anthropic description of a way to use the human intellect. It has no necessity other than that of providing a representation of a possible use of the human intellect that makes it consistent with our experience of the world at a level of consciousness which we refer to as the alert, problem-solving state of consciousness.

When the human intellect functions in a participatory manner (Table 3.2), its frame of reference is neither the part nor the whole. Part and whole are in fact the elements of a lexicon that we use to give an anthropic description of a participatory frame of reference. The English language—and in fact any language built on dichotomies such as subject-object—is not designed to convey participatory concepts in a participatory way, because it cannot resolve the epistemic tension created by the subject-object dichotomy.

Part	Relation	Similarity
It refers to the features of the whole, but the concept of whole is not needed *	*It is the archetype of the notion of change, not in time but in perception. Relation is used to reason about parts.*	*It is used to reason about the whole because it defines the relation of the part with the whole.*

Table 3.2: The components of the participatory worldview.

(*) Its presence is due to the fact that this is an anthropic description of a participatory frame of reference.

The Theme Of Part

Part is a concept that characterizes the whole as a manifestation of its parts. It does so in a mereological fashion, by describing how the parts form the whole as a result of a specific set of rules that determine what might be regarded as the evolution of a "system of parts". We will see an example of a dynamical system that illustrates a part-whole relation in Figure 3.4. I think that a good description of the difference between the participatory description of a participatory concept and the anthropic description of a participatory concept comes from the comparison between the notion of whole as an a priory concept, and whole as a property that emerges from certain relations between objects. Let us think of the difference between $2 + 2 = 4$ and the abstract definition of the sum of two integers. When we think about 4 as $2 + 2$, 2 acquires its ontological status from 4 because we conceptualize 4 as made of 2s. In contrast, when we define the sum of two integers abstractly, each integer acquires its ontological status from the definition of sum, which dictates how the elements—that we agree to call parts—can be combined to form new elements. In this case there is no need to have a concept of whole. What we mean by whole is: the interaction between objects according to a certain rule.

The Theme Of Relation

The theme of Relation captures the anthropic notion of change. Relation is a conceptual tool that we need to articulate concepts about parts. As seen in the description of the theme of Part, we need to construct concepts about multiple objects (or parts) because that is how we (appear to be able to) think about the whole as the manifestation of the interaction between objects. The theme of relation encodes a notion of change as a collection of parts.

The Theme Of Similarity

Similarity is a conceptual devices to reason about the whole. What this means is that to coin a notion of whole as that specific property of the interplay between objects, we need to objectify that property, and assign to it its own independent ontological status. This operation of objectification, which we find everywhere in mathematics and in philosophy, is the prototype of the theme of similarity. Another way to describe this is as the ability to think in a "many-as-one" and "one-as-many" fashion. It is the ability to assign to many the ontological status of one, and vice versa. The theme of Similarity is the opposite of the theme of Relation. Similarity assigns oneness to objects based on their mutual interaction, whereas Relation assigns the character of distinctiveness to an object in the form of the notion of change. Going back to the $2 + 2 = 4$ example, if we illustrate the operation

3.2 Two Ways To Grasp Concepts

of adding two numbers—two numbers becoming one—with the diagram $X \times X \to X$, and the operation of change, as in the theme of Relation, with the diagram $X \to X$, then we can visualize the connection between Similarity and Relation as in Figure 3.3-A.

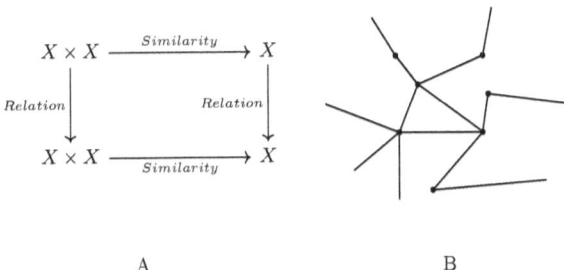

A B

Figure 3.3: (A) Connection between Similarity and Relation, and (B) illustration of the participatory mode of thought.

In Figure 3.3-B we use a graph to visualize the participatory mode of thought. The nodes of the graph represent the Parts, the edges represent Relations, and sets of nodes connected by edges represent the theme of Similarity. Examples of concepts and fields of study that reflect a participatory view are found in contemporary System Theory, in the study of dynamical systems, in the study of biological systems, in Chaos Theory, in certain interpretations of Quantum Mechanics and in cybernetics. Other notable examples are Fuller's Synergetics and Haken's Synergetics.

Phenomenology Of The Participatory Worldview

Examples of participatory mode of thought spring naturally in many physical systems. A physical system that epitomizes participatory thinking is the buzzer circuit in Figure 3.4. When the current from the battery C flows through A, it activates the electromagnet B, which draws A toward itself, breaking the contact at A. As the current stops flowing through B, the electromagnet becomes inactive, releasing A, and the cycle repeats itself. If we try to use causation to explain how a buzzer circuit works, we find that *if* there is contact in A, *then* there isn't contact in A, and vice versa, because the contact in A is destroyed as a result of the creation of the same. In this example the whole, that is, the sound produced by the buzzer, can be grasped only holistically rather than through the sharp knife of logic.

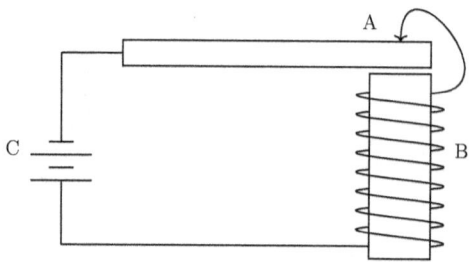

Figure 3.4: A buzzer circuit is an example of a circular cause-effect pattern.

3.3 Observational Structure Of Intentionality

Every intelligent creature on this planet has the ability to *observe* objects and states of affairs in the world. A hungry lion scanning the herd, looking for the weak, a basketball coach looking at the opponent's defensive and offensive tactics, or Davide in his car waiting at an intersection, are examples of *agents* observing *something*. But why do intelligent creatures observe things? The lion wants to eat, the coach wants to win, and Davide wants to reach his destination by car. There is a pattern here (Table 3.3): *observation is the gathering of data to achieve a well-defined goal through a certain strategy*. The lion wants to eat (goal) by preying (strategy) and it feeds its strategy by scanning the herd looking for the weak (data). The coach wants to win (goal) by getting a tactical advantage over his opponent (strategy) and he feeds his strategy by looking at his opponent's defensive and offensive tactics (data). Davide wants to reach his destination by car (goal) so he waits at the intersection (data) for the right moment to cross (strategy). There is no such thing as an observation without a purpose. Let's see this pattern more clearly. Table 3.3 shows what we call the *Observational Structure of Intentionality* of these three examples, namely: the intelligent agent that carries out the observation, the goal of the observation, the strategy to achieve the

3.3 Observational Structure Of Intentionality

goal, and the data to feed the strategy. The mechanism exemplified in Table 3.3 can be further summarized as in Figure 3.5.

Agent	Goal	Strategy	Data
Lion	Eat.	Prey on the weak.	Scan the herd.
Basketball Coach	Increase the chances to win.	Define a game strategy.	Look at the opponent's defensive and offensive tactics.
Davide	Reach my destination by car.	Determine when it's safe to cross.	Look at intersection.

Table 3.3: Examples of Observational Structure of Intentionalitys.

We might be led to think that this is how observations work, and that Figure 3.5 describes the cognitive circuitry that governs how we and other intelligent beings observe things. In part, this is true. Any observation carried out by an intelligent agent can indeed be broken down into the three components goal, strategy and data, and the analytical work of identifying these three components is by all means useful to begin to understand how we think about a goal, how our observations reflect and influence our own way of thinking, and how they determine what we see and what we don't see of a problem.

$$OBSERVATION = \{Data \xrightarrow{Strategy} Goal\}$$

Figure 3.5: The basic structure of observation.

However, if we examine Table 3.3 more carefully, we discover that goal and data are really just meaningless labels that denote the function that certain objects of thought or perception have or acquire in the implementation of a strategy, and we discover also that there are questions that we can ask about those goals, strategies and data that cannot be answered unless we make significant assumptions about each observation. Here are some of these questions: How does the coach select a strategy over another? and How does the coach know that it was the strategy he chose to increase his chances to win the game, and not something else? We

3 The Observation Of The Abstract Mind

might ask the same type of questions about strategy chosen by the lion and, I think, for whoever is driving my car—since I don't have one. In order to answer these questions, the coach would have to know with reasonable certainty that the likelihood to win over a certain opponent is affected by the strategy he chooses to counter his opponent's offensive and defensive tactics. Similarly, the lion would have to know that praying on the weak pays off over other hunting strategies. There is an intuitive way to express these ideas: the likelihood to achieve a goal changes with the strategy. And there is a precise way to put the same observations: there is a causal link between a strategy and the achievement of a goal. In this discussion, we do not consider the possibility of an agent taking a trial-and-error approach to achieve his goals, even though this can be a coherent technique conducive to a successful strategy. This position is consistent with successful observations, in which goal, strategy and data have reached a sufficient level of coherence that determines the sustainability of that particular method of observation, as opposed to random observations where the causal link between strategy and success is practically, and statistically, irrelevant.

It should be clear that causal relations like those just described between strategy and goal must hold also between strategy and data. That is to say: there must be a reason to believe that certain objects or states of affairs provide the data needed to activate a strategy that is believed to lead to the desired goal. What we have discovered in this analysis is that: *each observation encodes a belief system that determines the strategies that can be seen, evaluated and chosen to achieve the goal set by the observation, and that goal and strategy cannot be validated within the conceptual framework used to define the observation.*

Table 3.4 summarizes the structure of the belief system just described, which we call the *Validation System* of an Observational Structure of Intentionality. Given an observation, the process of identifying and describing its Validation System reveals the deep structure of its underlying belief system. The function of a Validation System's is to define a *model of truth* upon which a given Observational Structure of Intentionality bases the designation of success. The determination of an Observational Structure of Intentionality consists essentially in answering these questions:

- How do we decide when the goal of an observation is

3.3 Observational Structure Of Intentionality

achieved?

- How do we choose a strategy to achieve the goal of an observation?
- How do we choose the data that activates the strategy we chose?
- How do we measure the effectiveness of a chosen strategy?

Agent	How the goal is validated	How the strategy is selected	How the strategy is validated
Lion	*Prey killed and ready to eat.*	*To maximize the advantage over the prey.*	*Preying on the weak increases the likelihood to eat strategies.*
Basketball Coach	*Win or improve game.*	*To counter the opponent's offensive and defensive tactics.*	*Countering the opponent's offensive and defensive tactics increases the likelihood to win over the opponent.*
Davide	*Destination reached by car.*	*To drive safely in the city.*	*Paying attention at intersections increases the likelihood to reach the destination.*

Table 3.4: Examples of Validation System of observational structures.

When we can describe an Observational Structure of Intentionality and a Validation System consistent with the same, the network of causal relations that constitute its underlying belief system becomes evident. The description of a Validation System reveals the assumptions about how and why by doing or thinking about something in a certain way and in a certain context we get a certain outcome. Note that nothing in this line of reasoning leans toward a quantitative treatment of the Observational Structure of Intentionality: the *measurement* of the effectiveness of a strategy is, in other words, to be intended as part of the model of truth relative to the context where the strategy is defined.

3 The Observation Of The Abstract Mind

In the next section we apply these ideas to the Abstract Mind to obtain its Observational Structure of Intentionality.

3.4 What Does It Mean To Observe The Abstract Mind?

Let us recall from the introduction to Part I, that one of the assumptions of this research, is that we have direct access only to the Symbolic Intuitions of Reality.

A consequence of this premise, is that the observation of the Abstract Mind can only be an indirect observation: it can only be the observation of the Abstract Mind *through* a Symbolic Intuition of Reality. This conclusion leads to the question: What types of Symbolic Intuitions of Reality should be observe to study the Abstract Mind? To answer this question, let us recall from "The Structure Of The Symbolic Intuition Of Reality" in the introduction to Part I, that a Symbolic Intuition of Reality consists of two parts, intuitive and epistemic, and that:

- The *intuitive* part symbolizes a specific *type of mind-world transaction* by defining what is *intelligible*, and denotes the fundamental way in which we become aware of an object of thought or perception as a result of the interaction between the human mind and the World.

- The *epistemic* part contains a model of the cognitive structures at the foundation of human thought, which we identify with the meaning of the following 5 terms: *knowledge, comprehension, meaning, explanation* and *consciousness*.

Thus, the problem of defining the types of Symbolic Intuitions of Reality that are relevant to the observation of the Abstract Mind, is the problem of defining what *changes* in the intuitive and epistemic components of a Symbolic Intuition of Reality reveal or are conducive to an understanding of the inner workings of the Abstract Mind.

The changes in the *intuitive* part of a Symbolic Intuition of Reality are profound changes in the fabric of the mind-world transactions, whose significance and magnitude is defined by their implications on the fundamental structure of human thought. For this reason, these changes are not interesting at this early stage of the observation of the Abstract Mind, because they alter too much

3.4 What Does It Mean To Observe The Abstract Mind?

of what we intend to observe—we will have to wait until Parts III and IV to fully appreciate the magnitude and implications of these changes. As a result, the changes in the Symbolic Intuition of Reality we will focus on in this analysis are the changes in the *epistemic* part. There are two types of change in the epistemic part that are relevant to the observation of the Abstract Mind, and they are depicted in the diagrams of Figure 3.6.

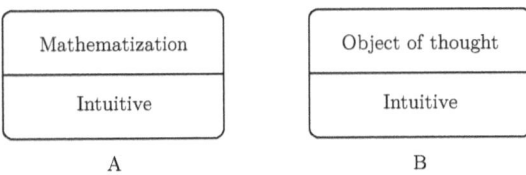

Figure 3.6: Two special types of Symbolic Intuition of Reality: phenomenology of mathematization (A), and awareness of an object of thought (B).

Phenomenology of mathematization

Figure 3.6-A denotes a Symbolic Intuition of Reality where the conceptual content of a mind-world transaction is transformed into *experience* by the action of an epistemic framework called *mathematization*. When a mind-world transaction is mathematized, its conceptual content is projected onto a zoo of mathematical objects that encode the features and properties of an overarching system of thought that we call Mathematics, where knowledge, comprehension, meaning, explanation and consciousness have a very specific meaning and significance. The peculiar features and properties of our mathematical experience of the World are generally referred to as our *mathematical knowledge of the World*, where the term knowledge is used as a mass noun to denote an array of conceptual hierarchies through which we comprehend the content of our symbolic intuition of the World according to the features and properties of a multitude of mathematical objects that represent various conceptualizations of quantity, order, pattern, symmetry, structure, abstract relation, correspondence, transformation, shape, infinity, and many more.

In Chapter 4 we will focus our attention on the observation of the phenomenology of this special type of Symbolic Intuition

3 The Observation Of The Abstract Mind

of Reality: the observation of *phenomenology of mathematization*. The name given to the observation of the phenomenology of mathematization is "The Architecture Of Mathematics", and is meant to emphasize that the basic structure of mathematization appears to display higher-order regularities that reflect, to a certain extent, some of the inner workings of the Abstract Mind. We will introduce a theoretical device based on the Observational Structure of Intentionality called The 4 Lenses, with which we will observe the cognitive and evolutionary view of the mental processes, the intellectual attitudes, and the styles of reasoning involved in the genesis of our mathematical knowledge of the World. The general Observational Structure of Intentionality of this type of observation is summarized in Table 3.5.

Goal	Strategy	Data
Explain the thought processes that give rise to the mind's ability to mathematize.	*Look for patterns in the way we define mathematical gadgets.*	*Look at the entire zoo of mathematical gadgets.*

Table 3.5: Observational Structure of Intentionality for the observation of the phenomenology of mathematization.

The Validation System for the observation of the phenomenology of mathematization is described in Table 3.6, and captures the basic coherence conditions we require for an observation to be consistent with the development of mathematical thought as we know it today.

3.4 What Does It Mean To Observe The Abstract Mind?

Goal validation	Strategy selection	Strategy validation
Mathematics as we know it today can be derived logically from the concepts identified by the strategy to explain the mind's ability to mathematize.	*A strategy to explain the mind's ability to mathematize must provide the minimum set of concepts needed to define any mathematical gadget in a way that is consistent with the current conception of mathematics.*	*Looking for patterns in the way we define mathematical gadgets is an inquiry method that is consistent with the current conception of mathematics, which is entirely based on the syntax of formal systems.*

Table 3.6: Validation System for Table 3.5.

There are more general and more thorough ways to describe a Symbolic Intuition of Reality, but we will have to wait until Part IV to introduce them. The characterization of the phenomenology of mathematization via The 4 Lenses is a first, general enough way to grasp the deep structures of thought involved in the creation of our mathematical intuition of reality that will help us comprehend practically the level of abstraction required in the study of the Abstract Mind.

Awareness Of An Object Of Thought

The diagram of Figure 3.6-B denotes a Symbolic Intuition of Reality where the epistemic part has a structure called *object of thought*, and where the conceptual content of a mind-world transaction is transformed into the most basic experience: the *awareness* of that object of thought. The awareness of an object of thought is the most elementary mechanism through which we humans are able to relate symbolically to reality, because the purpose and scope of the mind-world transaction from which the experience of awareness originates is fully defined by the metaphysical, irreducible correspondence between an object of thought, and the *intuitive knowledge* of that object.

The study of the awareness of an object of thought is the study of a special class of Symbolic Intuitions of Reality that mark the conceptual and epistemic boundary between the human mind and the World. This boundary is denoted in this research by the theoretical limit of what a human mind can grasp, and signals the separation between what's intelligible and what is incomprehensible in a given state of consciousness. A description of the awareness of an object of thought is, for this reason, an indispensable step in the exploration of the workings of the Abstract Mind because it signifies, so to speak, the contact surface that separates human consciousness from the World.

There is an astrophysical analogy to illustrate the relation between the study of the awareness of an object of thought and the workings of the Abstract Mind. Let us think of the Abstract Mind as the Sun. By studying the surface of the Sun and the radiations it emits, Astrophysicists can deduce some of the processes occurring in the interior of the Sun. The framework used to construct formal explanations about the interior of the Sun is of course what we call Physics, and is based on a worldview according to which everything that exists is made of matter and energy, and exists in space and time, and obeys certain universal laws with which it is possible to explain deductively the phenomenology of the known part of physical reality. Much like Astrophysicists need sophisticated tools to pry into the workings of the stars, and a worldview and a theory of Physics to construct formal explanations, to study the awareness of an object of thought we will need special conceptual tools, a worldview about the basic constituents of the mind-world interplay, and a framework to describe the physics of ideas that gives rise to the phenomenology of the Abstract Mind which we call Symbolic Intuition of Reality. We will develop a set of abstract tools to observe the genesis of an object of thought, and a framework to explain how an object of thought becomes the experience of awareness through which we are able to connect symbolically to reality. The study of the the inner workings of the Abstract Mind is essentially the study of what we can observe on its surface, and of the conceptual constructs that emanate from it: the phenomenology of the awareness of an object of thought.

Bibliographical Notes

This chapter is based on Lo Vetere 2008a, Lo Vetere 2013b, and contains a synthesis of several themes: from the Phenomenology of Perception, to Process Philosophy, Mathematical Ontology, Mathematical Epistemology, to the Philosophy of Physics, Philosophy of Mathematics, Philosophy of Time and Logic of Time.

The theme of the observation of mathematization stems from

3.4 What Does It Mean To Observe The Abstract Mind?

the search for a connection between intentionality and the prototypical structure of observation, and from the need to define a frame of reference to give a process-based description of the Symbolic Intuition of Reality. From the perspective of that privileged frame of reference, mathematization becomes an instance of an observational process, and the problem of describing mathematization becomes the problem of describing the phenomenology of an observational process.

The themes of Process Philosophy, and the discussion about the anthropic and participatory modes of thought were influenced, among others, by Baggott 2003, W. C. Salmon 1984, W. C. Salmon 2006, Stapp 2009, Stapp 2011, Benthem 1991, Whitehead 1967, Whitehead 1979, Rescher 1996, Rescher 2000, Fuller, Loeb, and Applewhite 1982, Haken 1984, Sherburne 1981, Feynman 1965 and Bateson 1979 for the buzzer circuit example.

Analogy With Quantum Entanglement And Nonlocality

We observe a phenomenon, perhaps, analogous to the one-dimensional world example in quantum physics: it's called quantum entanglement. Quantum entanglement refers to the physical phenomenon in which the state of a particle can be described only with respect to the state of a group of other particles. Entangled systems of particles have the property that events that change the state of one particle of the system—such as measurements—change instantaneously the state of all the other particles of the entangled system. The difficulty to explain how the information that the state of a particle has changed is instantly transferred to other particles is a direct consequence of a frame of reference in which particles are themselves described an distinct entities.

Chapter 4

The 4 Lenses And
The Architecture Of Mathematics

> The aim of mathematics is to explain as much as possible in simple terms.
>
> Atiyah 1978

This chapter introduces a theoretical device called The 4 Lenses, by means of which we can observe the ways in which the human mind mathematizes.

When we observe mathematics through The 4 Lenses, we see a cognitive and evolutionary representation of the mental processes, the intellectual attitudes, and the styles of reasoning involved in the genesis of mathematical ideas. By an evolutionary representation of mathematics, we mean the capacity to comprehend how certain structures of thought, certain styles of reasoning, certain belief systems, certain worldviews, and certain ways to come at the world with the demand to reveal deeper structures of reality, emerge as a result of crucial changes in our conception and understanding of mathematics.

The 4 Lenses is one of the first conceptual tools I created back in 1995 to begin to explore the problem of studying the human mind's ability to mathematize. The definition of The 4 Lenses was motivated by the need to have a method to systematically capture the mind's capacity to articulate thoughts—now and in

4 The 4 Lenses And The Architecture Of Mathematics

the future—and to classify the attitudes and worldviews that evolve with human intelligence as a result of the evolution of its metaphysical structures of thought. A detailed account of The 4 Lenses would require an entire book, but for the purpose of the study of the Abstract Mind, it is sufficient to describe the style of reasoning behind The 4 Lenses. To situate the discussion about The 4 Lenses, it is useful to review briefly the traditional approaches to the problem of defining the nature of mathematics. These traditional approaches are eloquently described by the questions that philosophers and mathematicians have been asking over several centuries in their quest to understand the nature of mathematics. We will use these traditional questions about the nature of mathematics as a frame of reference against which to define The 4 Lenses.

The Traditional Quest To Comprehend The Nature Of Mathematics

Traditionally, the quest for deeper understanding of mathematics has taken the form of a search for the practical and conceptual origins of mathematics. The questions that philosophers and mathematicians have been asking can be grouped into two main categories: the questions about the *structure* of mathematics, and the questions about the *nature* of mathematics.

The following questions give the gist of the type of problems raised by the study of the structure of mathematics:

- How is mathematics organized and what are the relations between its parts?

- Are the formalisms of mathematics based on the facts, or is mathematics just an entirely formal game which happens to be very effective in helping us understand the world?

The last question leads to

- If the formalisms of mathematics are based on the facts, what is the construction of mathematical formalisms based on?

- Why are there many foundations of mathematics, and what does it mean that there are many?

- What does motivate the development of mathematics?

Introduction

The questions about the structure of mathematics are all, not surprisingly, nuances of the fundamental problem of having a unifying view of mathematical thought that can be made consistent with the practice and the development of mathematics. These are very complex questions because they attempt to characterize mathematics from within the many horizons defined by the polymorphic appearances of mathematical concepts throughout the universe of mathematical thought, where it is very difficult to see the overarching cognitive structures from which those mathematical concepts originate. To understand why philosophers and mathematicians were asking those "questions from within mathematics", we need to consider the general philosophical positions from which those questions originate. Those general philosophical positions are seldom explicit, and pertain to a cosmology of thought in which mathematics is, in a sense, deprived of its mental origin, and treated like a cognitive skill.

To investigate the nature of mathematics, we ask:

- What is mathematics and what is its function?
- What is the origin of mathematics?
- What are the objects of mathematics and what is their ontological status?
- Is mathematics invented or discovered?
- ...and the omnipresent puzzle Why is mathematics so effective in describing the physical world?

These are questions about the phenomenology of a Platonic world, where the problem of the nature of mathematics is articulated as the problem of defining the relations between the presumed ontological status of mathematical concepts, and the reality we access through those concepts.

In contrast, with The 4 Lenses we *interpret* mathematics as a step in the evolution of the cognitive processes through which our species uses certain abilities of the human mind to relate to the world on a conceptual level. From the perspective of The 4 Lenses the human mind is an organ, mathematical thought is the primary manifestation of the metabolism of this organ, and concepts are the steps each metabolic process consists of. As such, mathematics reflects a moment in the evolution of human intelligence, and in the evolution of the metaphysical foundations of human thought, in much the same way in which the function and the metabolism of the organs in the human body reflect a moment in a process of biological adaptation. The questions we ask about mathematics from the perspective of The 4 Lenses are therefore not about the nature or structure of mathematics, but about the cognitive functions that give rise to the phenomenology of mathematics, and about the role that those functions play in the broader context of the evolution of the mind-world interaction.

The 4 Lenses are called Generative, Performative, Motivic and Methodic, and represent four basic and irreducible ways in which the human intellect seems to be able to articulate mathematical thought at the most fundamental level. These four modes of thought are reflected in the four directions along which we develop the cognitive structures that give rise to mathematical thought. Each lens is characterized by the 5 elements in Table 4.1.

Element	Description
Theme	*The theme characterizes the mind's ability to mathematize revealed by the lens.*
Suffix	*It may or may not be present, and characterizes the type of process codified by the lens.*
Spectrum	*The elements of mathematical thought revealed by the lens.*
Thought Process	*The set of cognitive processes involved in the workings of the lens.*
Observational Structure of Intentionality	*The observational structure that corresponds to the thought processes codified by the lens.*

Table 4.1: The structure of The 4 Lenses.

The *theme* describes the features of mathematical thought revealed by the lens. Almost all lenses have a *suffix* associated to the types of pictures of mathematical thought that they produce. The function of the suffix is purely mnemonic, and does not constitute a classification method of the views of mathematics produced by the lens. The image of mathematical thought produced by a lens forms a *spectrum*, in much the same way in which a beam of light dispersed by a prism into its spectral components forms an ordered sequence of colored lines or a rainbow. The workings of a lens are described by the specific *thought processes* responsible for the creation of the image of mathematics produced by the lens. To understand these thought processes we ask: How do we produce the image of mathematics described by the *theme*? An answer to this question explains how the human mind applies certain worldviews and belief systems to define and apply the type of conceptualization processes that produce that *spectrum*. The focus of the answer, though, is on the mechanism through which worldviews and belief systems influence or determine the genesis of those images of mathematics, not on the spectrum itself, which is meant to vary with our conception of mathematics. These thought processes are described via the *Observational Structure of Intentionality*.

4.1 The Generative Lens

When we observe mathematics through the generative lens, we see systems of ideas that allow us to formalize mathematics.

Theme

The theme of the generative lens is *modeling*. The generative lens makes us see the connections between a general conception of mathematics and a specific system of ideas to *realize* that conception, as depicted in Figure 4.1.

A formal system that can be used to formalize all or some of the central concepts of mathematics is called a *foundation of mathematics*. Crudely speaking, the quest for a foundation of mathematics can be thought of as the opposite of what we did in Section 1.1 to define a formal system. There, we defined four entities—Alphabet, Syntax, Axioms and Inference Rules—and used them to generate strings of characters of the Alphabet called theorems. In contrast, in the search for a foundation of mathematics, we choose a collection of theorems that represents a portion of the edifice of mathematics, and ask: What Alphabet, Syntax, Axioms and Inference Rules can be used to prove these theorems? The action of the generative lens on mathematics is twofold: on the one hand, it produces foundations of mathematics, and on the other hand, it reveals the metaphysical structure that

4 The 4 Lenses And The Architecture Of Mathematics

defines the cognitive patterns by means of which we seek and construct those foundations.

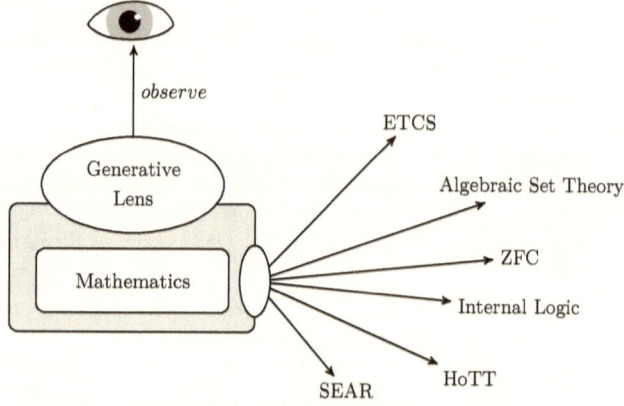

Figure 4.1: Part of the spectrum produced by the Generative Lens.

Suffix

There isn't a suffix associated to this lens.

The Generative Thought Process

The thought process at the basis of the generative lens is the construction of models of the concepts at the foundation of the modern conception of mathematics. In particular, the search for a foundation of mathematics has two objectives which can be categorized as follows:

1. Provide a description of what theorems can be formalized and proved by what formal systems.

2. Provide a description of the fundamental concepts such as number, set, function, shape etc. that make up mathematics as we know it today.

The first category is concerned with the proof-theoretic implications of a foundation of mathematics, and is, therefore, metamathe-

4.1 The Generative Lens

matical in nature—metamathematics is the study of mathematical theories with mathematical methods.

The second category is about the conceptual and practical implications of a model-theoretic view of the basic concepts of mathematics. This view has profound implications on the foundations of mathematics, because by redefining or eliminating or replacing certain basic concepts of mathematics, it redesigns a conception of mathematics. For example, this model-theoretic view of mathematics produces Topoi, which are mathematical objects in which mathematics itself can be defined, or produces mathematics not based on Set Theory.

Observational Structure

The Observational Structure of Intentionality of the generative lens defines the strategies that are suitable to characterize foundations of mathematics.

Goal	Strategy	Data
Characterize the theories that can be used as a foundation of mathematics.	*Construct a model of a basic concept of mathematics (e.g. Set), or replace it.*	*The structure of the basic concepts of mathematics.*

Table 4.2: The Observational Structure of Intentionality of the Generative Lens.

Examples

Notable examples of foundations of mathematics that we see through the generative lens (Figure 4.1) are various flavors of Set-theoretic foundations (e.g. Zermelo Fraenkel with the Axiom of Choice, Von Neumann-Bernays-Gödel), Category-theoretic foundations (e.g. Elementary Theory of the Category of Sets, Topos Theory, Algebraic Set Theory, Internal Logic) and Type-theoretic foundations (e.g. Homotopy Type Theory). The examples that follow offer a brief description of some these theories. I have chosen to describe some of these foundations for their historical, mathematical and philosophical importance, because they represent a fairly comprehensive spectrum of the contemporary approaches to the

4 The 4 Lenses And The Architecture Of Mathematics

foundations of mathematics, and for the insights they provide on the problem of the unification of the foundations of mathematics.

ZFC

The Zermelo-Fraenkel set theory with the axiom of choice, or ZFC for short, is one of today's most accepted foundations of mathematics. ZFC is a systematization of the naïve set theory, which is based in what is probably the most intuitive way to think about a set as a "bag of objects". However, as it turned out at the beginning of the twentieth century, while the notion of set as a "bag of objects" is useful in many practical applications, it is not suitable to construct mathematics.

ETCS

The Elementary Theory of the Category of Sets, or ETCS, is an axiomatic formulation of set theory in the language of category theory proposed by W. Lawvere in 1964. ETCS is an integral part of W.Lawvere's seminal work on the categorial foundations of Mathematics he started in the early sixties. The importance of ETCS is both mathematical and philosophical, like many of Lawvere's works, because it formulates in categorial language the idea that Set Theory should not be based on a membership relation but on isomorphism-invariant structures.

Algebraic Set Theory

Algebraic Set Theory, or AST for short, is a category-theoretic formulation of models of set theory developed by André Joyal and Ieke Moerdijk in 1988 and published in 1995. The insight at the basis of AST is that models of set theory are algebraic structures and that set theoretic conditions are related to algebraic ones.

Internal Logic

The intuition at the basis of what is called Internal Logic—also known as Categorial Logic—is that, since mathematics can be written in the language of logic, if we construct a mathematical object where we can define logic, then we can define the entire—or most of the—edifice of mathematics inside that object. This idea is implemented by Topos Theory. A Topos is a Category with "enough" structure that allows to define a logic and the notions of ordinary mathematics.

HoTT

As a constructive foundation of mathematics, Homotopy Type Theory, or HoTT for short, is a homotopy interpretation of Type Theory in which the notion of identity—the identity type—is interpreted in a homotopical fashion via the so-called axiom of univalence. Intuitively, HoTT is based on the observation that there is a correspondence between Dependent Type Theory and certain objects of Higher Category Theory such as groupoids, and uses this correspondence to define a generalization of Intensional Type Theory, also known as Intuitionistic Type Theory or Per Martin-Löf Type Theory, a type theory created by Swedish mathematician Per Martin-Löf in 1972. The interest in HoTT rests, besides its theoretical importance, in its application in the creation of automated proof assistants.

4.2 The Performative Lens

The performative lens makes us see how we deal with mathematical reasoning and practice.

Theme

The theme of the performative lens is *interpretation*. The performative lens makes us see how our *worldview* and *belief systems* determine, influence, magnify and undermine our ability to mathematize.

The name of this lens is borrowed from the Philosophy of Language, where the notion of performative utterance indicates those sentences that change what they describe. Indeed, the ways in which the various schools of thought define their conceptions of mathematics do modify the conception and practice of mathematics, given that each approach to the formulation of mathematical concepts is both descriptive and performative. Figure 4.2 illustrates some of the philosophical positions that we see through the performative lens.

Suffix

The action of the performative lens is captured by the suffix -*ism*, as in constructivism, nominalism, pragmatism, physicalism,

4 The 4 Lenses And The Architecture Of Mathematics

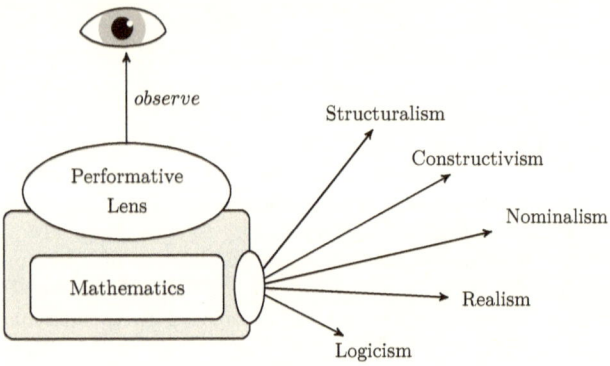

Figure 4.2: Part of the spectrum produced by the Performative Lens.

realism and logicism.

The Performative Thought Process

The performative lens reveals the basic belief systems and philosophical positions that influence and determine how we mathematize and how we interpret the nature, the function, and the practice of mathematics. The questions we ask when we observe mathematics with a performative lens are therefore of an ontological and epistemological nature, such as for example: What should the real mathematics be? Can we accept the existence of infinite mathematical objects? Does it make sense to prove the existence of a mathematical object without knowing how to construct it? The very act of asking these questions, the reasons to believe that asking these questions makes sense in the epistemic context in which we develop our conception of mathematics, and the arguments developed to explain what counts as an answer to these questions are all part of the picture of mathematics that we see through the performative lens. In essence, the structure of the performative lens is determined by the cosmology of thought through which we come at the problem of defining mathematical thought, and can be understood directly via the concepts that form each cosmology as described in the next section.

4.2 The Performative Lens

Observational Structure

The performative lens extracts the philosophical positions underlying the definition of mathematical concepts and the practice of mathematics. Its observational structure defines the views that are consistent with the styles of reasoning and the belief systems that characterize a given conception of mathematics.

Goal	Strategy	Data
Characterize the styles of mathematical reasoning and the views that define a conception of mathematics.	*Identify and classify the worldviews and belief systems underlying a conception and a formulation of mathematics.*	*The conceptions and the formulations of mathematical concepts, and the corresponding development of mathematics.*

Table 4.3: The Observational Structure of Intentionality of the Performative Lens.

To comprehend how the performative lens works, it is necessary to identify the fundamental concepts revealed by the lens, and the mechanisms by means of which those concepts shape mathematical reasoning. These concepts at the basis of a conceptualization of mathematics define and determine at the most fundamental level our conception of mathematics by marking out the boundary between what concepts and views should be construed as mathematical and what shouldn't. This lens functions by defining the ontological status of the concepts that form a conception of mathematics, and by characterizing the limits and the traits of their epistemology in relation to broader background philosophical positions, such as the nature of concepts, the nature of mathematical knowledge, a definition of world, reality, truth and many more. All of these background presuppositions constitute the arena in which the mind develops the ontological and metaphysical assumptions that lead to the various formulations of the nature of mathematics, and to the conceptions and development of mathematics in general. The overarching philosophical landscape in which the performative lens operates consists of a general cosmology of thought in which any argument is articulated in terms of the concepts of *truth, existence*, the *nature of concepts, reality, syntax* and *structure*. The choice and use of this cosmology of thought to define a conception of mathematics is a reflection of the intellectual evolution of our species, and represents the predominant understanding of the function and role of mathematics in our culture.

Examples

The nature and limits of mathematical reasoning as the art of creating abstractions of the world are the leitmotif of the performative views of mathematics. There is an important relation highlighted by the performative lens, between what we see of mathematics through this lens, and our ability to give an account of mathematics as the most successful cognitive technology created by our species. This relation emerges when we analyze the main themes in the Philosophy of Mathematics, such as the meaning of truth, the relation between mathematical truth and objects in the real world, the concepts of number, infinite and space, and the abstractions of algebra, logic, geometry and topology. Below is a list of what I think are the most interesting and influential approaches to the development of mathematical thought. Each presentation of mathematics given by the performative lens is characterized by the fundamental concepts used to motivate and construct a specific conception of mathematics. For example, a view on the nature of truth values gives rise to constructivism, a view on the nature of concepts gives rise to pragmatism, a view on the ontological status of concepts give rise to nominalism and so on.

Constructivism And The Nature Of Truth Values

The central idea of constructivism is to do mathematics without the principle of the excluded middle—the notion that every truth value is either true or false. Underlying this idea of mathematics is the belief that constructive mathematics is in many practical and theoretical ways the "real" mathematics. At the beginning of the 20th century, when the discovery of various paradoxes (e.g. Russell's paradox) triggered a collective search for consistent foundations of mathematics within mathematics itself—the so-called foundational crisis of mathematics—a number of mathematicians embraced a conception of mathematics which rejects the principle of the excluded middle and the axiom of choice—the notion that it is possible to select exactly one object from any given collection of sets–because these notions lead to nonconstructive mathematical proofs—proofs of the existence of mathematical objects that do not construct the objects whose existence they prove. With the advent of Topos Theory in the second half of the 20th century, our conception of logic had to be profoundly redefined to accommodate for objects that only partially exist, and to incorporate various alternative foundations of mathematics, and, as a result of this deeper and broader conception of logic, new types of constructivism emerged. Topos Theory taught us how to construct versions of mathematics in which most of the notions of logic are applicable only locally, and where the axiom of choice and the principle of the excluded middle are not valid. As observed by Lawvere and others, any topos with an extra structure called Natural Number Object—a structure that mimics natural numbers—has an internal logic which fails to satisfy the principle of the excluded middle and the axiom of choice, and can be used to describe most of mathematics. The implications of this discovery are that even non constructivist mathematician have to care about what can be proven without the principle of the excluded middle, because only constructive proofs can be interpreted in an arbitrary topos with the natural number object property.

Nominalism And The Ontological Status Of Mathematical Objects

Mathematical nominalism is an anti-realist and anti-platonist view of the ontological status of mathematical objects, according to which either mathematical objects do not exist at all, or they do exist in a non-spatial way and without causal powers. The aim of mathematical nominalism is to show that any interpretation of mathematics does not need or rely on any commitment to the existence of mathematical objects, that is: *there is no need to believe in the existence of mathematical objects to make sense of mathematics.* The first form of mathematical nominalism regards mathematical objects from a strong anti-platonist perspective, by denying the tenets of the platonic cosmology of concepts—the view that mathematical objects are ontologically objective, are not located in space and time, and have no causal connection with us. The first form poses two types of problems: either mathematics needs to be reformulated to avoid any commitment to mathematical objects, or it is necessary to explain how the use of a mathematics conceived in a platonic way does not involve any

4 The 4 Lenses And The Architecture Of Mathematics

commitment to the existence of mathematical objects as postulated by Plato. The second form of mathematical nominalism is a milder version of the first form, and replaces mathematical objects with other entities suitable for the application of mathematics, but results in a plethora of views on what these alternative objects should be, and on how they should be applied "mathematically"—these views are, among others, forms of mathematical fictionalism, deflationary nominalism and modal structuralism which we do not cover here. From the perspective of the The 4 Lenses, what matters in regards to the nominalistic attitude towards mathematization, is the ascription to mathematical objects of certain categories of existence, and the rationale and the worldviews supporting this process. This problem is less naïve than might appear at first glance. What we are suggesting, is that the type of problem raised by nominalism in regards to the interpretation of the ontological status of mathematical objects, is an expression of the specific structure of a cosmology of thought which does not fully define the role and the limits that a conception of existence should play in the development of mathematics. We will revisit this problem in Parts III and V where we will introduce the notion of objectivity in Interaction Theory (Section 10.6), which, together with the discussion about "Mathematical Physicalism" (Section 21.1) offers an answer to this conundrum.

Pragmatism And The Nature Of Concepts

The core idea of pragmatism is that concepts should be defined by means of their practical consequences. The pragmatic method of investigation consists in maintaining a clearly defined scope of the contents of concepts, and in clarifying and verifying each hypothesis by identifying its practical consequences. The focus of the pragmatist view is the mechanism by means of which concepts acquire their ontological status through the way they are used in reasoning. This approach is radically different from the platonic view in that it treats concepts as a technology rather than as abstract entities which happen to be useful to relate us to the world.

Logicism And Syntactical Nature Of Reality

Logicism is the idea that some or all mathematics is reducible to symbolic logic. The role of logic in logicism is not only technical, it is also metaphysical. At the very core of logicism, there is the belief that logic describes the laws of thought, and that as such, its role in describing the structure of human knowledge should be extended far beyond mathematics to cover philosophy and any other metaphysical dimension of the human intellectual activities. Probably the most important philosophical and mathematical implication of logicism is that is computationalizes most of the concepts we have created to relate us to our own existence and to the World. For logic means syntax, and syntax means computation. By ascribing to the horizon of human thought a syntactical nature, logicism essentially reduces the nature of thought and its manifestations to a process that can be mechanized and reproduced via computational methods.

Structuralism And The Holistic View Of Mathematical Thought

Mathematical structuralism is the view that the objects of study of mathematics are structures. This philosophical view, therefore, ascribes to all mathematical objects the ontological status of structure, and ties the discussion of the nature of mathematical objects to the broader philosophical context of structural realism, the philosophical view according to which either the only epistemic access to the World available to our mind is structure, or the World is itself made of pure structure. The term structure is used here to signify the relations that connect the parts to the whole, and the relations that define the parts via their being part of a whole, and the whole as the entity defined by its parts. The structuralist orientation of contemporary mathematics, propelled by the unifying views of the edifice of mathematics provided by a category-theoretical treatment of virtually any known mathematical concepts, and by the set of powerful analytical tools of categorification and decategorification, can be taken as the modern and most comprehensive model of mathematical structuralism.

4.3 The Motivic Lens

The motivic lens makes us see the different ways in which we mathematize.

4 The 4 Lenses And The Architecture Of Mathematics

Theme

The theme of the motivic lens is *process*. The motivic lens makes us see how different ways to *articulate a basic mathematical intuition* produce distinct but related pictures of that intuition.

Mathematics as we know it today consists of a collection of established procedures to produce mathematical representations of ideas, objects and states of affairs in the world. In the context of The 4 Lenses, the terms procedure and process are synonyms of Observational Structure of Intentionality in a way that will become apparent as we progress in the discussion. We have seen in Section 3.3 that the notion of Observational Structure of Intentionality encodes the purpose and strategy underlying the intentional act of directing one's attention at something. At a cognitive level, the Observational Structure of Intentionality of the motivic lens captures the dynamical interrelations between *observation* and *conceptualization*, and how these influence each other in a process that surfaces as directed thought. Figure 4.3 captures the most important ways in which we can look at the world mathematically. Algebra, Geometry, Topology etc. are not macro-categories for the classification of mathematical reasoning, they are peculiar styles of reasoning characterized by distinct ways of understanding, describing and solving problems.

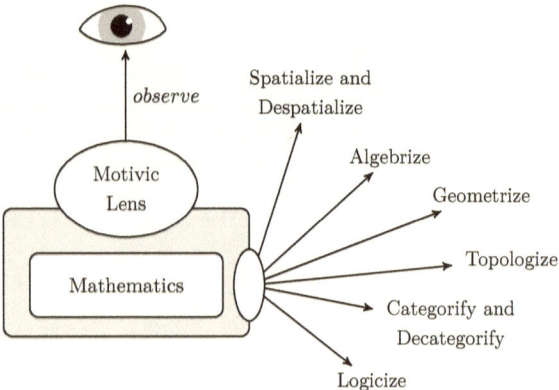

Figure 4.3: Part of the spectrum produced by the Motivic Lens.

4.3 The Motivic Lens

Suffix

The suffixes that capture the effect that the motivic lens has on mathematics are *-ize* and *-ify*, because they convey the general notion of process.

The Motivic Thought Process

The motivic lens conveys the distinct strategies with which ideas are translated into mathematical language. If we look carefully at the various styles of reasoning and at the dialectics of thought employed in the modern conception of mathematics, we recognize a handful of very specific ways in which the technology of formal systems is applied to create mathematical structures. The motivic thought process reveals precisely those peculiar ways of using formal systems, by deconstructing the relations between the ways in which the components of a formal system are used to define mathematical objects, and the description that that mathematical object gives of a certain reality.

Observational Structure

The Observational Structure of Intentionality is useful to explain the workings of the motivic lens, because it can be used to give a tactical description of the cognitive processes involved in mathematization.

Goal	Strategy	Data
Characterize the structure of the cognitive processes through which a concept is mathematized.	*Identify and classify the essential structures through which mathematical objects are constructed.*	*The elementary steps through which mathematical objects are constructed.*

Table 4.4: The Observational Structure of Intentionality of the Motivic Lens.

To characterize the structure of the cognitive processes through which a concept is mathematized via the Observational Structure of Intentionality (Table 4.4), we look at the elementary steps of the processes of mathematical construction, as defined by a specific

4 The 4 Lenses And The Architecture Of Mathematics

representation of a problem, in order to identify and classify the abstract patterns that form the structure of those processes. The outcome of this analysis is a description of the spectrum of the motivic lens, which outlines the key distinctions in the style of mathematical reasoning that give rise to the main branches of mathematics. Of course, the motivic lens can be used recursively on each line of the spectrum it produces, to reveal how further nuances of the same paradigm produce, for example, Algebraic Geometry, or Algebraic Number Theory, of Higher Geometry— Geometry in the context of Higher Category Theory—and many more. The motivic lens reveals the fundamental *processes* that characterize the emergence of various branches of mathematics, as in a sort of permutation game where the words algebra, logic, topology, geometry etc. are combined to name new disciplines.

Examples

The following examples summarize the main strategies used to mathematize concepts. Their description is, for typographical constraints, very brief and, inevitably, incomplete. But it captures, nonetheless, the mental attitude expressed by each approach to mathematization. Each strategy corresponds to a line of the spectrum of the motivic lens, and is characterized by a core *motif* that captures the goal of the Observational Structure of Intentionality associated to each mathematization.

Algebrize And The Motif Of Coherence

Algebraization is the interpretation of a problem or an idea as a system for the manipulation of symbols, without necessarily regard for the meaning of the symbols other than in relation to the translation itself. The key motif of algebraization is *coherence*. The core of any algebraic interpretation is encoded in the conditions that define the system for the manipulation of symbols, which consist of the symbols, the rules to manipulate those symbols, and the coherence conditions imposed on the manipulation rules in the form of equational properties. The purpose of algebraizing is to represent a problem or an idea so as to be able to express how the elements of that problem or that idea give rise to the phenomenology that seems to characterize what we *interpret* as a solution to that problem or as an insight into that idea.

4.3 The Motivic Lens

Goal	Strategy	Data
Coherence conditions imposed on the inference rules that capture a definition of truth or correspondence with the purpose of the algebraization process. (Axioms and equational properties)	*Rules for the manipulation of symbols that reflect the purpose of this mathematization.* (Algebraic Operations)	*A minimum collection of symbols descriptive of a problem or an idea.* (Alphabet)

Table 4.5: The Observational Structure of Intentionality of algebraization.

For example, when the set of natural numbers $1, 2, 3, \ldots$ is endowed with the ordinary binary operation of summation of integers, the identities such as 1+4=5 and 2+3=5 implicitly mean or encode a much deeper statement, as in the diagram below.

$$\begin{array}{ccc} 1+4 & \xrightarrow{+} & 5_a \\ =\updownarrow & & \updownarrow \alpha \\ 2+3 & \xrightarrow{+} & 5_b \end{array}$$

- There is a certain symbol 5_a that corresponds to 1+4 via +
- There is a certain symbol 5_b that corresponds to 2+3 via +
- There is an invertible morphism α between 5_a and 5_b which is *not* explicit in the definition of the binary operation of summation of integers +, for which *all* morphisms like α are regarded as equivalent, and for this reason we use the symbol 5 to indicate any morphism *equivalent* to α.

This rationale—which is a simplified version of an important sheaf-theoretic argument called *split-fibration*—is hidden and encoded in the sign =.

Geometrize And The Motif Of Relation

The geometric approach is somewhat similar to the algebraic approach, and describes a problem in terms of spatial relations between geometrical objects such as shapes or spaces. The key motif of geometrization is *relation* between spatialized objects. The similarity between geometrization and algebraization is more than syntactical. Geometrization, as we'll see later in this section, is in fact a *spatialization of algebraization* that consists in assigning to the dimensionless symbols of an algebraic system a set of geometrical objects. This process enriches each symbol, literally and metaphorically, with dimensions along which the symbol-object relates to other symbols spatially, and in which spatial relations replace syntactic rules.

Goal	Strategy	Data
Understand a problem or an idea in terms of relations between objects that codify certain features for interest of that problem or idea.	*Identity various kinds of relations between objects or spaces.*	*Geometrical objects or spaces descriptive of a problem or idea.*

Table 4.6: The Observational Structure of Intentionality of geometrization.

An example that illustrates the mechanism by which the Observational Structure of Intentionality of Table 4.6 operates is offered by the so-called geometries without points—other examples are the so-called incidence geometry and higher geometry and non-commutative geometry. Historically, the notion of geometrical point as an entity which has no parts, has been regarded by many philosophers and mathematicians—from Leibniz, to Husserl, to Whitehead, Lobachevsky, von Neumann, Tarski, Goodman and many more—as too abstract and idealized, and has triggered a number of proposals of alternative geometries in which the fundamental notion of point is replaced by the notion of region, which appears more natural and practical.

Logicize And The Motif Of Causation

Logic looks at the world as the arena of causality. The key motif of logicization is *causation*. The search for the causal structure of a problem or an idea has led to the invention of three types of logic: deductive, inductive and abductive, and to various generalizations of the basic ideas that these types of logic represent. Deductive logic is concerned with the study of the processes through which the truth of a conclusion necessarily follows from the truth of its

4.3 The Motivic Lens

premises: it is the study of how specific, always true conclusions follow from general principles. Inductive logic, instead, is concerned with assessing the plausibility of general propositions from the knowledge of individual instances of those propositions: it is the study of how general plausible, and possibly true conclusions follow from specific observations.

Goal	Strategy	Data
Coherence condition as model of truth values, often in relation to a correspondence theory of truth or equivalent theory.	*Determine the structure of the causal relations, typically via three types of inferences: deductive, inductive or abductive.*	*Elements descriptive of the causal structure of a problem or idea.*

Table 4.7: The Observational Structure of Intentionality of logicization.

The third type of inference, called abductive, is concerned with propositions based on their ability to account for what is observed: it is the study of how plausible, possibly true conclusions follow from incomplete observations. These three types of inferences are further specialized and refined when we introduce modal operators in rational reasoning—such as in modal logic—that expand the framework of causality with other dimensions such as belief, necessity, possibility, and with events that are obligatory, prohibited or permitted.

Topologize And The Motif Of Change And Invariance

Topology looks at the world through the lens of the concept of continuity (and other related concepts). The key motif of topologization is *change and invariance*. The topological study of a problem consists in translating certain features of the problem into the features of space and form, and in generalizations of these concepts, such as, for example, in the context of Higher Category Theory, where higher-dimensional objects encode the types of invariance that define a topological object. Topology can be regarded as a spatialization of Logic for reasons similar to why Geometry can be regarded as a spatialization of Algebra. In a topological argument, the data is a space or the shape of an object, the strategy is the study of the continuous deformations of that space or shape that correspond to the features of the problem being topologized, and the goal is the determination of the conditions that make those deformations continuous, because the conditions for continuous deformations correspond to the features of the problem being topologized.

Goal	Strategy	Data
Describe mathematical objects via their properties of continuity or smoothness.	*Characterize the transformations between mathematical objects that preserve continuity and smoothness.*	*Models of continuity and smoothness.*

Table 4.8: The Observational Structure of Intentionality of topologization.

Two examples of the connections between Logic and Topology are provided by Stone Spaces—which are topological representations of Boolean algebras (the algebra of propositional logic)—and by Internal Logic—which we have already mentioned in the discussion about the Generative Lens (Section 4.1), which is a formal construction of logic, and of mathematics itself, inside an ambient mathematical object with sufficient structure—these objects are *categories*.

Categorify And The Motif Of Universality

Category Theory is a set of tools to describe mathematical objects via the structure-preserving transformations of those objects (Table 4.9). An informal definition of Category is the following

Definition 4.3.1. (Category – *informal*)
A *category* can be thought of as a diagram—vertices connected by arrows—where the vertices denote mathematical objects of the same kind, and the arrows denote sets of transformations—called *morphisms*—between those objects. The arrows of a category satisfy certain composition rules and coherence conditions, and each mathematical object has a special arrow associated to it called identity.

4.3 The Motivic Lens

In the diagram below, the arrow $f = h \circ g$, where $\circ$ is a binary operation to compose morphisms.

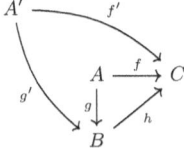

For example in **Set**, the category of sets, the objects denote sets and the arrows denote functions between sets. Similarly in **Top**, the category of topological spaces—a topological space is a set enriched with a structure to define continuity—the objects denote topological spaces and the arrows denote continuous functions between them. The key motif of categorification is *universality*. Universality in Category Theory denotes a vast range of constructions that generalize the following general fact: *"[m]any properties of mathematical constructions may be represented by universal properties of diagrams"* (MacLane 1998). What makes a mathematical construction universal is its capacity to capture the fundamental mechanism by means of which a given mathematical gadget functions. For example, if in the diagram above we consider a new object A' and a new pair of arrows from A' to B and C, say $f' : A' \to C$ and $g' : A' \to B$, the condition that $f' = h \circ g'$—read f' *factors* uniquely through h—makes that diagram universal, because h, in that particular configuration subsumes *any* factorization of arrows.

Goal	Strategy	Data
Describe mathematical objects via structure-preserving morphisms.	*Characterize the relations between mathematical objects.*	*Morphisms between mathematical objects.*

Table 4.9: The Observational Structure of Intentionality of categorification.

Category theorists see the world, as well as mathematics, as a place made of pure structure, and translate problems and ideas in

4 The 4 Lenses And The Architecture Of Mathematics

the language of Category Theory via a process called *categorification*, which consists in finding a representation of mathematical object in terms of certain structure-preserving transformations involving that object. There are two types of categorification: vertical and horizontal. Vertical categorification generalizes a concept from a set-theoretic context to a category-theoretic context, by defining a correspondence between set and category. Very roughly, vertical categorification regards the elements of a set as objects of a category, functions between sets as morphisms of categories—called functors. Horizontal categorification, also known as oidification, is the process of defining a category with a single object that describes a certain concept, and then generalize that category to a category with multiple objects. Decategorification is the opposite process: given a category, it forgets some of the information encoded by the category in the form of arrows, coherence conditions between them, as well as the properties that certain diagrams might carry.

Spatialize And The Motif Of Interaction

Spatialization is the process of ascribing to objects of thought or perception an arena of quantitative or qualitative becoming. We will examine spatialization in greater detail in Parts III and IV. For the moment, it suffices to emphasize that the process of studying an idea in terms of its quantitative and qualitative evolution is a general strategy to define the purpose of our interaction with that idea. The key theme of spatialization is *interaction*.

Goal	Strategy	Data
Describe the genesis of mathematical concepts as the spatialization or despatialization of other mathematical concepts.	*Characterize and classify the abstract transformations that spatialize and despatialize mathematical concepts.*	*Entities that share a common organizing principle.*

Table 4.10: The Observational Structure of Intentionality of spatialization.

The diagram of Figure 4.4, which I call the *propeller diagram* illustrates the main relations between four branches of contempo-

4.3 The Motivic Lens

rary mathematics. In the diagram, the transitions from left to right denote *spatializations*, the transition from right to left denotes *despatialization*, and the transitions in the up and down directions denote *respatializations*. The four quadrants are wrapped by a dashed box to signify that the same movements up, down left and right, can be iterated in all four directions, a bit like folding a piece of paper to construct an origami. The processes of folding the mathematical structures inside the dashed box correspond to *categorification* and *decategorification*, each contributing to an additional step of spatialization and despatialization.

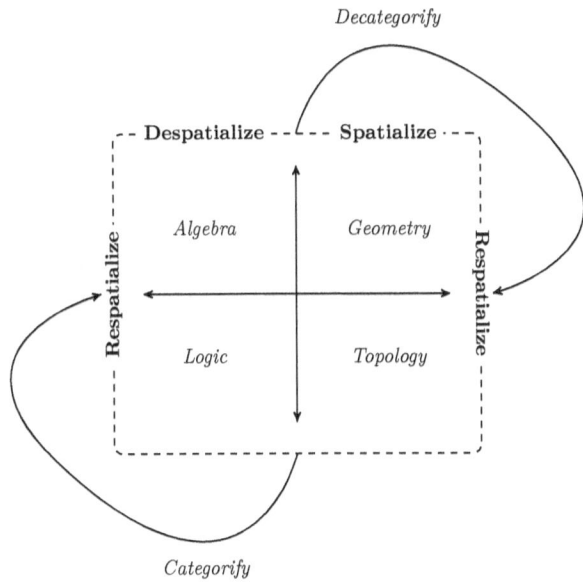

Figure 4.4: The propeller diagram illustrates the spatialization processes between some of the components of the motivic spectrum of mathematics.

In the description that follows, the term *higher* refers to generalizations of mathematical structures in the context of Higher

4 The 4 Lenses And The Architecture Of Mathematics

Category Theory.

Algebra And Geometry

These are related by transformations that assign geometric objects and higher structures to symbols and higher algebraic structures respectively, geometric relations and higher morphisms to algebraic ones and higher morphisms respectively, and geometric forms of identity and higher congruences to equations and higher equivalences respectively. The spatialization of symbols consists in defining a correspondence between a dimensionless symbol and a dimensional one, such as a geometrical figure. This strategy applies to higher structures as well, and is captured by a general correspondence between Algebra and Geometry called Isbell duality or Isbell conjugacy, after American mathematician John R. Isbell. In higher structures, spatialization (despatialization) takes the form of categorification (decategorification) and of various definitions of basic algebraic and geometrical structures in the context of Higher Category Theory.

Logic And Topology

The transition from Logic to Topology consists in constructing a space or an higher structure which carries properties that correspond to a type of logic. Similarly, the transition from Topology to Logic forgets the underlying space or higher structure. The general framework to reason about both types of transitions is Internal Logic—but there are other mathematical constructions that topologize logic, such as the already-mentioned Stone Spaces and their generalizations. Internal Logic is the category-theoretic process of reproducing parts of the edifice of mathematics in a category with enough structure to codify the basic mathematical constructions needed to create mathematics. One of Lawvere's groundbreaking contributions to modern mathematics is to have discovered a general scheme to construct these versions of "embedded mathematics" inside categories—originally called Categorial Logic (Lawvere 1969a). In these embedded mathematics, for example, the logical operators $\top$ (truth), $\bot$ (false), $\wedge$ (conjunction), $\vee$ (disjunction), $\Rightarrow$ (implication), $\exists$ (existential quantification) and $\forall$ (universal quantification) correspond to certain constructions inside a category—some of these constructions being also universal—such as objects ($\top$), unions ($\wedge$), intersections ($\vee$) and generalized forms of equivalence called adjunctions ($\exists, \forall$)—more

4.3 The Motivic Lens

about adjunctions in Section 15.3.

Algebra And Logic

The relations between Algebra and Logic are characterized by the general schema according to which algebra can often represent a deductive system. This basic intuition, first suggested by the fact that set-theoretic operations model Boolean algebra, has been generalized in many directions. One of those directions is Internal Logic, which we have already discussed. Other are specific algebraizations of logical systems, such as Heyting algebras and combinatory logic. Another important example of this general paradigm is found, once again, in the work of Lawvere (Lawvere 1963), in which he introduces what are today called Lawvere theories, which are a general method for doing universal algebra—the algebra of fundamental mathematical constructions—in the context of a category with a certain structure.

Geometry And Topology

In (higher) Geometry, spatialization emphasizes the spatial relations between geometrical objects (higher geometrical structures) and the structure of geometrical congruences (higher equivalences). In Topology, spatialization emphasizes the relations between spaces (higher topological structures), and the transformations (higher morphisms) that preserve certain properties of interest of those spaces. Roughly speaking, the respatialization process from Geometry to Topology, forgets the objects and remembers only their structure, and forgets the geometrical congruences and remembers only the geometrical transformations that preserve the properties of those congruences; the same happens in a similar fashion in higher contexts. Conversely, the respatialization process from Topology to Geometry sees in Geometry as a powerful language to represent certain general facts described in a topological context via peculiar relations (higher morphisms) between geometrical objects (higher structures), typically to emphasize or to study properties that would be difficult to explain or compute in a topological context.

Categorification And Decategorification

We have already described categorification in the section dedicated to the corresponding line of the motivic spectrum. Here we want to highlight briefly another important aspect that these two constructions emphasize about the deep structure of the edifice of mathematics. The classic mathematical constructions—the constructions that mathematicians have been using prior to the invention of Category Theory in the mid 40's—are *all* the decategorifications of higher structures which are, in many philosophical and technical ways, more natural. The point is, when we decategorify a higher structure, for example a category—recall the informal definition of category we gave a few sections ago—we destroy some of the information codified by that category. For example, in a decategorification of *Set*—the category of sets—may eliminate all the arrows, and remain with a mathematical construction made of objects only, namely, only sets. A consequence of this process is that many of the nice universal constructions and coherence conditions codified by the diagrams and the equations of a category are destroyed. This is the reason why categorification, being an attempt to recover the information destroyed, is an intrinsically difficult process, because it demands to rediscover or guess the correct equational properties and the coherence conditions that define higher version of that mathematical construction.

Spatialization As Dualization

Another concept that characterizes the ways to mathematize related via spatialization (or despatialization) is *duality*. Duality is a form of philosophical and mathematical equivalence that expresses the notion that two objects can be replaced by one another in a set of defined contexts. This makes equivalence, and thus duality, a generalization of *identity*.

4.4 The Methodic Lens

The methodic lens makes us see the various approaches to the formulation of a mathematical theory.

Theme

The theme of the methodic lens is *mode of thought*. A mode of thought determines the approach to the formulation of a mathematical theory; it is the specific style of reasoning with which we

4.4 The Methodic Lens

mathematize. The methodic lens makes us see the formulation of mathematics, or of a mathematical concept, either as a synthetic process—a process in which the role of a theory is to codify *higher structures of thought*—or as an analytic process—a process in which the role of a theory is to encode the *detailed relations between concepts*.

When we construct mathematical ideas, we seem inclined to use predominantly the anthropic mode of thought, as seen in Section 3.2. As a result of this general mental attitude, the directions along which the human intellect seems to be able to articulate, understand and explore ideas boil down to two, which can be called: up and down, or general and particular, or basic and derived. In the parlance of the formulation of mathematical theories, the up and down directions of thought are called respectively synthetic and analytical. This is to say, we can either look for an axiomatization of a system of ideas, such as in the synthetic and syntactic approaches to the formulation of concepts, or we can look for a substrate of concepts upon which to build a theory, such as in the analytic and semantic approaches.

Suffix

The action of the methodic lens is represented by the suffix *-tic*, as in analytic, synthetic, syntactic and semantic.

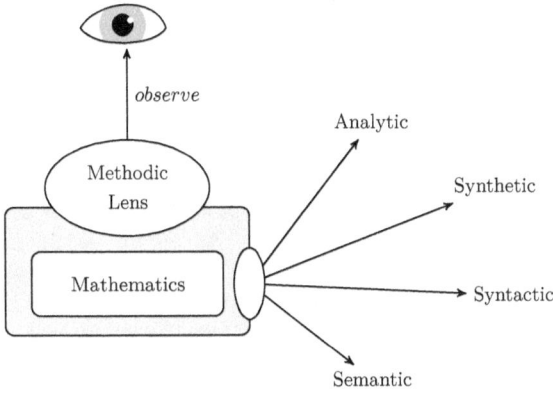

Figure 4.5: The spectrum produced by the Methodic Lens.

4 The 4 Lenses And The Architecture Of Mathematics

The Methodic Thought Process

At the basis of the methodic thought process, there is a belief system that defines how a theory can or should be formulated to be in full accord with our current conception of mathematics. This belief system is a reflection of the anthropic orientation we described in Section 3.2, and gives rise to the two known ways to articulate the structure of a mathematical theory, which are called analytic and synthetic. These two types of formulations codify the observational structure underlying any known process of mathematical abstraction by means of which we seem to be able to translate ideas into mathematical language. The analytic method comprises the analytic and the syntactic approaches to the formulation of a theory. The synthetic method comprises the synthetic and the semantic formulation of a theory. At the core of the methodic lens there is, therefore, the view that the ways to articulate concepts are essentially two. This is a rather peculiar fact about the human mind's capacity to think. We will see in Chapters 5 and 10 how to situate the nature of concepts within the broader context of the metaphysical structures of thought, and the implications that this perspective has on our ability to conceive concepts. For the moment, what's important about this lens is that it reveals that we seem to be able to construct hierarchies of concepts either by subsuming primitive concepts (analytical), or by abstracting other concepts (synthetic).

Observational Structure

To comprehend the difference between the analytical and the synthetic approaches to the formulation of a concept, we need to examine the role that the particular cognitive strategy codified by the formulation of the concept acquires in the reasoning that yields to the—analytical or synthetic—definition of a concept. The role played by the cognitive strategy is signaled by the definition of an ad-hoc type of computation. As we will see in greater depth and detail in Chapter 11, any formulation of a concept, and a fortiori, any formulation of a concept based on the formal systems methodology, codifies a conception of computation. When we examine a problem with the intent to construct an analytic representation of it, we focus on the representations of the problem that provide the substrate of basic concepts upon which we make computations about that problem. In this case, the computational

4.4 The Methodic Lens

rules that hold together a theory's definitional structure are the manifestation of the relations that *tie the substrate concepts to the concepts of the theory*. In contrast, when we examine a problem with the intent to give a synthetic description of it, we focus on the features of the problem that characterize the problem in itself, and *the computational rules derived from a model of a synthetic presentation of the problem, express the rules through which the model is constructed*. These two observations about the relations between a conception of computation and the two known modes of thought are illustrated in the examples.

Goal	Strategy	Data
Characterize the nature of a mathematical concept.	*Define a concept based on its intrinsic nature, or in relation to other concepts.*	*The structure and the features of mathematical concepts.*

Table 4.11: The Observational Structure of Intentionality of the Methodic Lens.

The goal of the methodic lens is to characterize what aspects of the nature of a problem a theory should emphasize. The nature of a problem is distinct from the problem itself, and is strictly related to our ability to observe that problem and to the quality of our observation. Based on our current understanding of mathematics, problems can be regarded from two complementary viewpoints: from an axiomatic viewpoint, which emphasizes the irreducible elements that define a problem, and from an analytic viewpoint, which emphasizes how a computational formulation of the problem encodes and structure of the problem itself, in a way consistent with our experience of that problem. The strategy to articulate the analytic description of a theory, is centered around the notion of which equational and computational properties should characterizes a theory. In contrast, the synthetic approach to the formulation of a theory follows a strategy centered around the notion of structure, and of the structure-preserving changes for the theory at hand. The synthetic strategy attempts to define a theory per-se via its own structure, and without reference to a particular representation of that structure that is conducive to a computational view of it. In the synthetic approach the computational aspects of a theory are always a byproduct, and never at

4 The 4 Lenses And The Architecture Of Mathematics

the conceptual foundation of the theory. Another useful way to describe the analytic and synthetic strategies to the formulation of a theory is to say that they give two distinct characterizations of definitional identity (see also Chapter 5). The analytic approach defines a new identity—a theory—via the fixed identities of the substrate concepts. The synthetic approach defines a new identity dynamically, via the models of the theory it defines—and in this regard it is, I think, more natural.

Examples

Analytic

The analytic way to formulate a theory relies on the way in which the manipulation of the elements of a problem, for example via equations or transformations, codifies the nature of the problem described by the theory. For example, Analytic Geometry makes use of formulas and systems of coordinates as a way to describe geometry, and dispenses entirely with any geometric representation of the mathematical entities it describes. The forms of computation available in Analytic Geometry are directly tied to the formulation of the theory itself—namely, systems of equations—because that is precisely the way the theory is constructed from a set basic concepts. The analytic methods are often the source of a syntactic interpretation of a problem, where the syntactic rules correspond to the relations between substrate concepts.

Synthetic

The synthetic way to formulate a mathematical theory relies on the elements of a problem that define the nature of the problem, and does not focus or does not give much prominence to the computational nature of the problem as a way to describe the same. The synthetic approach is often referred to as axiomatic in that it consists in the formulation of axioms that encode the key features of the intended mathematical concept. For example, Synthetic Geometry studies geometrical figures as such, without recourse to formulas. The computational methods of a synthetic theory are necessarily derived from a model of the theory, because the construction of a model ties a synthetic theory to a specific representation system. The interpretation of the synthetic definition of a theory gives rise to a semantic for that theory.

4.4 The Methodic Lens

Bibliographical Notes

The ideas underpinning The 4 Lenses were introduced in a number of unpublished papers I wrote in the 90's, Lo Vetere 1996a, Lo Vetere 1996c, Lo Vetere 1996b, and appear in the form they have in this book in Lo Vetere 2014. I have used The 4 Lenses throughout my research more or less informally and implicitly as a thinking tool to navigate the quirky abstractions at the basis of the deep structure of mathematical thought. In these papers there is, between the lines as a background idea, and never explicitly stated, a first rudimentary concept of a phenomenology of the Abstract Mind model. But in the 90's I wasn't thinking in terms of Symbolic Intuition of Reality. The general philosophical aspects of the problem of a holistic and evolutionary description of mathematics were inspired by numerous authors. It is impossible to cite all of them. The following list should give the flavor of the many views, schools of thought and syntheses embraced by this chapter: MacLane 1986, Connes, Lichnerowicz, and Schutzenberger 2001, Hartimo 2010, Lucas 2000 and Whitehead 1968.

The images of mathematical thought provided by the generative lens belong to the vast literature on the foundations of mathematics. The examples I present in the generative lens can be found in or were inspired primarily by Benacerraf and Putnam 1964, Kunen 2009 (ZFC) Lawvere 1963 and Lawvere 1964, Lawvere 2005 (ETCS), Steve Awodey 2008a, Joyal and Moerdijk 1995, Berg and Moerdijk 2007 (Algebraic Set Theory), MacLane and Moerdijk 1994, J. L. Bell 2008, Goldblatt 2006, Jacobs 2001, Peter T. Johnstone 1977 (Internal Logic), Steve Awodey and Warren 2009, Steve Awodey, Garner, et al. 2011, Steve Awodey, Gambino, and Sojakova 2012, Steve Awodey, Pelayo, and Warren 2013, Univalent Foundations Program 2013, Jacobs 2001 (HoTT).

The views of mathematics given by the performative lens can be found in or were inspired by Troelstra, Dalen, and Ranis 1999, M. J. Beeson 1985, Bishop and M. Beeson 2012, Bourbaki 2006 (constructivism), Burgess and Rosen 1999, Azzouni 2007, Azzouni 2009, Azzouni 1997, Benacerraf and Putnam 1964, Benacerraf 1973, Lewis 1986, Resnik 2000 (nominalism), Benacerraf 1965, Feferman 1999, Russell 1908, Whitehead and Russell 1925 (logicism), Maddy 1992, Steven Awodey and Carus 2007, Steve Awodey, Pelayo, and Warren 2013 (structuralism).

4 The 4 Lenses And The Architecture Of Mathematics

The discussion about the motivic lens was inspired by Borceux and Janelidze 2001, Whitehead 1979, M. A. Shulman 2008, Hodges 2009, Lawvere 1972, Lawvere 1992, Peter T. Johnstone 1986, Grothendieck 1983, McLarty 2011.

The discussions about the analytic and synthetic structures of thought described by the methodic lens inspired by Kock 2006, Hodges 1997, Hodges 2008, Yanofsky 1999, Hartimo and Okada 2016, Girard 1995.

PART II

Inside The Abstract Mind

What does mathematics tell us
about the workings of the mind?

Introduction

> We have to remember that what we observe is not nature in itself but nature exposed to our method of questioning.
>
> Heisenberg 1971

Here we define a framework to describe how the Abstract Mind creates the intuitive part of a Symbolic Intuition of Reality.

In the introduction to Part I, we explained that the Abstract Mind creates Symbolic Intuitions of Reality, and that a Symbolic Intuition of Reality is made of two parts called *intuitive* and *epistemic*.

> *"The intuitive part symbolizes a specific type of mind-world transaction by defining what is intelligible, and denotes the fundamental way in which we become aware of an object of thought or perception as a result of the interaction between the human mind and the World. "*

Our desire to sacrifice abstraction for clarity, suggests two reasons for beginning to study the Abstract Mind by describing how it creates concepts rather than a complete Symbolic Intuitions of Reality. First, focusing on the genesis of concepts makes the description of the workings of the Abstract Mind a lot more intuitive, because it gives us the opportunity to isolate much more efficiently the nature of this deep metaphysical problem. Second, the definition of epistemology used in the Symbolic Intuition of Reality depends on a special type of concepts called **concrete** (Chapter

Introduction

10), which we use in this research to characterize Abstract Mind states. It seems therefore more natural to describe epistemologies in the context of a specific model of Abstract Mind rather than abstractly, which is what we do in Part III where we introduce a model of Abstract Mind called Interaction Theory.

To describe how the Abstract Mind creates concepts, we begin by explaining how we regard concepts in this research (Figure 4.6-A), and then tackle the problem of studying how the Abstract Mind creates them (Figure 4.6-B and C). We regard concepts as having a *dual nature*, interferential and intentional, for reasons that will become clear as we progress in this discussion. Chapter 5 examines the metaphysical structure of concepts, and introduces the important notion of *definitional interference* to characterize their *interferential* nature. Chapters 6, 7 and 8 describe the intentional nature of concepts through a framework called Synthetic Structure of Intentionality in two steps (Figure 4.6-B and C).

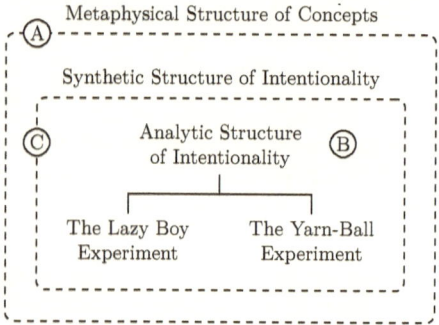

Figure 4.6: Logical structure of Part II

We can think of an Abstract Mind that produces *only* concepts—as opposite to an Abstract Mind that produces full Symbolic Intuitions of Reality—as a theoretical model of a human mind incapable of looking at the World through the concepts it produces. This simplified model of Abstract Mind produces the Symbolic Intuition of Reality of Figure 3.6-B, which we copy here for convenience.

The Structure Of Appearance

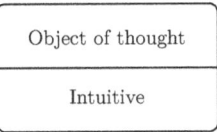

An imaginary mind that observes the World through the concepts it produces, but that has no access to, or does not develop, the cognitive processes necessary to create any kind of knowledge from those observations, does not *see* the World: it is just aware of it. It creates mental impressions that correspond solely and entirely to the experience of awareness of an object of thought or perception. This imaginary mind does not relate to the World in any meaningful way because it is unable to ascribe meaning to its conceptualizations of the World. It has memory but no experience. Its Symbolic Intuition of Reality is, so to speak, *empty*.

In this extended introduction we want to explain the techniques we will use in the next four chapters to study this simplified model of Abstract Mind.

The Structure Of Appearance

Our sensory experience of the world is largely created by the brain. It is not made of the raw data we gather through the five senses, but of what the human brain makes of that data. If we ignore the structure of appearance, things really *are* as they appear. It is when we begin to comprehend the structure of appearance, and deconstruct how we relate to reality conceptually and sensorily, that we begin to see things as they are.

To comprehend the structure of sensory appearance, we have discovered, and, in some cases, explained, sensory illusions. Sensory illusions are sensory stimuli that reveal how the discrepancy between reality and appearance originates at a physical, physiological or cognitive level. There is a strong analogy between the mechanism through which sensory illusions are used to study perception, and the way we will use carefully constructed illusions to study the genesis of concepts in the Abstract Mind.

To get a feel for how sensory illusions work, let's examine briefly two important types of illusions: optical illusions and illusions that trick the hierarchical structure of perception.

Introduction

Optical Illusions

An optical illusion (Locher and Escher 1978, Buonarroti and Cencetti 1989) occurs when the perceived image differs from the image causing the perception. This phenomenon reveals how the brain—and, in some circumstances, the mind—integrate or rewrite the data we perceive through our senses, and creates a picture of reality that does not correspond to what causes the perceptual experience (Gregory 1997). Broadly speaking (Gregory 1991), visual illusions can be organized into three categories: physical, physiological and cognitive. Physical illusions are caused by the environment. For example, a straight stick half immersed in water appears bent. Distant objects appear blurred when observed behind hot air, such as the heat haze created by jet engine exhaust, or by asphalt in a hot summer day. Physiological illusions are caused by the way the physiology of perception adjusts itself to cope with intense stimuli. Afterimages are a common example of physiological illusions. Cognitive illusions are thought to be caused by the way perceptual stimuli are organized conceptually. The cognitive superstructures we unconsciously superimpose on perception may give rise to alternative interpretations, to distortions or paradoxes, and to fictional images altogether.

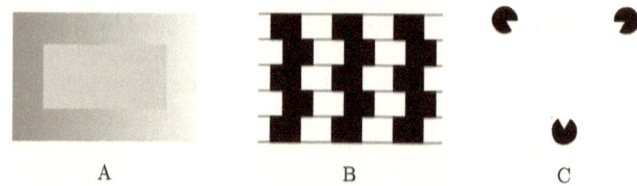

Figure 4.7: Three examples of optical illusion: simultaneous contrast illusion (A), café wall illusion (B), and Kanizsa's triangle (C).

In Figure 4.7-A the rectangle in the center seems to fade from light gray to dark gray, but is in reality just one color. This phenomenon occurs because the mind emphasizes the contrast between slightly differing shades of gray, producing the visual experience of a color gradient in the rectangle in the middle. In Figure 4.7-B, named after a nineteenth-century café in Bristol, U.K., the horizontal lines are parallel, but they appear sloped—

various explanations of this phenomenon have been proposed, there isn't one I lean towards to. Figure 4.7-C creates the illusory contour of a white triangle covering three black circles—it is believed that the explanation is to be sought in the structure of the visual cortical regions.

McGurk Effect

The McGurk effect, or the effect of "Hearing lips and seeing voices" (McGurk and MacDonald 1976) occurs in speech perception as a result of the priority of vision over hearing, and of the interaction between hearing and vision that produces speech perception. Speech perception needs both hearing and vision, but because vision has higher priority, if what we see doesn't match the sound that reaches our auditory system, what we *actually hear* is what we see, not what's being said: vision overrides hearing if what we see clashes with what we hear. To hear what's actually being said in a McGurk experiment, you need to close your eyes.

Metaillusions

We measure the quality of a concept by how reliably and faithfully and meaningfully we connect to reality through that concept. This is primarily an indirect and utilitarian measure of how much we can trust the cognitive processes that underpin the creation of the conceptual apparatus that connects us to reality symbolically. Roughly, we can describe this measure as a sort of tacit agreement between the mind and the senses that stipulates the following duties:

- the mind is responsible for portraying a symbolic representation of the sense data, and that representation must have a quality and a complexity consistent with the kind of correspondence between concepts and sense data needed to acquire knowledge and make sound decisions,

- the senses are responsible for providing to the mind a sufficient and reliable amount of sense data in response to an intention.

Seeing a car parked on the sidewalk would be meaningless if we believed that our mental experience of that car on the sidewalk, what we call *the seeing of that car parked on the sidewalk*, was a hallucination, or an otherwise unreliable experience: an experience that has little or no meaningful relation with the objective fact that we mean when we say "there is a car parked on the sidewalk". The belief that our mental experience of reality has an intrinsic degree of veridicality, even if that experience is not or cannot be thorough, is a fundamental assumption about the nature of the conceptual relations that we form in our mind, and through which we relate to the world symbolically. This fundamental assumption about the way we conceptualize is, therefore, unrelated to the *content* of the mental experiences it produces. It is entirely and solely and exclusively about the very mechanism through which we *construct* that experience, consciously and unconsciously.

Through concepts—which do not necessarily need to have a linguistic representation to carry out their cognitive function—the human mind is able to grasp reality symbolically, in much the same way in which through perception, the mind constructs a sensorial map of the physical world. There are concepts that can trick the mechanism through which the human mind conceptualizes, in much the same way in which there are sensorial stimuli that give rise to sensory illusions that trick the physiological or the cognitive mechanism through which we perceive reality through our senses.

In sensory illusions, especially in optical illusions, the fundamental content of a visual experience is unchallenged; only some secondary features of the visual experience give rise to the optical illusion. In an optical illusion, the mental experience of reality can still be trusted because its degree of correspondence with reality is substantially intact. We may see images that are not there, as in the Hermann grid illusion, or misjudge lengths and colors, as in the Ponzo and in the Checker shadow illusions respectively, but the fundamental content of the sensory experience, unless we hallucinate, is substantially reliable. In an optical illusion, the correspondence between the conceptualization of what we see, no matter how distorted by physical, physiological or cognitive factors, and the stimulus that causes the illusion, can still be trusted because the fundamental structure of that experience is fundamentally intact.

This changes radically in a metaillusion. A metaillusion does not challenge the content of a mental experience: *a metaillusion challenges the correspondence between concept and reality, while remaining consistent with perceptual experience.* Metaillusions cast doubt on the pact between the mind and the senses by showing that the very process of conceptualization cannot be trusted, and that therefore the degree of veridicality of the correspondence between concepts and reality is, or can be turned into, a fundamentally unreliable instrument of knowledge. To add insult to injury, a conceptual illusion shatters the concept-reality correspondence while remaining consistent with perceptual experience, which explains why most metaillusions are *perceived* or *interpreted* as paradoxes. They are regarded as paradoxes precisely because they shatter the concept-reality correspondence, but reducing metaillusions to paradox machines does not do any justice to how insidious and vicious and sophisticated metaillusions really are.

Definition 4.4.1. (Metaillusions)
A *metaillusion* is an explanation or a demonstration that constructively challenges the structure and nature of a Symbolic Intuition of Reality.

The keyword in Definition 4.4.1 is the adverb *constructively*, which is used in its mathematical sense, and denotes the fact that a metaillusion consists in the construction of a notion that shatters the concept-reality correspondence. In contrast, a non constructive statement is existential: it simply asserts the existence of a paradoxical notion without showing how that notion is rooted in a web of known concepts through which we know how to relate reliably to reality. The requirement that a metaillusion must explicitly construct the paradox it gives rise to is essential, as we will see, because there is a fundamental correspondence—known as Synthetic Structure of Intentionality (Chapter 8)—between the construction of a metaillusion and the deconstruction of the Symbolic Intuition of Reality from which a paradoxical experience originates.

Example Of Metaillusion

We will see a metaillusion in detail in Chapter 7. Here, to complete this long introduction to metaillusions, we want to mention an example of metaillusion called the Banach-Tarski paradox: there exists a decomposition of a solid three-dimensional sphere into a finite number of pieces that can be recombined in a different way to yield two identical copies of the original sphere. The Banach-

Introduction

Tarski paradox is a theorem that shatters a basic assumption of the common conceptual correspondence with reality: you can't create stuff out of thin air. The paradox raises many questions. Since the known laws of Physics suggests that we cannot create stuff out of thin air, why can we demonstrate mathematically that we can create two identical copies of a sphere? How can we relate the concept of a sphere to an object in the physical world, knowing that we can deduct logically a statement that does not correspond to any known physical law? This question conveys the pain of realizing that "sphere" is not necessarily a safe reference to a physical object, and this realization alone is enough to undermine the very purpose of concepts. The Banach-Tarski paradox, once explored thoroughly, is an example that forces us to question whether we can trust the concept of "that object out there" as such, regardless of its shape, which is a basic notion upon which we reference physical reality conceptually for practical purposes. Even if it is quite clear that the paradox is deeply rooted in certain mathematical abstractions which most people would rightly regard as distant from everyday experience, it still shakes the very structure of the fundamental belief in how we use our ability to mathematize to relate to reality.

A Technique To Probe The Abstract Mind

The technique we use to study how the Abstract Mind creates concepts is based on the following strategy. Let **Q** be a simplified version of a certain problem **P**. Suppose we want to find the solution to **P**, knowing only a solution to **Q**. If we could add to **Q** some of the features that would make its solution closer to the solution of **P**, then we would be in a position to spot the structure of the solution of **P**. This is, intuitively and practically, the strategy we will follow to study the Abstract Mind. But this is also a well-known strategy, widely used in Physics and Mathematics, at the basis of what is known as Perturbation Theory, which gets its name precisely from this incremental way of exploring the solution to a problem.

To study the Abstract Mind we apply a strategy analogous to Perturbation Theory. We construct two versions of the same concept, an elementary type of space, and a *metaillusion* based on the same concept, and then we extrapolate from the patterns that the genesis of the two concepts have in common, the underlying structure of the mechanism by means of which the Abstract Mind constructs *any* concept. Figure 4.8 summarizes this idea.

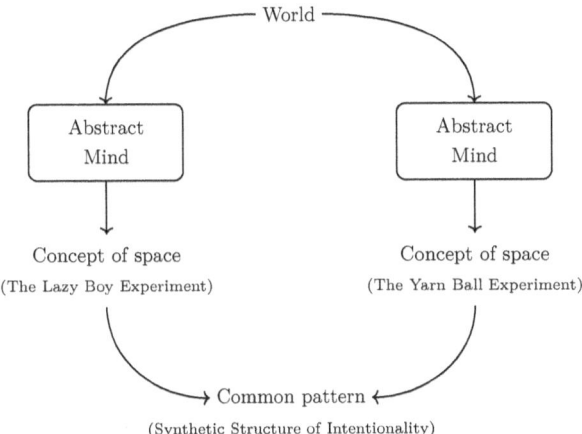

Figure 4.8: Schema to deconstruct a metaillusion with perturbation probes.

The choice of space to probe the Abstract Mind is motivated by the fact that space can be easily constructed in many ways while remaining consistent with the ordinary perceptual experience of the World, which makes the use of metaillusions easier and more intuitive. The feature of remaining consistent with perceptual experience simplifies the analytical work to reveal the anatomy of the underlying cognitive processes that produce space, and makes space a good candidate to probe the Abstract Mind.

Chapter 6 presents the first thought experiment (Figure 4.8, The Lazy Boy Experiment), where an elementary concept of space is produced by the Abstract Mind as a result of a perceptual experience of some objects on a desk. The construction of the concept of space is described in a step-by-step fashion to reveal gradually each layer of abstraction added in the process. Chapter 7 is the perturbation probe. It is a thought experiment and a metaillusion (Figure 4.8, The Yarn Ball Experiment) designed to challenge each individual element of the cognitive processes involved in the conceptualization of our physical experience of space. Chapter 8 deconstructs the two thought experiments and examines a pattern that emerges from this analysis called Analytic Structure of Intentionality. The Analytic Structure of Intentionality is a first rudimentary way to describe the key elements of what appears to be the Abstract Mind's inner structure. A synthetic description of the Abstract Mind is given by the Synthetic Structure of Intentionality model (Figure 4.8), which is the conceptual framework we use to make hypotheses about the inner structure of the Abstract Mind. What makes the Synthetic Structure of Intentionality model a good model to describe the Abstract Mind, is that it offers an accurate description of the human mind's ability to mathematize in terms of the workings of the Abstract Mind, that is consistent with the observational structure of the same (Sections 3.3 and 3.4).

My philosophical morals are lax: I use improperly the term *intentionality* to define the structure that the two thought experiments—The Lazy Boy Experiment and The Yarn Ball Experiment—in have in common. The abuse consists in using a term which has a precise philosophical meaning out of context, thus implying or suggesting or assuming a correspondence between the two contexts that justifies the use of the term. My motivation for using intentionality to explain these thought experiments is that the pattern that emerges from this analysis has a strong correspondence with the philosophical definition of intentionality given by Searle (Searle 1983).

Key Concepts Of Part II

1. The notion of definitional interference characterizes the nature of concepts via the transient mutual relations between their definitions.

2. It is possible to describe the essential structure of the Abstract Mind by analyzing a class of spatial concepts characterized by the property of producing conflicting descriptions of the same perceptual experience while remaining sensorily consistent with it.

3. The Synthetic Structure of Intentionality is a theoretical model of Abstract Mind that allows us to describe the Symbolic Intuition of Reality as a conscious, organic process driven by intentionality.

Chapter 5

A Metaphysical Structure Of Concepts

> I was in New York yesterday, today is Sunday isn't it, and I had some business in the morning and I did two shows and then I grabbed the cab and I raced the airport and I try to catch the 11 o'clock and I figured I missed it and I did, so I laid over in Kennedy for three hours to catch Eastern out to Atlanta Georgia, I was there for two hours, called Delta into Dallas Texas, and I was two hours there to get TWA into LA. I know it's not a very interesting story, I only mentioned it because I wanted to point out that, in regard for whether or not I wanted to be here, I was relatively indifferent.
>
> Peter Falk,
> 24th Primetime Emmy Awards acceptance speech.

In this chapter we introduce the notion of *definitional interference* to characterize the nature of concepts.

There are two primary reasons why we want to characterize the nature of concepts:

1. Because the Abstract Mind produces Symbolic Intuitions of Reality, which are made of concepts.

2. Because concepts are the *only* observable manifestation of the Abstract Mind's activity.

5 A Metaphysical Structure Of Concepts

The first reason is practical. We want to comprehend the nature of the entities produced by the Abstract Mind because concepts are the entities we use to think. The second reason is foundational and theoretical. To model the Abstract Mind, we need to understand the elements of its phenomenology.

Any philosophical debate about the nature of concepts, reflects necessarily an approach to the study of the mind, because concepts are generally regarded as the constituents of thought. My position on the nature of concept is codified by the process-centric view underpinning the definition of the Abstract Mind model, and by my proclivity to think of the ontological status of concepts from a pragmatist angle. I think that, in the context of the Abstract Mind model, asking What are concepts the constituents of? is as meaningful and accurate as defining a chisel as an item sold in a hardware store.

The concepts we examine in this chapter form a very special class of concepts called **concrete** concepts. We will define the class of **concrete** concepts in Chapter 10, and in the same chapter we will explain why it is sufficient to focus on the class of **concrete** concepts to study the nature of concepts. What we need to know about **concrete** concepts to situate the discussion about the nature of concepts, is that in any given state of consciousness, **concrete** concepts are fully defined by their property of being *irreducible*. For example, at the alert, problem-solving state of consciousness, the notion of identity is a **concrete** concept because the human intellect is not able to reduce identity to more elementary notions. **concrete** concepts have, among other properties, the property of having a particularly simple metaphysical structure by virtue of being irreducible. As a result of having simple metaphysical structures, **concrete** concepts have also simple definitions, and it is precisely their definitional spareness that will come in handy in the study of the nature of concepts we present here.

The project of characterizing the nature of **concrete** concepts begins with the observation that we *define concepts into existence*. Thus, to comprehend how **concrete** concepts come into existence, it is necessary to understand the process through which definitions acquire their ontological status: we want to comprehend their deep nature, how they arise and how they disintegrate. The structure of this chapter reflects precisely this rationale, and culminates with the exact statement of the nature of concepts codified by the notion of definitional interference. As illustrated in Figure 5.1,

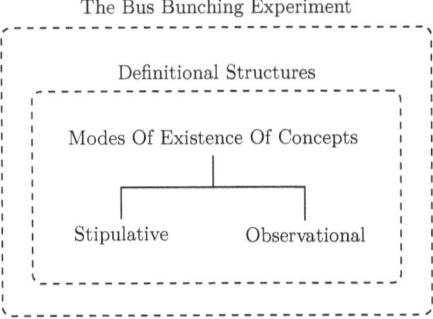

Figure 5.1: Construction of the notion of definitional interference.

since concepts are defined into existence, the study of definitions is also, and, necessarily, the study of the concepts' modes of existence. The knowledge of the concepts' modes of existence allows us to abstract and synthesize the universal structure of definitions, which we capture with The Bus Bunching Experiment (Section 5.4). The The Bus Bunching Experiment is a thought experiment designed to introduce the notion of definitional interference in an extremely practical manner. It is, like the other thought experiments we present in Part II, my attempt to convey highly abstract concepts in practical terms, and through the metaphysical content hidden in everyday experience.

The notion of definitional interference situates the discussion about the structure of the Abstract Mind we develop in the next chapters, and permeates the entire book. Wherever in this book we talk about concepts, the reader should interpret my use of the term concept according to the principle of definitional interference.

5.1 Stipulative And Observational Modes Of Existence

Definitions are a manifestation of the creative power of language: *any* language. Here we use the term language as synonym of form of expression. A painter translates the emotional and metaphysical structure of her experience into lines, colors, shapes, volumes, brushwork and subject. Her canvas is made of light. A musician's

5 A Metaphysical Structure Of Concepts

canvas is time: she paints with rhythm, pitch, melody, timbre, texture and dynamics. A dancer's canvas are space and time. She paints with her body, with her movements, with the space she fills, with the timing and the energy of her movements. A chef's canvas are the senses of taste, sight and smell. She paints with flavor, texture, shape, color and aromas. A writer's canvas are thoughts and emotions. She pains with words, rhythm, sentence structure and figures of speech.

No matter what the language, any expression in any language defines a lens through which we can look at the world and at ourselves: this is why language is often seen as a tool of creation, because it defines realities into existence. If we look at the fundamental structures that definitions—and thus concepts—have in common, we discover that there are two types of structures, which we call *stipulative* and *observational*. These two structures symbolize the modes of existence in which we seem to have epistemic access to concepts. Rather than focusing of the relationship between definiendum and definiens, we describe concepts through their modes of existence, and through the concept of definitional interference we introduce later in this chapter. The idea we want to convey here is that, although we will be describing the stipulative and observational modes of existence as distinct and separate structures for explanatory purposes, they are meant to be understood as two sides of the same coin.

With these ideas in mind, we can begin to examine how concepts *seem* to exist, bearing in mind that this is not how they *really* exist, as we shall see in the "The Bus Bunching Experiment" (Section 5.4), but it is a convenient way to begin to scrutinize their deep origin.

A concept is defined into existence in a *stipulative* manner if *first* we give an abstract definition of the concept and *then* construct or present an object that matches that definition. Of course, one can define a unicorn and never find one in the physical world, but the concepts we are interested in are references to objects or states of affairs that we can relate to with our mathematical technologies. As an example, let's pick the concept of identity. An example of *definitional identity* is the concept of prime number, which consists of a formal definition of prime number—a natural number greater than 1 that has only trivial divisors, namely, 1 and itself—followed by examples of prime numbers such as 2, 5 and 587.

5.1 Stipulative And Observational Modes Of Existence

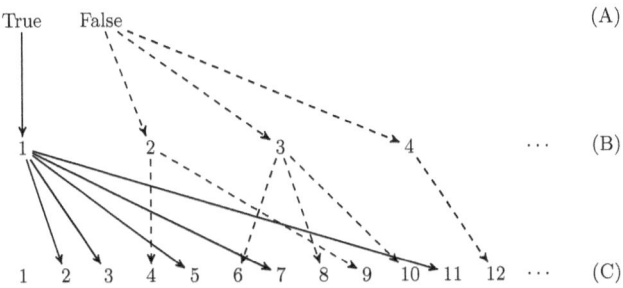

Figure 5.2: Illustration of the stipulative mode of existence of prime numbers.

Figure 5.2 illustrates how a concept acquires its stipulative mode of existence as a three steps process.

A. The stipulative process starts with a definition of prime number, which characterizes what a prime number *is* (True) and *isn't* (False).

B. To construct an instance of prime number, we enumerate all the natural numbers, and look for those with only one divisor greater than 1.

C. The definition in A acts as a contract between natural numbers. The dashed arrows in Figure 5.2 indicate the composite numbers—the numbers with non trivial divisors. Those are the numbers to which the prime number contract in A does not apply. The solid arrows indicate the natural numbers to which the prime number contract applies; the prime numbers.

In contrast, a concept comes into existence in an *observational* manner if *first* we consider a collection of objects or states of affairs, and *then* apply to those objects or states of affairs a certain Observational Structure of Intentionality (Chapter 3.3). Continuing with the concept of identity applied to prime numbers, an *observational identity* of prime numbers is a three-step process (Figure 5.3) where:

A. We consider the collections of all natural numbers.

B. We apply to the collection of all natural numbers a certain Observational Structure of Intentionality defined as follows:

5 A Metaphysical Structure Of Concepts

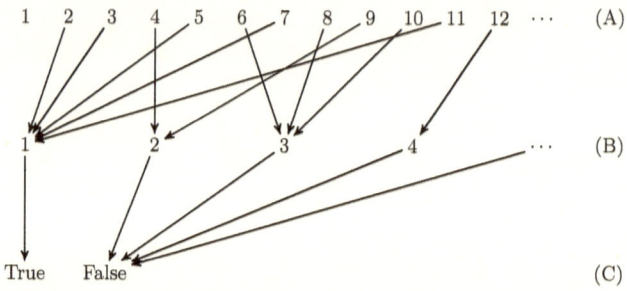

Figure 5.3: Illustration of an observational construction of the concept of prime number.

Goal	Strategy	Data
Identify the natural numbers with the minimum set of divisors.	*Group the natural numbers by their number of divisors.*	*The set of natural numbers.*

C. We focus on the group of natural numbers with only one divisor—1 is not being counted as a divisor.

In this case, a set of natural numbers is endowed with the identity of prime number as a result of the Observational Structure of Intentionality described in Figure 5.3, which revolves around a strategy to detect and categorize certain regularities in the number of divisors of natural numbers. This example shows why this definition of prime number is observational, because it ascribes to natural numbers an ontological status that is a reflection of a specific Observational Structure of Intentionality that we call *prime number*. The transition from B to C in Figure 5.3 assigns two values (1 or 2) to the elements of a set of numbers (B), thus acting as a truth function, or, in the parlance of Set Theory, acting as a membership function.

5.2 Observational Identities

Of the two modes of existence we introduced in the previous section, the stipulative is probably the most common and the most

5.2 Observational Identities

intuitive, because it consists in the application of a definition to objects and states of affairs. The process of applying a definition can be very complicated. For example, consider the following definition: a real number x is of type δ_k, with $k = 0, 1, 2, 3, \ldots 9$ if it is irrational—cannot be written as $\frac{a}{b}$ with a,b integers—and if for every prime number p, its pth decimal digit is k. For example, for us to say that the number $x = 2.155857534153\ldots$ is of type δ_5, and assuming that its $2^{nd}, 3^{rd}, 5^{th}, 7^{th}, 11^{th}, \ldots$ digits are 5, we should be able to prove that there aren't two integers a,b such that $\frac{a}{b} = x$, which is not trivial. Stipulative definitions are everywhere in Mathematics and Physics, and are the purest expression of our understanding of concepts, because they condense our deep understanding of an idea, and of how the interconnections of that idea with other ideas creates meaning in a larger context of our knowledge of the realities we access through that idea.

But where do stipulative definitions come from? Although we stated in the previous section that the stipulative and observational modes of existence are two sides of the same coin, which would make the question Which concepts' mode of existence comes first? a chicken-and-egg problem, the intentional and creative nature of knowledge seems to suggest that the stipulative nature of concepts is the result of that very creative process of knowledge, which, as such, is triggered by observation. Thus, the observational nature of concepts subsumes their stipulative nature. I think that this conclusion, notwithstanding the limitations of a language that is clearly inadequate to describe a participatory phenomenon like the nature of concepts in which the subject-object duality becomes meaningless, can be taken as a first, cursory interpretation of the interplay that gives rise to concepts as they appear to our intellect.

In this section we want to give two important examples of the observational mode of existence, and to this end, we will continue to use the concept of identity as a reference for the observational mode of existence—recall that this discussion about the metaphysical structure of concepts focuses on **concrete** concepts only due to their generative nature.

As we have seen, in the observational mode of existence, there is no a priori stipulative definition, there are no definitional structures upon which to articulate any thought at all, except the overarching conceptions that define intelligibility and knowledge. In the absence of stipulative definitions, the objects of thought or perception are given an identity to signify that they become *temporarily* part of a the specific Observational Structure of Intentionality through which we access those objects of thought or perception. Thus, a concept's observational identity functions as a reference to a general interim stipulative definition of identity, used merely to mark that a certain object becomes distinct from other objects as a result of activating a certain strategy part of an Observational Structure of Intentionality. We observe that objects are or become distinct from other objects in two ways: by virtue of carrying an independent ontological status, and by activating certain observational structures. Two important examples of observational identities are the *relational identities* and *pattern identities*. Relational identities are identities assigned to objects or states of affairs by virtue of displaying certain regularities, typically with respect to a causal frame of reference. Pattern identities are identities assigned to objects or states of affairs by virtue of displaying certain geometrical or topological regularities, with respect to a mathematical frame of reference.

5.3 Relational And Pattern Identities

In Figure 5.4 the circles represent processes, and the arcs connecting them represent various relations of influence and conditioning. If there exist a subset of processes that influence and condition each other in a tight recursive way—the gray circles in Figure 5.4—while remaining coupled to the wider context in terms of relations of mutual influence, then those processes are endowed with an identity by virtue of displaying a kind of organizational closure. This is an example of relational identity.

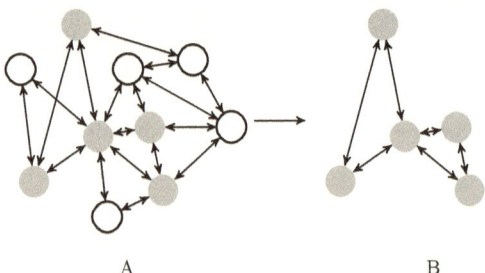

Figure 5.4: Illustration of relational identity.

5.3 Relational And Pattern Identities

Relational identity emerges in biological and in artificial systems, and is often associated with features of self-organization and self-generation. It is well known that most, if not all natural systems are self-organizing, autonomous systems. The way we capture the intrinsic nature and structure of their properties of autonomy and self-organization is via specific observations aimed at isolating what we think are the manifestations of those properties in the system's phenomenology and evolution. We find another example of relational identity in Nutritional Sciences and Dietetics. Certain diets are designed according to the rationale that the effects of a nutritional pattern are determined by the quantities of certain foods. Other diets are built on the belief that what matters the most to achieving a target body fat ratio are the ratios of certain foods or other relations between individual nutrients.

In all these cases where a system evolves naturally from an unintelligible state, to a state that we are able to interpret as a partially ordered state as a result of a certain observation, we characterize that state by stipulating its identity to signify the emergence of a certain, intelligible order. The part of the system that displays order is the part of the system to which we ascribe an identity by virtue of matching a general paradigm case in which *regularity is seen as a manifestation of an objective, intrinsic identity.*

In the case of pattern identity, the geometrical patterns play the role that causality plays in the relational identity example. In Figure 5.5 the five vertical lines at the center of the image look like a square as a result of a certain observational structure with which we may examine that image.

Figure 5.5: Illustration of pattern identity.

In this case the concept of identity is used to signify that a system displays regularity in the form of geometrical patterns that reflect the patterns of the Observational Structure of Intentionality through which we look at that system.

5.4 The Bus Bunching Experiment

In the previous sections we discussed the two modes of existence of concepts, and how these shape the way we define concepts into existence. In this section, we are going to revisit and challenge those ideas, reframe them in a much more accurate and powerful frame of reference, and show that the metaphysical nature of concepts is not definitional: it's *interferential*.

The notion we are going to introduce to characterize the metaphysical nature of concepts is called *definitional interference*, and seems to have several advantages over the traditional philosophical accounts of concepts:

- It eliminates the need of a fixed frame of reference to get epistemic access to concepts: there is no need to think about concepts as mental representations, or abilities, or Fregean senses etc.

- It eliminates the need of a notion of identity to reference concepts, in fact, the notion of definitional interference shows how elusive and unnecessary the notion of identity is altogether.

- It eliminates the phenomenon of ontological continuity that accompanies many traditional philosophical accounts of concepts, in which concepts become bound to a definition or to an underlying philosophical view which dictates what a concept is and how it acquires its identity through the changes allowed by this or that philosophical view.

The underlying theme of definitional interference can, perhaps, be described by the following metaphor:

> *Concepts do not exist as ontologically independent and immutable entities, rather, their nature is akin to the nature of interference patterns on the surface of a lake that are continuously created and destroyed.*

To introduce the notion of definitional interference, we need a thought experiment. I have been thinking about how to explain

5.4 The Bus Bunching Experiment

definitional interference in simple terms for quite a while, when finally in December 2015, I was on a bus from Dublin to Dun Laoghaire[1], when something interesting happened: my bus stopped to load and unload passengers, and another bus running along the same route overtook my bus. This is not a rare phenomenon in public transportation, it's called bus bunching. Bus bunching occurs when two or more buses running along the same route are in the same place at the same time. This phenomenon is a problem for travelers because it breaks the bus headway regularity and results in an abnormal distribution of passengers across buses, but it is also an extraordinary example that illustrates a deep metaphysical and mathematical problem: it is an example of *definitional interference*. To understand why bus bunching is interesting from a philosophical and mathematical point of view, we must understand why it is problematic, and to understand why it is problematic, we must ask a simple question: How do we identify a bus? The answer should be straightforward: to identify a bus we need a notion of *identity* for buses.

Any bus has a route number, which indicates where the bus stops and when it's scheduled to run. Let's organize what we know about buses with two definitions that we'll call: Route Number and Stop Condition.

Route Number

A route number consists of a timetable and a nonempty set of bus stops. Each bus is assigned to one route number only, and multiple buses can be assigned to the same route number. The buses are scheduled to be evenly spaced, and each bus may run multiple times a day.

Stop Condition

A bus must stop at a given bus stop, if the bus stop belongs to its route number, if there are passengers on the bus that have requested the bus to stop, or if at the bus stop there are passengers that want to get on the bus, unless the bus is running at full capacity and no passenger on the bus has requested the bus to stop.

---

[1] A beautiful town on the East coast of Ireland, about 10 miles south of Dublin.

5 A Metaphysical Structure Of Concepts

The Route Number describes how a bus service operates, and sets the context for the Stop Condition. The Stop Condition describes what it takes for a road vehicle that can carry many passengers to be an actual bus. Let's use now these two definitions to determine when two buses are *identical*. When two buses have the same route number they stop, by definition, at the same bus stops, but this condition alone is not sufficient to determine that the two buses are identical because two buses with the same route number might run at different times. So for two buses to be identical they must run along the same route number at the same time. Let's call this the *identity principle for buses*:

> *Two buses are identical if and only if*
> *they run along the same route at the same time.*

With this data in mind, let's imagine this scene: some people are waiting at a bus stop and two buses running along the same route arrive at that bus stop at the same time causing a bus bunching situation. Bus **A** didn't keep to its schedule, and is the first to reach the bus stop. When it reaches the bus stop, it stops to load passengers. Bus **B**, which is following bus **A**, is on time, but when it reaches the bus stop it doesn't stop like it would normally, because none of its passengers requested the bus to stop. This scenario is possible if the bus **A** can load all of the passengers that want to get on the bus, so that the bus **B** doesn't have to stop to load the rest of the passengers. When two buses give rise to the bus bunching situation just described, they satisfy the identity principle for buses at the particular point in time in which the bus bunching occurs, because they run along the same route, and because they are at the same bus stop at the same time, so we might expect them to become temporarily indistinguishable. However, this is not exactly what happens.

When two buses give rise to the bus bunching situation just described, the simultaneous presence of the bus that didn't keep to its schedule (bus **A**), and the bus following it (bus **B**), *suspends* the Stop Condition for the bus **B**, and this causes the bus **A**— the bus that stops to load passengers—to *take temporarily **B**'s identity*. Here is another way to put what happens. In a bus bunching situation, the presence of a bus loading passengers (bus **A**) *prevents* the Stop Condition to apply to the bus following it (bus **B**). As a result of the temporary suspension of the Stop Condition for bus **B**, bus **B** does not stop to load passengers as it

5.4 The Bus Bunching Experiment

would if there wasn't a bus in front of it loading passengers: thus temporarily *ceasing* to be a bus. The bus that doesn't stop to load passengers ceases to be a bus at the bus stop where bus bunching occurs, because the Stop Condition that defines "bus" *cannot* apply to it, as a result of the presence of a second bus that didn't keep to its schedule. In other words, bus bunching disintegrates **B**'s definitional integrity. Note how the number of passengers in the bus bunching situation just described, signals when we can think of the concept of identity in a classical, a non interferential way. Bus **B** doesn't stop because the number of passengers waiting at the bus stop is finite—there aren't enough passengers to fill two buses or n buses in general—and if the number of passengers at a bus stop was infinite, there would have to be an infinite number of buses serving that route, in which case bus bunching would not occur. This observation, essentially, situates the notion of definitional identity—the identity of a concept ascribed to it by a definition—as a limiting case of definitional interference.

The mechanism by means of which bus bunching breaks the definitional structure of a bus, becomes even more apparent when we generalize bus bunching. In a generalized bus bunching situation: multiple buses run along multiple routes that share a set of stops. What happens when two or more buses arrive at the same bus stop at the same time? When bus bunching occurs in a scenario in which multiple buses run along multiple routes that share a set of stops, the suspension of the definitional structure of a bus involved in bus bunching, can be caused in more ways than in the simplified bus bunching scenario—two buses running along the same route—because more buses compete to pick up passengers at every bus stops shared by distinct route numbers. When this combined bus bunching occurs, passengers have more options to travel between two stops shared by distinct route numbers, and this results in an increased chance to break the definitional integrity of one or more of the buses that cause bus bunching.

What's important about this generalized bus bunching scenario is that it gives us a picture of identity far richer than the static and absolute definition of bus given by the Route Number and the Stop Condition. This thought experiment shows us that buses acquire and lose their identity as a result of the presence of other buses competing for passengers. It shows us how buses and passengers form a dynamical system in which buses running along a given route continuously share their identity.

The picture of identity that emerges from the generalized bus bunching experiment, is that identity is not defined by a single entity, but by the interplay of several actors, namely: passengers, cars, road works, and other agents causing traffic jam or other delays, and of course by multiple buses running along distinct routes that share some stops. In this dynamical picture of identity, questions like "What bus is that?" make sense only in relation to a given definitional interplay within which we situate our reference to identity. The very raising of a question about the identity of a bus, without specifying the definitional interplay that gives rise to that particular instance of identity, indicates that there is something basic that we are misunderstanding of what it is that we are asking a question about. It means that we ignore the dynamic nature of the ontology of concepts, it means that we ignore that concepts are like interference patterns that emerge and disappear as a result of the continuous interplay of systems of definitional structures in a phenomenon we have called *definitional interference*.

5.5 Definitional Interference

The general principle illustrated by The Bus Bunching Experiment is that definitions, and thus concepts, are to be understood not as static entities but through the dynamical nature of the cognitive processes in which multiple definitions converge and evolve. We generalize the ideas we described in The Bus Bunching Experiment in the direction of definitional interference, because the logical structure of that thought experiment is entirely built on the mechanism through which definitional integrity—the conditions that hold a definition together—depend on instances of other definitions. The metaphor we used to convey this idea is that of the interference patterns formed by waves on the surface of a lake. Asking What is a concept? according to this view of the nature of concepts, is like asking where is a wave: the very raising of this question is incompatible with the nature of the things the question is about.

What we can say, in light of these observations about the nature of concepts, is that definitions stipulate the existence of certain relations between entities in a context in which those relations are conducive to a meaning. Any other consideration about the nature of concepts, their identity, their ontological continuity, seems to

5.5 Definitional Interference

be incongruent with the nature of concepts. I have no difficulty in going even further in this direction and say that concepts are merely a linguistic tool to reference the phenomena of definitional interference that emerge locally as a result of the mind-world interactions. If we confine a conception of concepts to the domain where they seem to emerge from a definition, we rediscover a traditional conception of concepts which is based on one's own philosophical preferences: I regard those not as conception, but as *descriptions*. Nonetheless, the argument illustrated by The Bus Bunching Experiment seems to indicate that a localized view of concepts does not quite convey their true nature, which emerges much more vividly within the larger arena of definitional interferences.

A rigorous treatment of definitional interference is beyond the scope of this book. However, a way to formalize the intuition of definitional interference could be the following. Let's indicate with D_i and D^i the stipulative and observational forms of a definition D, with d_i and d^i instances of those forms, with $d_{i,j}, d_i^j, d^{j,k}$ the transformations $d_i \to d_j$, $d_i \to d^j$ and $d^j \to d^k$ respectively. We can identify d_i with $d_{i,i}$, d^j with $d^{j,j}$. These three types of transformations are what holds together a definition's definitional integrity: they define the dynamics that govern the transition from the observational to stipulative forms and back—recall the discussion at the beginning of Section 5.2 about how stipulative definitions seem to originate from observational ones, and how the interplay between the two forms gives rise to the dynamical nature of concepts captured by definitional interference. Intuitively, by D, we mean a certain mathematical object consisting of $d_{i,j}, d_i^j, d^{j,k}$— i,j,k varying in a set of indices—and of transformations between them, characterized as follows:

Definition 5.5.1. (Definitional Interference)
A *definitional interference*, is a mathematical object consisting of three distinct entities $d_{i,j}, d_i^j, d^{j,k}$ and transformations between them, in which it is possible to define an internal first order logic, such that $p \Rightarrow \neg q$, with $p, q = d_{i,j}, d_i^j, d^{j,k}$ and $i \neq j \neq k$.

The negation $p \Rightarrow \neg q$ (read *not q follows from p*) in Definition 5.5.1 is itself based on a notion of identity, which defines a mode of being of the things defined by D. For example for a bus, its mode of "being a bus", applies for as long as the bus operates as a bus, as we have seen in The Bus Bunching Experiment.

5 A Metaphysical Structure Of Concepts

Typically, a mode of being is based on an implicit assumption of ontological continuity. Thus, for example being a bus while in the context of a bus operating as a bus means that the being a bus is without interruption. It is this feature, embedded in the implicit assumptions of a definition of bus, that support the notion of identity for buses. Note, however, that the suspension of the definitional integrity is not related to the ontological continuity of a given identity: *it doesn't matter that a bus was not being a bus while its definitional integrity was suspended.* This suggests that definitional integrity has its own internal logic, which allows for the superposition of multiple cancellations of a definitional integrity through multiple concurrent occurrences of $p \Rightarrow \neg q$. The very suspension of definitional integrity counts as definitional interference. Thus $\neg D_i$ is the suspension of—or the negation of—D_i's definitional integrity, whereas D_i means that its definitional integrity is intact.

In the Motivic Lens (Section 4.3) we situated the term spatialization as one of the key processes through which the human mind seems to mathematize. We will dig a lot deeper into the finer structure of spatialization in the chapters dedicated to $\mathcal{M}$-*signatures*. Here we want to use the notion of definitional interference to revisit what we have discovered so far about spatialization with the following

Definition 5.5.2. (Spatialization – *via definitional interference*) Spatialization is the process of revealing the intrinsic nature of an object of thoughts via the technology of definitional interference.

Bibliographical Notes

In Lo Vetere 2015b I describe an embryonal form of definitional interference called *bounded identity* via The Bus Bunching Experiment. I introduced the notion of bounded identity to explain the counterintuitive existence of concepts that cannot be cloned or copied: concepts defined by the bizarre property of not admitting multiple instances of themselves. Later I realized that there was a more abstract and somewhat more natural frame of reference to reason about the nature concepts, which is what I call *definitional interference*, of which bounded identity is a special case.

The notion of identity and its interpretation via definitional

5.5 Definitional Interference

structures remains central in the ordinary mode of reasoning and in many of the ideas I present here because the human intellect appears to rely on identity to articulate thoughts. Several works on the notion of identity in Category Theory have inspired my research of the nature of concepts: Bénabou 1967 sets the scene for the development of modern category-theoretic methods to reconstruct identity, Lawvere 1996a and Rodin 2013 expand on the manifestations of identity and its category-theoretical implications and nature.

There are several examples of relational identity in mathematics and in biology. The concept of relational identity is somehow present, but only implicitly, in Maturana and F. J. Varela 1980 and Francisco J. Varela, Thompson, and Rosch 1993 but is assumed or implied or hidden in the context of the kind of system thinking they develop in their works. The same observation about the use or emergence of implicit notions of relational identity applies to Cybernetics (Wiener 1949, Wiener 1952, Wiener 1961, Arbib 1970, Arbib and Hesse 1986), System Theory and Chaos Theory (Thom 1972, Bateson 1972). There, relational identity takes the form of a variety of generalizations or manifestations of the basic idea of Galois Theory, or of generalizations or manifestations of Fixed Point arguments and their connections to cohomology.

Strangely, when I was studying the deep nature of concepts, way before the argument I begun to develop in Lo Vetere 2015b, I had not connected that topic to the Symbolic Intuition of Reality. Relational and observational identity, and the overarching archetypical framework of definitional interference are, perhaps, what we should be using as a foundation of modern System Thinking.

Chapter 6

The Lazy Boy Experiment

> The role of space as arena of becoming has as one
> consequence a quite specific form of the
> transformation of quantitative into qualitative; the
> seemingly endless elaboration of varied cohomology
> theories is not merely some expression of
> mathematicians' fanatical fascination for fashion,
> but flows from the necessity of that transformation.
>
> Lawvere 1992

The exploration of the Abstract Mind begins here with a thought experiment to reveal the layers of abstraction involved in the construction of a concept of space.

Some objects on a desk and a set of arm positions will help us deconstruct a set of thought processes by means of which we are able to construct a symbolic representation of our experience of those objects. The aim of this elementary thought experiment is to probe the structure of the Abstract Mind to begin to reveal its internal structure.

We begin by establishing a correspondence between objects of perception and objects of thought via a set of arm positions, and use this correspondence to develop a concept of identity for the symbolic process by means of which we reference those objects. Then we show that this correspondence between identities is universal, and use this fact to define a conceptual gadget that encodes the entire symbolic process described in this chapter: *space*.

6.1 The Thought Experiment

Imagine the following scene: a boy is sitting at his desk. The boy is so lazy that all he does is sit on his chair and move his right arm to grab the objects on the desk. In front of him on the desk there are his computer, his mug, a pencil and a stapler. When he looks at his desk he can see all the objects on it, but he is lazy, very lazy, so he always counts to ten before moving his body more than what is strictly required for his survival. When the lazy boy wants to write something with the pencil, all he has to do is leave his lazy right arm bent at a 90 degree angle and grab the pencil that's beside the computer mouse. When he wants to drink coffee, things get a little bit more adventurous because he has to stretch out his lazy right arm to grab the mug that's in front of him. It's when he wants to use the stapler that things get complicated. The problem is, even if he stretches out his arm, the stapler is too far for him to grab it while sitting. He would have to get off the chair to get the stapler, but he is too lazy to do that, so he never uses the stapler. What is interesting about this lazy boy sitting at his desk? Let's summarize what happens in this scene.

Object on the desk	Arm position to grab the object on the desk.
Pencil	*Arm bent at 90 degrees.*
Mug	*Arm fully stretched out.*
Stapler	*No arm position.*

Table 6.1: Correspondence between arm positions and objects on the desk.

If we read Table 6.1 from left to right, we basically reconstruct the scene with the lazy boy. But if we want to understand what happens in the lazy boy's mind, we need to read Table 6.1 from right to left. If we do so, we discover that we can think of objects as arm positions, that is to say: Table 6.1 functions like a dictionary to translate arm positions into objects on a desk. So for example, this dictionary tells us that if the lazy boy keeps his right arm bent at a 90 degree angle, all he can do is grab the pencil beside his mouse, and if his right arm is fully stretched out, the only thing he can grab is his mug. The deceptively simple structure of the technology described in this example, contains the seed of a powerful machinery to construct concepts: because it exemplifies the ability to formulate thoughts about something by thinking about something else.

6.2 Object Classifiers

Let's now examine in more detail the effect of the dictionary of Table 6.1 on the lazy boy's ability to conceptualize the objects on the desk. If he stretches out his right arm, he discovers that there are other objects on the desk that he can grab other than his mug. Similarly, if he keeps his right arm bent at a 90 degree angle, he discovers that there are other objects that he can grab other than the pencil beside the mouse. Note that when his right arm is stretched out, the lazy boy cannot grab the pencil, so we can think that the lazy boy cannot grab the same object on the desk with two distinct arm positions. As a result of this observation, we could upgrade the dictionary of Table 6.1 with more objects associated to the same arm position, in such a way that no object is associated to two distinct arm positions.

A convenient way to describe this observation about the effect of the arm position on the objects is to say that each arm position acts as an *Object Classifier* for the objects on the desk. For, every object on the desk is classified by one and only one position of the lazy boy's right arm. What this means is that some objects on the desk become "like the mug" if the lazy boy can grab them by stretching out his right arm, some objects become "like the pencil" if the lazy boy can grab them without extending his arm, and some objects become "like the stapler" if the lazy boy cannot grab them while sitting in his lazy boy chair, either because they are too far from him, or because they are too close to his body for him to grab them, even when his right arm is bent at a 90 degree angle. With this upgraded dictionary, the arm positions become a conceptual lens through which the lazy boy can observe what's on his desk.

However, there is a more accurate way to describe the thought process enabled by an Object Classifier: an Object Classifier enables a crucially important cognitive architecture in the lazy boy's mind, it allows him to *identify* the objects via his arm positions by means of a notion of sameness. For, when two or more objects become like the mug as a result of being classified by the same arm position, they acquire a quality of sameness which makes them indistinguishable from an arm position viewpoint.

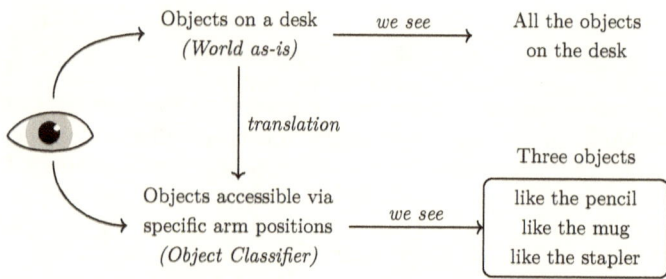

Figure 6.1: Illustration of the abstraction process that uses a set of arm positions to translates sense data into an Object Classifier.

The cognitive architecture that transforms an ideal image of the World as-is, into the world that we see through an Object Classifier is summarized in Figure 6.1. When the lazy boy looks at the desk without any conceptual lens, he sees the objects on the desk. In contrast, when he looks at the desk through the lens of the Object Classifier, he sees three objects: the objects *like* the pencil, *like* the mug and *like* the stapler. The construction of an Object Classifier hinges on the mode of access to the objects on the desk, and in this thought experiment we have chosen a lazy boy because he can produce only three types of arm positions, which result in a simple Object Classifier.

6.3 A First Concept Of Space

There is, I think, a central question we should ask at this point: What is the relation between the objects out there on the desk, and the abstract dictionary constructed by the lazy boy in this thought experiment? The creation of a dictionary to translate arm positions into objects on a desk is an abstraction process that uses arm positions to characterize the spatial relations between

6.3 A First Concept Of Space

the objects on the desk and the lazy boy's body. It is important to note that what we call "spatial relations between the objects on the desk" do not exist independently of the notion of "moving the right arm to grab an object on the desk". Spatial relations are indeed only a convenient way to reference the correspondence between types of arm movements and objects on a desk based on certain verifiable success criteria, such as the possibility to grab an object by positioning the right arm in such and such way. So there are no spatial relations between objects as such: what there is, is a universe of events, some of which validate the strategy of positioning an arm in such and such way so as to grab the intended object.

This observation about the ontological status of spatial relations between objects, clarifies the cognitive function provided by the ability to create dictionaries. The function of a dictionary, and, consequently, the function of an Object Classifier, is defining a way to validate the abstraction process from objects to arm positions.

The Object Classifier enables the lazy boy to know the objects on his desk via the notion of space, which makes space a cognitive technology. The space created by the lazy boy is a technology to deal effectively with the stuff on his desk for a certain set of purposes such as writing and drinking. Let us then describe briefly this technology with these three questions and answers:

What problem does the lazy boy space solve?

The lazy boy space is a technology to solve the problem of relating the lazy boy's intention to do something (write, drink) with an environment in which that intention can be fulfilled (his desk).

How does space solve the problem it is designed to solve?

The lazy boy space uses a method of access to the objects on the desk—that in this experiment consists in grabbing the objects on the desk via three types of arm positions—to define an abstract correspondence between the objects on the desk and the perceptual events that validate that method of access. This correspondence between objects of perception and objects of thought constitutes the syntax of a synthetic experience of the physical objects that takes place in the lazy boy's mind.

What knowledge is needed to construct the lazy boy space?

These elements are needed to construct a lazy boy space:

- the definition of the intention that motivates the construction of the space (such as the intention to write or drink)
- the definition of a strategy to fulfill the intention (such as grabbing an object with the right arm in such and such way)
- the definition of an environment that validates the strategy (there must be objects on the table with such and such characteristics)

For the moment we will not elaborate on the assumption that there must be also the knowledge of how to conceptualize an Object Classifier, this will be explained in greater detail later in the book.

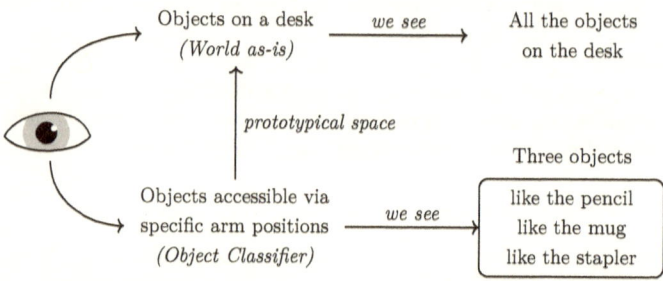

Figure 6.2: Illustration of the correspondence that gives rise to a prototypical concept of space.

Intuitively, what we call *space* is the process of thinking about the objects on the desk from the Object Classifier viewpoint. This process is illustrated in Figure 6.2 by the arrow that goes from the Object Classifier to the World as-is. It is a prototypical notion of space because it relies on a particular choice of a method to access the objects on the desk, namely, the three arm positions that the lazy boy can produce. Space is the process of ascribing the ontological structure of the objects of thought that the lazy boy forms through an Object Classifier to the objects on the desk. It is an interpretation process built on the notion of identity provided by an Object Classifier.

6.4 The Dynamic Structure Of Space

If we look carefully at Figure 6.2, we notice that there is a pattern. When the lazy boy observes the objects on the desk, he sees the objects as they are. In contrast, when he decides to use his right arm to interact with the objects on his desk, he sees three types of objects: the objects like the pencil, like the mug and like the stapler. If we compare the mere observation of the objects on the desk with what the lazy boy sees through the movement of his right arm, a pattern begins to emerge. There are the objects on the desk. There is a way to interact with those objects, which in this thought experiment consists in looking at them, or by grabbing them, through a set of arm positions. And there is a mental image of those objects that manifests itself as a result of a specific way to interact with those objects. The objects seen as they are appear as objects, whereas the objects seen through the Object Classifier generated by the three arm positions appear as three types of entities that we labeled "like the pencil", "like the mug" and "like the stapler".

There is no special reason why the lazy boy should interact with the objects on his desk by looking at them or by grabbing them with his right arm. As a matter of fact, these two ways to interact with his environments are completely arbitrary. A direct consequence of this observation is that we can think of the objects on the desk not as an unmediated presentation of what *is* on the desk, but as the objects on the desk that we see through the Object Classifier generated by a specific mode of interaction that we call "looking at those objects". So what the lazy boy sees on his desk, is *always* the product of a certain Object Classifier. If he has a lazy day, a day in which even moving his right arm is too much to ask, what he sees by just looking at his desk is a bunch of objects. If he feels particularly energized and decides to move his right arm, what he sees on his desk is a set of three objects.

6 The Lazy Boy Experiment

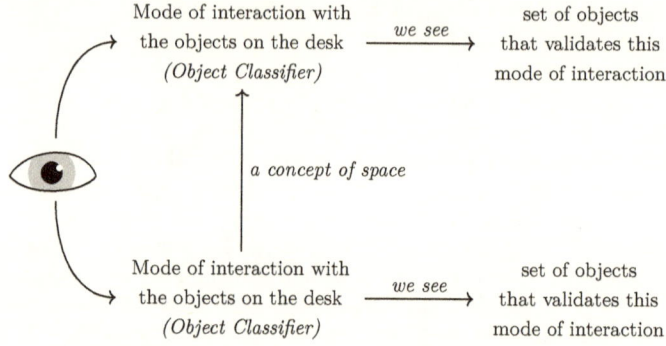

Figure 6.3: A presentation of the concept fo space as transformation between Object Classifiers.

These observations reveal a deeper nature of what we called space. Since everything we see is presented to our intellect by an Object Classifier, it is more accurate to think about space as a transformation between Object Classifiers (Figure 6.3).

Bibliographical Notes

These reflections on the deconstruction of a basic concept of space are based on my work to give an elementary presentation a fundamental structure of mathematical thought Lo Vetere 2009a and on my interpretation of the category-theoretic version of the underlying ideas Baez and Dolan 1998, Rodin 2005, Lawvere 1969a.

Chapter 7

The Yarn-Ball Experiment

> The [other] immensely important point to grasp is that "real" is not a normal word at all, but highly exceptional; exceptional in this respect that, unlike "yellow" or "horse" or "walk", it does not have one single, specifiable, always-the-same meaning.
>
> Austin 1962

> Where conscious subjectivity is concerned, there is no distinction between the observation and the thing observed.
>
> Searle 1992

In this chapter, we continue the exploration of the Abstract Mind with an argument and a thought experiment in which a counterintuitive, one-dimensional representation of the ordinary three-dimensional space, begins to reveal the inner workings of the Abstract Mind.

Our perceptual experience of space is three-dimensional. Our imagination and our spatial intuition are three-dimensional, and despite the amazing power of the human mind, we are not able to think in more than three dimensions. This means that, regardless of what empirical coordinate system we use to conceptualize space, our intuitive, unmediated way to understand space is three-dimensional. This is a very peculiar feature of the human intellect which has many subtle repercussions in the way we seek knowledge and construct concepts. There is probably an evolutionary explanation for this. A way to verify how the human mind seems to be wired to think in three dimensions, consists in trying to imagine a two-dimensional world, such as a sheet of paper, where the only way one can travel is in two directions inside the sheet of paper. Most people who try to do this thought experiment notice that they have to think about it for a while before they can actually imagine what a two-dimensional experience of space would be like. The reason why it is not intuitive to think about a two-dimensional space is that we instinctively try to use our three-dimensional experience of space to imagine what it would be like to live in a two-dimensional space. Our inability to think in more than three dimensions is one of the many reasons why mathematics is such a powerful cognitive technology. Mathematics is a spectacularly powerful cognitive technology—probably the most effective cognitive technology ever created by our species—because it allows us to reason about things that we are not able to imagine. The ability to use mathematical conceptualization to go beyond the limits of imagination, amplifies human intelligence in ways that we do not always fully understand. For example, we can reason about problems in a space with 26 dimensions by simply learning some mathematical rules to manipulate symbols that represent the algebraic structure of that 26-dimensional space.

7.1 The Thought Experiment

A method to identify the position of a point in a three-dimensional space is described in Figure 7.1. Let P and Q be two distinct points in the ordinary three-dimensional space. In order to identify the position of Q with respect to P we take a clothespin, a wool hank and a tape measure and follow these steps:

A. Wind the wool hank into a ball around P.

B. Keep winding the yarn around the ball until the yarn touches Q, then stop winding and clip the clothespin where the yarn touches Q.

C. Unwind the ball of yarn.

D. Measure the length of the yarn between P and Q.

We define the *yarn-number* of Q with respect to P as the length of the yarn between P and Q that we measure in step D. What does this result mean? What does it mean that we are identifying

7.2 How To Wind A Ball Of Yarn

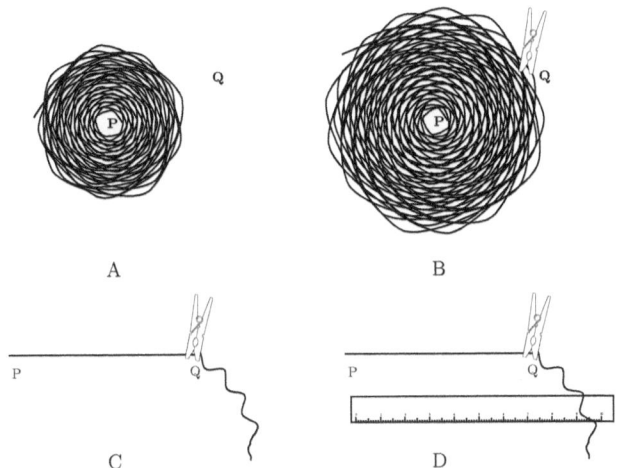

Figure 7.1: Steps to identify the position of a point in space with a ball of yarn.

the position of a point in the three-dimensional space with only one measurement instead of three, as in the Cartesian coordinate system? Before we can answer these questions, we need to look at this experiment in more detail. The method we have followed to find the position of Q with a ball of yarn does not depend on the choice of Q, so for us to determine whether this is an actual coordinate system we need to find if different points in space have different yarn-numbers.

7.2 How To Wind A Ball Of Yarn

One objection to the yarn-ball argument could be that there are many ways to wind a ball of yarn, and that each way results in a different yarn-number for the same point Q, because it requires a different amount of yarn to reach Q. The existence of multiple ways to wind a ball of yarn becomes a problem if we cannot repeat the same measurement and find each time the same yarn-number for a given point Q. The problem this objection raises is therefore: How we can make the yarn-ball method a repeatable method of measurement? The solution to this quandary is simple: we need to define a convention. Today, for example, we define 1 meter as

7 The Yarn-Ball Experiment

the distance traveled by light in vacuum in $\frac{1}{299792458}$ of a second. The assumption behind this definition is that the speed of light is a universal constant of Nature, and therefore the distance traveled by light in vacuum in $\frac{1}{299792458}$ of a second is constant. Similarly, to define a convention to wind a ball of yarn, we must find a way to systematically visit each point on a sphere. So how can we standardize a method to wind a ball of yarn?

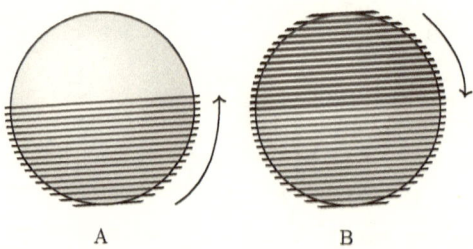

Figure 7.2: Stylized illustration of a method to wind a ball of yarn.

There are many solutions to the problem of winding a ball of yarn: the solution we present here is probably the simplest. The way to wind a ball of yarn we introduce is called "up and down", and consists in winding the yarn around the ball of yarn from the bottom to the top and back again. Figure 7.2 shows the yarn being wound around the ball (A) from the bottom to the top and (B) from the top to the bottom. This technique to wind the yarn around the ball, makes it easy to construct the exact same ball of yarn each time, and to use the same amount of yarn each time. Other ways to wind a ball of yarn employ more sophisticated techniques that rely on the symmetries of regular polyhedra.

7.3 Precision

Another objection to the yarn-ball argument could be the following: How can we find the position of a point located in the empty space between crossing threads? There is empty space between crossing threads because the yarn thickness is not zero, so if a point Q is in the empty space between crossing threads its yarn-number is an approximation which indicates the minimum distance between Q and one of the threads traversing the region of space Q is in.

In the metric system, for example, to increase the precision of a measurement, it is sufficient to switch to a smaller decimal sub-multiple: from centimeters ($\frac{1}{100}$ of a meter) to millimeters ($\frac{1}{1000}$ of a meter) and so on. What is the equivalent of the decimal sub-multiple unit in the yarn-ball system of measurement? To answer this question, we need to observe that the precision of a yarn-number increases as the empty space between crossing threads decreases, and that the empty space between crossing threads decreases with the yarn thickness. So the thinner the yarn, the smaller the empty space between threads, the greater the precision.

7.4 Taming Infinity

There is yet another objection to the yarn-ball argument. This one is a direct consequence of the need to reduce yarn thickness to increase the precision of the yarn-numbers. When we reduce yarn thickness, the amount of yarn needed to identify the position of a given point Q increases because the yarn has to be wound around the ball more times to fill the same volume of space. This observation suggests that to increase indefinitely the precision of the measurement of a yarn-number for a given point Q, the amount of yarn needed to carry out the measurement increases indefinitely, and so does the yarn-number. A way to visualize this peculiar phenomenon that occurs in the measurements of yarn-numbers is to imagine that the increase in the precision of a measurement results in the stretching of space. Imagine a point Q of an imaginary three-dimensional space with this "stretching" property, and a Cartesian coordinate system in this space. If we measure the three Cartesian coordinates of Q in meters, we find that they are, say, $(2, 7, 4)$ meters, but if we switch from meters to millimeters to increase the precision of the measurement, we find that the Cartesian coordinates of Q are $(3000, 8000, 5000)$ millimeters, that is, respectively 3, 8 and 5 meters.

The problem of the yarn-number increasing indefinitely with the precision of the measurement can be solved with a trick. Instead of defining the yarn-number L of Q as the length of the yarn between P and the clothespin (see Figure 7.1 (B,D)) we can define a new yarn-number L' of Q as $L' = L \times \delta$, the product of the old yarn-number L and the yarn thickness δ. Intuitively, the effects on the new yarn-number L' of L increasing indefinitely as

δ decreases indefinitely cancel each other out when we multiply them together. Think about these products: $1 = 1\frac{1}{1} = 10\frac{1}{10} = 100\frac{1}{1000} = \frac{1}{1000}\ldots$, where $1, 10, 100, 1000\ldots$ are the old yarn-numbers L, and $\frac{1}{1},\frac{1}{10},\frac{1}{100}\ldots,\frac{1}{1000}\ldots$ are the yarn thickness.

7.5 What Have We Discovered?

In this thought experiment, we have identified the position of a point in the three-dimensional space with one number, the yarn-number, and we have done so starting from the ordinary perceptual experience of the physical world. In contrast, we know that starting from the same perceptual experience of the physical world, we can construct a concept of space with a Cartesian coordinate system, which requires three numbers to identify the position of a point in a three-dimensional space. What does this result mean? What does it mean that we can construct concepts of space so different from each other that are consistent with the same perceptual experience?

Here "consistent" refers to two distinct features of this construction of space.

The one-dimensional concept of space, obtained from unmediated sense data via the yarn-ball method, is consistent with the traditional way in which we perceive the physical space around us. Consistency, here, means consistency with the immediateness and unity of physical experience, and of our consciousness of it. It means that the physical space that we see through the concept of space developed in the yarn-ball method, is a space that we can immediately recognize as the ordinary, three-dimensional picture of the physical world, in which physical objects correspond to objects in our mind without the mediation of algebraic manipulation of symbols. This immediateness is not a given. In mathematics, concepts of space where the points are transformations between mathematical gadgets, or where points are entire mathematical structures, or even other spaces, are very common and far from intuitive. In these cases where mathematical abstraction overpowers completely human imagination, the correspondence between the "stuff" in the physical world and the abstractions in our mind is lost, and the only way to reason about objects and states of affairs in the physical world is via the manipulation of algebraic symbols that we know how to interpret. In this thought experiment, we

7.5 What Have We Discovered?

didn't have to resort to any degree of abstraction that would have inevitably created a gap between the objects in the physical world and the abstractions of our mind.

The concept of space obtained with the yarn-ball method is consistent with a reference method to construct concepts of space, namely, the Cartesian coordinate system, which suggests that there is a correspondence at the structural level between the two methods.

On the one hand, we have the practical, constructive nature of The Yarn Ball Experiment, where we can literally use our hands to build a concept of space that we can use to relate with the physical world around us. On the other hand, we see a correspondence between the steps that we follow to conceptualize space with the Cartesian method, and the steps that we follow to conceptualize space with a ball of yarn. Both constructions of a concept of space, Cartesian and yarn-ball, share the same haptic nature: they both require certain tools and they both rely on certain procedures to use those tools. Both constructions of a concept of space require the same level of abstraction, that is: the association between points in space and numbers is the result of direct measurements of physical objects. The immediateness of both ways to construct a concept of space is what makes them structurally similar.

	Cartesian	Yarn-Ball
Tools	*Tape measure.*	*Tape measure, yarn, and clothespin.*
How the points in space are associated to numbers	*Definition of a unit of measurement, and one measurement per each coordinate.*	*Definition of a standard way to wind a ball of yarn, and one measurement.*

Table 7.1: Comparison between the Cartesian and the Yarn-Ball methods to construct a concept of space.

Now that we have made this correspondence between these two methods manifest, we ask: How can we interpret the fact that two methods to construct a sensorially based concept of space give rise to two distinct concepts of space, where the element of distinction is what appears to be a fundamental, ontologically objective feature of physical space such as dimensionality? What

7 The Yarn-Ball Experiment

is it that determines this difference in the type of concept of space produced by these two methods? The first thing we must observe, is that what we have found is a feature of the concept of space, and not a feature of physical space. The point is to establish the implications of this feature for the way we use mathematics to relate to objects and states of affairs in the world.

There is a specific point in the process of conceptualization of space in this experiment, in which we divert from the Cartesian path and create the path that leads to the construction of a one-dimensional concept of space. This bifurcation occurs precisely when we decide to interface with the points in the physical space in a different way. It is the decision to use a strategy that involves using a yarn that determines the new path that leads to the one-dimensional picture of the physical space. Other strategies, such as for example using spherical coordinates—used in Geography and Astronomy—would lead to other three-dimensional pictures of space. So what seems to emerge from this analysis is that there is a relation between the strategy chosen to interface with space, and the concept of space constructed as a result of that choice.

Bibliographical Notes

This thought experiment is based on the notion of space-filling curve. A space-filling curve is a continuous line that "fills" the entire space contained in a square, or in a cube, or in a 4-dimensional cube and so on. [1] Space-filling curves (Sagan 1994) were discovered by Italian mathematician Giuseppe Peano (1890), and are based on a counterintuitive result due to German mathematician Georg Cantor on the measurement of the size of a set—a concept known as *cardinality*. Cantor discovered that the unit interval $[0\ldots 1]$—which contains numbers such as 0.05, $\frac{\sqrt{2}}{3}$, $\frac{3}{4}$, $\frac{\pi}{4}$, ...—has the same cardinality as the unit square, or the unit cube, and in general as any n-dimensional unit cube, and that therefore the unit interval can be put in a 1-1 correspondence with. Peano discovered that it is possible to construct the correspondence Cantor proved theoretically between the points on the unit interval and the points in a unit square as a continuous line (Figure 7.3). Similar, counterintuitive constructions based on the same result are Banach-Tarski paradox—a geometric paradox in which a solid

---
[1] A technical definition of space-filling curve is that of a continuous map whose range contains a finite-dimensional manifold.

7.5 What Have We Discovered?

ball in three-dimensional space is divided into a finite number of disjoint subsets, which rearranged in a different way yield two identical copies of the original ball.

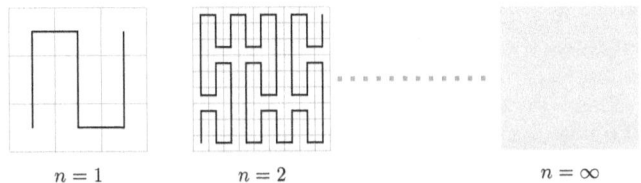

Figure 7.3: Illustration of the first two iterations of a Peano curve.

The Yarn Ball Experiment is my own work (Lo Vetere 2008b), and shows how to construct a three-dimensional space-filling curve without any notion of recursion or fractal geometry as in the conventional mathematical treatment of the subject. This thought experiment provides also an example of a general pattern to deconstructs the conceptualization of space which is the focus of Chapter 8.

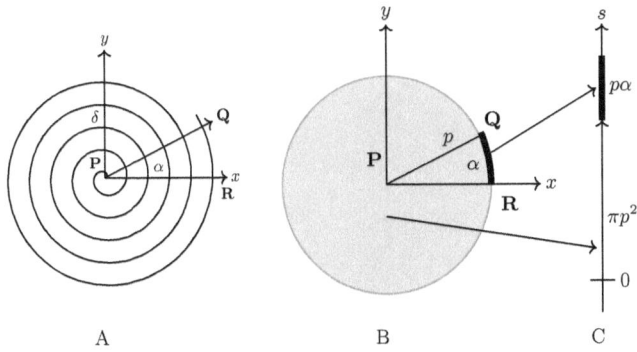

Figure 7.4: Illustration of the construction of a yarn-number in 2 dimensions.

Here I sketch the construction of a yarn-number in the two-dimensional case for pedagogical purposes with only elementary, high school math concepts. Let p be the length of the vector

$\overline{PQ}$, α the angle that $\overline{PQ}$ forms with the x-axis, and δ the yarn-thickness. We can approximate the length of the arc of $\overparen{PQ}$ along the Archimedean spiral centered at P with step δ (Figure 7.4-A), as the sum of n circumferences of radius $\delta, 2\delta, 3\delta, .., n\delta$, that is, $L = \sum_{i=1}^{n} 2\pi\delta i = \pi\delta n(n+1)$. Furthermore, we can approximate n with $\lceil \frac{p}{\delta} \rceil$ which gives $L = \pi\frac{p^2}{\delta} + \pi p$. Thus, by defining $L' = L\delta$ we get $L' = \pi p^2$—plus a term that vanishes as δ tends to zero. By adding to L' the length of the arc $\overparen{RQ}$, which does not depend on δ, the length of the path $\overparen{PQ}$ along the Archimedean spiral centered at P, for arbitrarily small values of δ becomes $L'' = \pi p^2 + p\alpha (mod\ 2\pi)$.

This calculation sends a point in polar coordinates (p, α) to a non negative real number in two steps (Figure 7.4-B). First it sends the area of a disc of radius p to a segment of length πp^2 and then it adds to this segment the length of the arc $\overparen{RQ}$. The segment on the s-axis (Figure 7.4-C) that measures $\pi p^2 + p\alpha$ is a two-dimensional yarn-number.

Chapter 8

A Synthetic Structure Of Intentionality

> ... [P]erception is an intentional and causal
> transaction between mind and the world. The
> direction of fit is mind-to-world, the direction of
> causation is world-to-mind; and they are not
> independent, for fit is achieved only if the fit is
> caused by the other term of the relation of fitting,
> namely the state of affairs perceived.
>
> Searle 1983

In this chapter, we introduce an interpretation of a fundamental property of mental states called *intentionality* to model the human mind's ability to conceptualize.

As Searle puts it (Searle 1983), intentionality is

> *"that property of many mental states and events by which they are directed at or about or of objects and states of affairs in the world".*

This chapter revolves around two themes which can be regarded as two sides of the same coin. There is the theme of how intentionality explains what appears to be the universal pattern through which the human mind *constructs* concepts. And there is the theme of how the human mind is able to *connect* to reality symbolically through concepts. These two themes are connected through the intentional nature of the mind-world interplay. The account

8 A Synthetic Structure Of Intentionality

we present here of how intentionality explains conceptualization constitutes a frame of reference to reason about the basic mental functions modeled by the Abstract Mind. The explanation of how concepts connect us to reality is a stepping stone to understanding and generalizing the computational nature of our conception of knowledge, and the reasons why that nature seems to decode physical reality.

The theoretical model of Abstract Mind we present here is called Synthetic Structure of Intentionality, and is designed to capture the basic structure of the intentional mental processes underpinning the construction of concepts. The argument to introduce the Synthetic Structure of Intentionality is developed in two steps.

1. We analyze how The Lazy Boy Experiment and The Yarn Ball Experiment produce a concept of space, and show that they have the same structure. The insight into the anatomy of thought that emerges from this introductory analysis is condensed in a theoretical model called Analytic Structure of Intentionality. It's called analytic because it rests on a substrate of concepts out of which intentionality is built.

2. We examine the constituents of the Analytic Structure of Intentionality, and explain how their interplay gives rise to the mind's ability to conceptualize. This second step reveals the exact relations between the types of abstractions that produce concepts, and leads to the definition of a more sophisticated model of Abstract Mind called Synthetic Structure of Intentionality.

The ideas presented in this chapter constitute the frame of reference for the creation of formal models of the Abstract Mind that we will see in Part III.

8.1 An Analytic Structure Of Intentionality

Let us recall that the technique we employed to explain the workings of the Abstract Mind consists in probing the Abstract Mind with two versions of a concept of space to unearth the structure of the cognitive processes that create concepts. To this end, we presented two thought experiments to construct a concept of space, The Lazy Boy Experiment (Chapter 6), and The Yarn Ball Exper-

8.1 An Analytic Structure Of Intentionality

iment (Chapter 7). Those two thought experiments were designed to accentuate the features of the thought processes that appear to make up the human mind's ability to mathematize.

Here, we begin to dissect the two constructions of space we saw in The Lazy Boy Experiment and in The Yarn Ball Experiment to reveal their common structure. Let us first briefly recall what we observed in the two thought experiments.

In The Lazy Boy Experiment, the lazy boy wanted to grab the objects on his desk with his right hand. He discovered that there were three types of arm positions to achieve his goal, and this realization led him to the formulation of a concept of space via the definition of a dictionary to translate arm positions into objects on the desk. Similarly, in The Yarn Ball Experiment, we wanted to identify a point in the three-dimensional space by measuring its position. Our strategy relied on a clothespin, a wool hank, a tape measure, and on a strange method to use those three objects to assign yarn lengths to points in the three-dimensional space. This strategy led us to the formulation of a one-dimensional representation of the three-dimensional space.

There is a pattern here. There is the *Will* to interact with the objects in the physical world. There is the *Intention* to act upon that Will in a specific way. There is a set of *Actions* that characterize how the Intention to act upon that Will is executed. And there is a mechanism by means of which the *Object Classifier* (Section 6.2) generated by the set of Actions gives rise to a *Concept* of space. Will, Intention, Action, Object Classifier and Concept—capitalized to indicate that they are the steps of a process—are a way to deconstruct the processes of conceptualization we saw in the two thought experiments, and to reveal the different types of abstractions involved in the same. These five elements represent the Analytic Structure of Intentionality, in that they describe intentionality in terms of a collection of underlying primitive notions.

8 A Synthetic Structure Of Intentionality

Step	Lazy Boy Experiment	Yarn-Ball Experiment
Will	*Identify the objects on the desk.*	*Identify the position of points in space.*
Intention	*Interface haptically with the objects on the desk.*	*Measure the position of a point in the three-dimensional space.*
Action	*Extend the right arm to grab the objects on the desk.*	*Wind a ball of yarn, clip the clothespin, unwind, and measure the yarn length.*
Object Classifier	*The arm movement classifies the objects on the desk.*	*A standardized method to wind a ball of yarn classifies the points in the three-dimensional space.*
Concept ontological	*The interaction with the objects on the desk via the movement of the right arm creates a concept of space with three points (pencil, mug and stapler) each corresponding to a type of arm movement.*	*Winding the yarn creates a correspondence between the points in the three-dimensional space and the points of a one-dimensional space.*
Concept mathematical	Space as transformation between Object Classifiers.	

Table 8.1: Comparison between The Lazy Boy Experiment and The Yarn Ball Experiment

In Table 8.1, the Will element characterizes the goal, thus it answers the question: What is the desired outcome? The Intention characterizes a general course of action to cause the outcome defined by the Will, thus, it answers the question: What is needed to cause the desired outcome? The set of Actions characterizes a strategy to implement the goal defined by the Intention, thus it answers the question: How can we cause the desired outcome? The Object Classifier characterizes the goal that a certain set of Actions is directed at, thus it answers the question: How can we validate the goal against the desired outcome? The Concept is the culmination of the chain of cognitive acts that starts with the Will. The structure of the Concept encodes every step of this process—Will, Intention, Action, and Object Classifier—and is the lens through which we look at the world. Thus, the Concept answers

8.1 An Analytic Structure Of Intentionality

the question: How can we experience the desired outcome? where the term "experience" refers to the entire reference framework created by the interplay between Will, Intention, Action and Object Classifier.

From a philosophical perspective, the Concept defines the structure of human experience: the lazy boy's experience of the objects on his desk is created by the movement of his right arm, and our abstract connection with the points in space is created by a certain way to wind a ball of yarn. From a mathematical viewpoint, we want to construct a mathematical object that codifies the philosophical structure of a given concept of space: we want to create a *model* of that concept. That model—which is one of infinitely many—is a transformation between Object Classifiers.

Step	Purpose	Question
Will	*Define the goal.*	*What is the desired outcome?*
Intention	*Characterize a general course of action to cause the outcome.*	*What is needed to cause the desired outcome?*
Action	*Characterize a strategy to implement the intention.*	*How can we cause the desired outcome?*
Object Classifier	*Characterize the target the action is directed at.*	*How can we validate the goal against the desired outcome?*
Concept	*Define a concept that codifies this entire process.*	*How can we construct our experience of the wanted outcome?*

Table 8.2: The constituents of the Analytic Structure of Intentionality

These questions are summarized in Table 8.2 for convenience. What begins to emerge from this cursory analysis, is that each step of the Analytic Structure of Intentionality leads to the next step as a result of some kind of selection process. For example, the Intention to act upon one's Will according to a specific plan is the result of a process that identifies a certain course of action among a universe of possibilities. Similarly, there are many ways in which the plan described by an Intention can be executed, which indicates that there must be a mechanism by means of which a specific

8 A Synthetic Structure Of Intentionality

Action is selected—or elected, if we want to give an evolutionary twist to this discussion—among a universe of possibilities.

With these broad ideas in mind about the Analytic Structure of Intentionality, we want to dig a lot deeper into the core aspects of the structure of intentionality, and to this end, we have to distill what characterizes intentionality at a higher level of abstraction. As a general rule, a method to reveal the more rarefied nature of a concept, consists in dissecting the relations between that concept and other related concepts. In this case, the analytic representation of intentionality consists of an ordered sequence of concepts, namely, Will, Intention, Action and Object Classifier—omitting Concept, which is the manifestation of the superposition of the other four concepts. Thus, there are three relations we need to examine: Will-to-Intention, Intention-to-Action and Action-to-Object Classifier. The analysis of these three relations is what we call the Synthetic Structure of Intentionality (Figure 8.1), where synthetic is to be interpreted in the mathematical sense, that is, *axiomatic*.

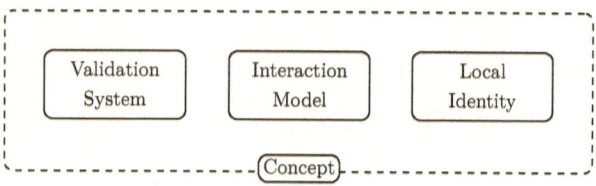

Figure 8.1: The Synthetic Structure of Intentionality.

The Synthetic Structure of Intentionality is a theoretical model that encodes the distinct cognitive architecture necessary to explain what we perceive or observe as the deliberate genesis of a concept, as a result of the intention to achieve a specific goal. The synthetic character of this theoretical model of intentionality is that it recasts the frame of reference of Will, Intention, Action and Object Classifier into a structure consisting of three elements that codify higher cognitive functions, namely: a Validation System, an Interaction Model and a Local Identity.

8.2 Validation Systems

A Validation System codifies that which is intelligible to an Abstract Mind at an archetypical level, and defines an Abstract Mind state.

The plan to fulfill a want is the manifestation of a distinct Intention. Will and Intention share the property of being directed at or of or about objects and states of affairs in the world. For if I have a desire, it must be a desire that such and such thing happens, or it must be a desire of something, or to do something. Similarly, if I have an Intention, it must be an Intention to do something, or to act upon something. However, the distinctive feature that characterizes Intention is that Intention is directed at something in a structured way that involves some idea or projection about a future in which the goal is part of a subjective reality. Intention is about planning for the goal set by one's aim. Thus, there is a causal relation between Will and Intention: Intention explains Will. This causal relation determines the additional structure of an Intention: it is the extra structure that encodes the belief system that makes an Intention consistent with the effect that it brings about.

But there is more than just causality that connects Will to Intention. There is a sense in which intentionality establishes a tension towards a goal, and stores in that tension the potential for action. There is no such thing as intentionality without this tension and there is no such thing as intentionality without an aim. Intentionality without an aim is just an aspiration which can vary only in intensity. So the next question we need to ask is: How can we characterize the aim encoded by intentionality?

Validation System : $Will \rightarrow Intention$

A concept that describes the transition from Will to Intention is the notion of *Validation System*. A Validation System characterizes what qualifies a specific Intention that satisfies the want symbolized by the Will element of the Synthetic Structure of Intentionality. In this sense, a Validation System represents the capacity to articulate thoughts. The concept of Validation System is more complex than causality because it encodes a variety of coherence conditions that are needed to make causality a consistent and epistemically accessible concept upon which we build explanations. For example,

8 A Synthetic Structure Of Intentionality

the goal to have a sip of coffee (Will) can be achieved by getting a cup of coffee (Intention), because the notion of sipping coffee (effect) as a result of getting a cup of coffee (cause) is consistent with the ordinary frame of reference in which we can *explain* the use of the ideas of sipping coffee and getting a cup of coffee to deliberately interface with the objects in the physical world. In this case the causal link between Intention and Will is evident and even trivial, and relies upon our knowledge of how to cause the desired effect and how we can use this causal relation to explain our Intention.

The notion of *explanation*, in conjunction with the *knowledge* of a causal relation is crucial here. There is a subtle distinction to make about the dynamics that involve causal relations and the construction of explanations. Explanations are what makes new data consistent with what we already know to be true. We use causality to construct explanations. However, the knowledge of a causal relation is not sufficient to produce an explanation. What's needed to produce good explanations, is the knowledge to make new data consistent with the system of ideas that receives the new data, and this knowledge has no special relation with causality.

The next example illustrates the difference between the knowledge of a causal relation between facts or ideas, and the knowledge needed to construct an explanation of those facts or ideas based on causality. In The Yarn Ball Experiment, we asked how to identify the position of a point in the three-dimensional space. We know that a Cartesian coordinate system (cause) provides a way to locate the position of points in space (effect), however, as we saw in the experiment, what constitutes the actual content of the causal relation can be far from intuitive. For, finding the position of a point in the three-dimensional space as a result of the application of set of instructions that involve a clothespin, a wool hank, a tape measure, and a peculiar way to wind a ball of yarn is not that obvious.

A Validation System defines the conditions of satisfaction that an Intention needs to meet to be consistent with the goal it causes. However, as we saw in this discussion, the structure of the Intention element is richer than causality because it encodes the belief system that characterizes *how* Intention causes the desired effect: to comprehend this mechanism we need the notion of Interaction Model.

8.3 Interaction Models

An Interaction Model codifies how an Abstract Mind can think when it is in a state defined by a given Validation System.

Let us now examine the *structure* of a Validation System. From a functional viewpoint, a Validation System provides a new frame of reference in which the achievement of a goal is consistent with the frame of reference in which the goal is defined. This means that: the definition of the cause (Intention) extends the definition of the effect (Will) as a result of the data encoded in the Validation System. Crudely speaking, a Validation System defines the horizon inside which every plan to achieve a goal can be imagined. It is the arena where any set of actions towards a goal takes place. But how do we decide that a goal has been achieved? For example, how do we know that we are sipping coffee from a mug as a result of acting upon our desire to have a sip of coffee, and according to our intention to achieve that goal in a specific way? The notions that capture the answers to these questions are **success** and **success condition**.

- **success** defines a Validation System by marking its ontological boundary: it defines the meaning of *is* and *isn't* by characterizing when a Validation System ceases to exist—to cease to exist in this context means that the definitional structure of the Validation System loses its integrity and thus its applicability.

- **success condition** is a model of **success** that defines when a goal is achieved.

In the context of the Synthetic Structure of Intentionality, the definition of a **success condition** is functional to the definition of the Validation System's status. For, Intention and **success** are contiguous and mutually exclusive notions from an ontological viewpoint. When there is **success**, the Intention that propels towards the state of **success** ceases to exist, and when there is Intention, **success** is not yet achieved: this is how any given **success condition** is instrumental to the genesis of the ontological status of a Validation System. Any given **success condition** provides the conceptual boundary between the Intention to act upon a desire, and the objects and states of affairs *after* the goal is achieved. The function of a **success condition** is to signal when the Validation System, intended as the epistemic domain in which any intentional acts takes place, ends.

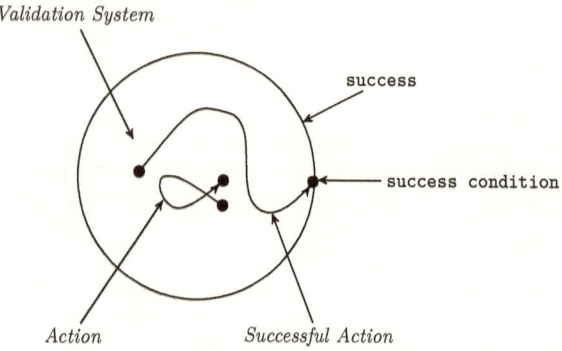

Figure 8.2: Illustration of the concepts of **success** and **success condition** in a Validation System.

The function of **success** is to axiomatize **success conditions**. The notion of **success** distills the basic, irreducible features that are necessary to conceive any intelligible **success condition**. A way to visualize the concepts of **success** and **success condition** is via the metaphor of a circle (Figure 8.2) that separates two regions of space: the space inside the circle, where any conceivable action to reach a goal takes place, and the region outside the circle, which represents the domain where no set of actions is consistent with any intelligible definition of goal. The Validation System is represented as the region of space *inside* the circle, and the actions are represented as paths entirely contained inside the Validation System. The successful actions are those that result in the achievement of the desired goal set by the Will, and are represented as paths inside the circle that end on the circumference. A practical way to interpret what a **success condition** does, is to regard it as a cognitive mechanism that defines what is *intelligible*. In the language of the Synthetic Structure of Intentionality model, intelligible means "whatever can be described by a Validation System" and denotes, for this very reason, the structure of thought. Thus, intelligibility and Validation System are interchangeable notions in the context of the Synthetic Structure of Intentionality.

Interaction Model : *Intention → Action*

A convenient way to capture the transition from Intention to Action is to think of it as the definition of a certain *Interaction Model*. The notion of Interaction Model encodes how we decide to come at the project of attaining a goal, what strategy we use and why. When we add a *model* of **success condition** to a Validation System we can define an action plan. What we are gradually constructing here is a layer cake where each layer enriches the layer below with a feature that enables more abstract cognitive processes to surface and develop. For example, the transition from Will to Intention enables *framing* via the notion that characterizes the realization of the Will, and the transition from Intention to Action enables *scoping* via the notion of **success**, and via **success conditions**, being models of a given notion of **success**. In summary

An Interaction Model is a mechanism to get epistemic access to the world in a way that is consistent with a given notion of **success**.

There is one more observation to make about the concept of **success** within the context of Validation Systems: **success**, and thus any **success condition**, must be decidable and epistemically accessible. For, the concept of **success** must be such that it must be possible to decide when it is achieved, and consequently, a representation of **success** must be decidable within a system of thought consistent with the Validation System to which that particular representation of **success** refers to. The claim that a Validation System works on inaccessible assumptions, or based on an inconsistent system of ideas disintegrates any notion of **success** the Validation System may encode. Thus, we discover that the conditions of satisfaction of a Validation System define goals that allow for an epistemically accessible definition of **success**. In the next section we'll see systems of ideas where the notion of success isn't always epistemically accessible.

8.4 The Limits Of Formal Systems

Formal systems are an example of Interaction Models, and a particularly important one for two reasons: because they are used to construct mathematics, and because when formal systems are

8 A Synthetic Structure Of Intentionality

non-trivial, they can be used to study Interaction Models that depict a Validation System where the notion of **success** isn't always epistemically accessible—recall the discussion in Section 8.3 about the requirement for a notion of **success** to be decidable and epistemically accessible. So there are two reasons to dig deeper into the structure of formal systems from the Synthetic Structure of Intentionality viewpoint.

First, let us describe the correspondence that allows us to use formal systems to model Interaction Models. The interpretation of formal systems we present here is based on a correspondence between the notions of well-formed formula and formal proof, which are an integral part of the technology of formal systems, and the Actions that take place in a Validation System as described in Figure 8.2. Let us recall briefly a few points from the discussion about formal systems of Section 1.1. A well-formed formula is a sequence of symbols or words obtained by applying the rules of a formal system. A formal proof is the last well-formed formula of a sequence of well-formed formulas. Table 8.3 introduces the correspondence between formal systems and Interaction Models. This correspondence suggests that we can think of a Validation System as the universe of all the well-formed formulas that can be generated by the language of the formal system.

Interaction Model	Formal System
Validation System	*The set of all well-formed formulas.*
success	*The set of all provable theorems.*
success condition	*A theorem.*
Action	*A sequence of well-formed formulas.*
Successful Action	*A formal proof.*

Table 8.3: Correspondence between Formal Systems and Interaction Models.

With this data in mind, let us explore the most interesting part of this correspondence: the construction of successful Actions. As we saw in Section 8.3, the definition of **success** is vital to the physiology of a Validation System, because it provides a true-false mechanism to determine the achievement of the goal codified

8.4 The Limits Of Formal Systems

by the particular Synthetic Structure of Intentionality that that Validation System is part of. In the context of the mathematical study of formal systems, there are two key notions that we must recall before we can proceed in this analysis: completeness and consistency. A formal system is *complete* if any well-formed formula in the language of the formal system can be proven from the set of axioms of the formal system, and it is incomplete otherwise—proving a formula means constructing a formal proof of it. A formal system is *consistent* if it does not contain a well-formed formula such that the well-formed formula and its negation are provable from the axioms. To illustrate the idea of a notion of **success** that is not epistemically accessible, we want to use the correspondence of Table 8.3, and some important results about formal systems, to find examples of undecidable theorems—theorems that cannot be proven in the minimum formal system needed to state them.

Based on what we know about formal systems, an inexhaustible supply of undecidable theorems is given by the so-called *self-referential paradoxes*. Self-referential paradoxes are the manifestation of the following fact: *if a formal system F is non-trivial, then it has properties that cannot be described in the language of F*—where non-trivial means with sufficient symbolic complexity to express arithmetic. A famous self-referential paradox is the Liar paradox (generally credited to Epimenides, circa 600 BC), which in natural language takes the forms "This sentence is false", or "I am lying", and in symbols $P \Rightarrow \neg P$ (P implies its negation). The Liar paradox shows that natural languages cannot reliably talk about their own truthfulness. These sentences describe themselves because they make a claim about their own features. The paradox consists in the fact that every assumption about those sentences contradicts itself—if the sentence is true then it is false, and vice versa. Note that not all self-referential sentences are paradoxical. For example "This sentence contains 9 vowels" is decidable (we can *verify* that it is true) and provable (we can *prove* that it is true). Another famous paradox of the same kind is Russell's paradox. Russell's paradox shows that naïve set theory is inherently flawed because sets cannot express reliably their intrinsic property of membership. Consider the set S of all sets that are not members of themselves, and ask: What set is S a member of? If S is not a member of itself, then its own definition imposes that it must contain itself, but if S contains itself, then it contradicts its own definition of being the set of all sets that are not member of themselves. Again, Russell's paradox emerges as a result of a statement about sets to describe a self-referential feature of a certain set. These are only two of a long list of historically important examples of self-referential paradoxes—such as Tarski's undefinability theorem, Cantor's theorem, Richard's paradox, Turing's halting problem and many more.

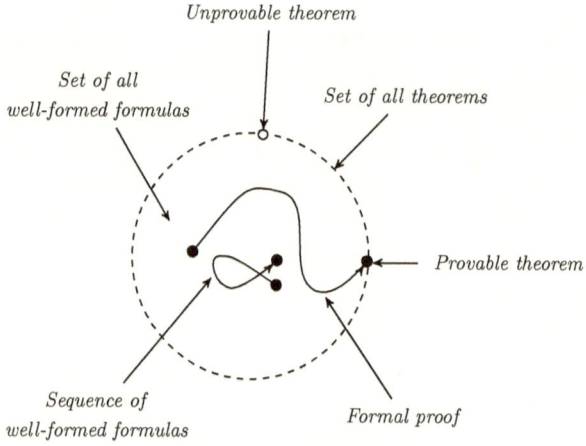

Figure 8.3: Illustration of the notion of success in a non-trivial formal system.

8.5 Gödel's Argument

Is there a way, given a formal system, to systematically construct theorems that cannot be proven? Given a sufficiently complex formal system—where sufficiently complex means with sufficient symbolic complexity to express arithmetic—a method to construct infinitely many theorems that cannot be proven is given by Gödel's first incompleteness theorem (1931). In his famous work on incompleteness, Gödel showed a method to construct infinitely many self-referential statements in the language of a formal system that are, by design, not provable. Gödel's incompleteness results shows that arithmetic cannot adequately express the provability of infinitely many theorems of arithmetic. Figure 8.3 uses the same visual metaphor used to describe the notion of **success** in Section

8.3, to visualize the set of all theorems of a given formal system. The circumference is a dashed line to signify the fact formalized by Gödel's first incompleteness theorem that there are theorems which we cannot prove from within the formal system that generates them—an important consequence of Gödel's work is that to prove a theorem $\mathcal{T}$ that is not provable in a given formal system F, we would have to "extend" F; however, in the new formal system F' that contains F and in which we can prove $\mathcal{T}$, there would still be infinitely many theorems that cannot be proven, by virtue of Gödel's first incompleteness theorem, and this goes on ad infinitum. In the formal systems parlance, Gödel's incompleteness theorems express the relation between completeness and consistency: any formal system in which you can do at least elementary arithmetic cannot be both consistent and complete.

8.6 Lawvere's Fixed Point Theorem

To fully appreciate the deep nature of the property of incompleteness revealed by self-referential sentences, it is indispensable to interpret the correspondence of Table 8.3 with the help of a set of conceptual tools that allow us to study incompleteness within a much broader set of problems. Incompleteness is the manifestation of an important property of non trivial formal systems that has deep mathematical and philosophical implications because of its significance for the nature and limits of mathematical knowledge.

A first step in this direction consists in recognizing that Gödel's incompleteness, self-referential paradoxes and fixed point theorems share a common structure—fixed point theorems are statements about the conditions under which a function f satisfies the equation $f(x) = x$, where x is called a fixed point of f. In 1969 Lawvere described in category-theoretic terms (Lawvere 1969b) the conditions for incompleteness and paradoxes, and subsequently showed that they are the same. Thanks to his pioneering work, we have today a profound insight into the deep structure of formal systems, and a very sophisticated technology to describe the universal pattern underlying self-referential paradoxes: this new description is contained in Lawvere's Fixed Point theorem, which recasts self-reference in the language of Category Theory.

8.7 Local Identities

A Local Identity codifies an object of thought when an Abstract Mind is in a state defined by a given Validation System.

Let's now turn our attention to the role played by the Object Classifier in the Analytic Structure of Intentionality. In The Lazy Boy Experiment, the lazy boy used a dictionary (Object Classifier) to translate arm positions into objects on a desk, and, as a result of that translation, he discovered that the objects on the desk became like the mug, like the pencil and like the stapler. Similarly, in The Yarn Ball Experiment, when we standardized a method to wind a ball of yarn (Object Classifier), we discovered that the points in the three-dimensional space could be replaced by the length of a piece of yarn: the three-dimensional points became one-dimensional when we transformed them into the length of a piece of yarn. So if we compare one more time these two thought experiments, we notice that the action of the Object Classifier in the Synthetic Structure of Intentionality is twofold:

1. It performs a specific cognitive function that provides a correspondence between the Action—which is a strategy to implement the Intention—and the object that the Intention is directed at.

2. As a result of 1, it allows us to *identify* the object the Intention is directed at with our Action to interface with that object. This is what we mean by "grasping a concept". The act of grasping is both literal and metaphorical, and is based on the analogy between the physical contact necessary to grab an object, which occurs through, for example, a hand, and the peculiar type of mental contact necessary to grab a concept, which occurs through identity.

The function of an Object Classifier is to supplement the translation of Intention into Action with a correspondence between the Action, and the object the Intention is directed at.

This special correspondence between Action and Object Classifier is what we call *an identity*. Any given *identity* constitutes the mechanism by means of which we reference the object the Intention is directed at in the context of that specific Analytic

8.7 Local Identities

Structure of Intentionality, in much the same way in which, for example, the seeing is the experience of vision.

Local Identity : *Action → Object Classifier*

In The Lazy Boy Experiment, an identity is given by a dictionary to translate arm positions into objects on a table, and in The Yarn Ball Experiment it is given by the method to wind a ball of yarn. We call this correspondence Local Identity to signify that it functions like a specialized concept of identity within the context of that specific set of cognitive processes that transform the Will to interface with the World into a Concept that enables us to look at the world for the specific purpose encoded by that Will and that Will only. But by introducing the notion of Local Identity we do a lot more than just codify the relation between Action and Object Classifier. In a way that will become clear in Part III, the concept of Local Identity gives us a very practical definition of what we commonly refer to as *thought*.

It is only through the specific structure of thought provided by a Local Identity that we are able to create the connection between mind and world. Those connections are our *experience* of reality: they *are* reality. It is precisely by enabling this fusion between the thinking and the thought, between mind and reality—no dualism intended—that we *think*. Note that reality is anything we are or become aware of. Reality is any change in our perception of the world and of our own thoughts because changes in awareness are what we commonly refer to as consciousness.

Bibliographical Notes

The philosophical content of this chapter is inspired by the work of many authors in the fields of the Philosophy of Mind, Process Philosophy and Category Theory. The theme of intentionality and its role in the manifestation of rationality was influenced by Searle 1983, Searle 2002, Dretske 1980, Fodor 2008. The theme of intentionality in relation to experience and consciousness was influenced by Strawson 1994. The deconstruction of The Lazy Boy Experiment and The Yarn Ball Experiment, and the notion of Synthetic Structure of Intentionality were developed by the author in Lo Vetere 2009b, Lo Vetere 2009a. The themes of phenomenology, in particular analytic phenomenology, bodily awareness, perception and consciousness were influenced by Searle

8 A Synthetic Structure Of Intentionality

1990, Searle 1992, Searle 1990, Dennett 1991, Merleau-Ponty 1958. This list is, as usual, incomplete.

The interpretation of self-referential paradoxes via the notion of interaction model is in Lo Vetere 2009b, and the discussion about a unified framework to describe self-referential paradoxes is in Lawvere 1969b and Yanofsky 2003. The architecture of the mind, the notion of a causal mind, the structure of reason and the formulation of causal links as the basis of the process of explanation were influenced, among others, by Putnam 1975, Ryle 1949, Dretske 2000, W. C. Salmon 1984, W. C. Salmon 1998, Hardy 1940. The reductionist views of the mind such as Fodor 2008 were also incorporated in some initial investigations and then abandoned. Some historical themes about the structure of mathematical speculation were inspired by Hadamard 1945, Goodman 1951, Hintikka 1969, Hintikka 1996, Feferman 2000, Feferman 2012, Minsky 2006, Bohm 1992. The theme of mental reality and cognition were influenced by Searle 2001, Searle 2004, Strawson 1994, N. U. Salmon 2007.

PART III

An Introduction To Interaction Theory

How can we explain our
symbolic intuition of reality?

Introduction

> The search for hard-to-vary explanations is the origin of all progress.
>
> David Deutsch

This third part has two main goals. As an introduction to Interaction Theory, its objective is to outline the structure of a theory that redefines the foundations of human thought based on a *unifying system of metaphysical principles* that describe the fundamental structure of the mind-world interplay. As an introduction to a *style of reasoning*, its objective is pedagogical. The concepts we introduce in this part are in fact necessary to comprehend the examples of Symbolic Intuition of Reality we examine in Part IV, where we apply the ideas and methods of Interaction Theory to describe three tightly intertwined foundations of human thought corresponding to three distinct types of intelligibility. In this scene-setting introduction we review the worldviews and background presuppositions at the basis of Interaction Theory, and begin to couch the main themes of this theory.

Interaction Theory is a model of Abstract Mind based on the Synthetic Structure of Intentionality. To comprehend how Interaction Theory is constructed, it is therefore necessary to comprehend what counts as a model of Abstract Mind, and specify the fundamental correspondence used to represent this assumption in the language of the Synthetic Structure of Intentionality.

The description of Abstract Mind we are interested in in this research, is a self-contained, self-consistent system of thought that we can carry around in our heads like we carry around in our heads the modern scientific explanation that the Earth orbits around the Sun, or the recipe for a tuna sandwich, or the project of a tree house. The framework we want to be able to carry around in our heads, is a multidimensional system of thought that allows us relate symbolically and practically to the World at the epistemic level revealed by each Symbolic Intuition of Reality, and at the higher epistemic levels revealed by the dynamic properties of the Abstract Mind.

Within the intellectuality defined by a specific Symbolic Intuition of Reality, we want to be able to define an epistemic project, and a prototypical structure of mathematization from which to derive and develop mathematical descriptions of the World.

Within the higher order intellectualities defined by the dynamical properties of the Abstract Mind, we want to be able to redefine our symbolic and practical knowledge of the World in the broadest context of the levels of reality revealed by each individual Symbolic Intuition of Reality, and by their mutual correspondences and interactions.

In the introduction to Part I, we said that the Symbolic Intuition of Reality consists of two parts, *conceptual* and *epistemic*:

> "*[t]he epistemic part contains a model of the cognitive structures at the foundation of human thought, which we identify with the meaning of the following 5 terms: knowledge, comprehension, meaning, explanation and consciousness.* "

The choice of those 5 terms as the foundation of human thought is, of course, an assumption. We could have chosen other terms to conceptualize an archetypical intellect, such as for example a prototypical notion of *communication* or *computation* or *artistic intuition*. The reason why we *identify* the archetypical structure of the human intellect with the meaning of the terms knowledge, comprehension, meaning, explanation and consciousness we give in Interaction Theory, is dictated by a variety of reflections and worldviews that influenced and inspired this research, and by the requirement of Interaction Theory to remain consistent with the modern conception of mathematics. We won't elaborate on these reflections, secure in the knowledge that the reader will pick up the metaphysical and mathematical sensitivity of the author as we progress in this discussion, and from the examples we present in Part IV.

Postulates

The postulates of Interaction Theory explain how the concepts and worldviews used in the construction of a model of Abstract Mind based on the Synthetic Structure of Intentionality are believed to correspond to reality. Hence, each postulate consists, in general, of two kinds of claims: ontic and epistemic. Ontic claims are about the things that are believed to exist in the World. Epistemic claims are about the relevant knowledge we assume to have about those things. The postulates of Interaction Theory are of two kinds, postulates about the invariance and universality of the elements that form the cosmology of thought in which Interaction Theory is developed, such as for example consciousness and the structure of knowledge, and postulates about the mind-world transactions.

Postulates About Invariance And Universality

This group of postulates states that the fundamental cognitive structures that preside over the genesis of knowledge and that give rise to a Symbolic Intuition of Reality do not change with the states of consciousness.

Postulate 8.7.1 (The Synthetic Structure Of Intentionality Is Constant).
The ontic content of this view that the Synthetic Structure of Intentionality is an *invariant* of human consciousness. This postulate is a coherence condition needed to justify the search for the universes where the three components of the Synthetic Structure of Intentionality can vary.

The epistemic content of this postulate is that the conception of *intelligibility* codified by the interplay of the three components of the Synthetic Structure of Intentionality is an integral part of the phenomenology of consciousness.

An important consequence of this postulate is that the Synthetic Structure of Intentionality should be considered part of the basic features of consciousness, together with the unity of the experience of consciousness and the flow of consciousness.

Introduction

This postulate is used extensively in Interaction Theory to define models of the Synthetic Structure of Intentionality parametrized by models of intelligibility.

Postulate 8.7.2 (The Modes Of Consciousness Are Invariant).
The ontic content of this view is that the modes of consciousness are an invariant of the mind-world interplay. This postulate is used as a coherence condition in Interaction Theory to describe consciousness and to model the Symbolic Intuition of Reality.

The epistemic content of this postulate is that the modes of consciousness correspond to types of *spatialization*, as described in Section 14.1.

Postulate 8.7.3 (The Universal Structure Of Knowledge).
The ontic content of this view is the existence of a universal cognitive mechanism based on a conception of intelligibility through which we humans define knowledge.

The epistemic content of this postulate is that this universal cognitive mechanism is fully described by what in Interaction Theory is called the **reconstruction of the success condition**, which is a theoretical device used to characterize the genesis of a conception of knowledge.

The **reconstruction of the success condition** is used in Interaction Theory to define the archetypical ways in which cognitive technologies are defined in a given Validation System.

This postulate captures the observation that knowledge has always the form
$$A = B$$
where A and B are certain entities defined in a given validation system, and where the meaning of the symbol $=$ is defined via the underlying notion of *intelligibility*. For example, this postulate states that in a mode of thought defined by a conception of intelligibility based on the notion of identity, knowledge has necessarily the form $A = B$ where $=$ denotes a mathematical *model* of identity—e.g. equality, isomorphism, equivalence, adjunction, fibration, homotopy, cobordism, etc.

Postulate 8.7.4 (The Knowledge-Reality Interface Is Invariant).

The ontic content of this view is that the cognitive mechanism through which the we relate symbolically to reality, that in Interaction Theory is called Knowledge-Reality Interface, does not change with consciousness: it is an *invariant* of the mind-world interplay.

The epistemic content of this view is that the Knowledge-Reality Interface is fully described by the diagram of Figure 8.4, which we will examine in detail in Parts III and IV.

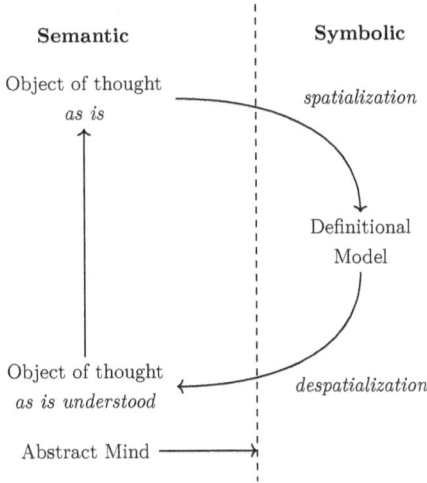

Figure 8.4: Structure of the Knowledge-Reality Interface.

Postulates About The Mind-World Transactions

These postulates describe the core features of the mind-world interplay, and set the scene for the definitions Interaction Theory is based on.

Postulate 8.7.5 (The Mind-World Transactions Are Real).

Introduction

The ontic content of this postulate is that the mind-world transactions are real, they are not theoretical abstractions.

The epistemic content of this postulate is that the mind-world transactions can be described by the *language of process*, like we describe digestion, photosynthesis and any other natural process.

This postulate is used to justify the definition of the Abstract Mind as a model of those archetypical cognitive functions that give rise to a Symbolic Intuition of Reality.

Postulate 8.7.6 (The Nature Of Concepts Is Interferential).
The ontic content of this view is that the nature of concepts is entirely described by the notion of *definitional interference* (Sections 5.4 and 5.5). This postulate is used to justify the use of the Synthetic Structure of Intentionality to model the Abstract Mind.

The epistemic content of this postulate is a procedure to generalize the correspondence between the mental and the physical domains via the notion of definitional interference. The underlying belief is that the nature of definitional interference itself conveys a correspondence between the mental and the material realities.

The debates about the nature of concepts often reflect contrasting approaches to the study of the mind, to our conception of knowledge and of Philosophy itself. The interpretation of the phenomenology of concepts codified by this postulate is that concepts always manifest themselves in the mental and the perceptual domains: *it is always possible to identify a given concept as the extension to the mental domain of perceptual events, or, vice versa, as the extension to the perceptual domain of mental phenomena.*

An important consequence of this postulate is that it points to the logical necessity of the existence of *hierarchies of conceptions of concepts*, each defining the term "concept" in a fashion different but interrelated with the other conceptions. The most basic form of conception of concept is of course that of a mechanism by means of which we construct a correspondence between the mental and the material domains. It is a device that constitutes the irreducible foundation of our perceptual experience of the world. Higher conceptions of concepts define the correspondence between higher mental domains and lower mental domains.

To get a feel for what this means, let us call this basic type of concept a *type*-0 concept. Examples of *type*-0 concepts are identity. As already observed, identity extends to concepts (mental domain) the physical act of grabbing an object (material domain). From this definition, it follows that *type*-1 concepts are defined to grasp *type*-0 concepts, *type*-2 concept are defined to grasp *type*-1 concepts, and in general *type*-$(n+1)$ concepts are defined to grasp *type*-n concepts.

$$\textit{type-n} \xrightarrow{\textit{type-}(n+1)} \textit{type-n}$$

We can picture a *type*-n concept as a node in an infinite graph where the arcs that connect two nodes represent a *type*-$(n+1)$ concept.

Postulate 8.7.7 (Definitional Closure).
The ontic content of this postulate is that the set of postulates of Interaction Theory is complete, and fully describes the structure of the mind-world interplay modeled via the Abstract Mind model.

The epistemic content of this postulate is that it is possible to explain the mind-world interplay by focusing solely and exclusively on the action of intentionality on consciousness as modeled by the Synthetic Structure of Intentionality.

This postulate is used to define the **Action** element of the Fundamental Correspondence—which we present later in this introduction to Part III.

Postulate 8.7.8 (The Structure Of Intelligibility).
The ontic content of this view is that at any given level of consciousness, there exist certain concepts that fully define intelligibility by virtue of being or becoming irreducible at that level of consciousness. In Interaction Theory these irreducible concepts are called `concrete`, and define the conceptions of intelligibility used to model Abstract Mind states.

The epistemic content of this postulate is that each Abstract Mind state is fully defined by a single `concrete` concept.

Introduction

Postulate 8.7.9 (Foundation Of Human Thought).
The ontic content of this postulate is that, at an archetypical level, the architecture of human thought is fully defined by the meaning of the terms knowledge, comprehension, meaning, explanation and consciousness.

The epistemic content of this postulate is that it is possible to explain the genesis and evolution of the intellectuality of a Symbolic Intuition of Reality solely and exclusively from a model of knowledge, comprehension, meaning, explanation and consciousness consistent with the other postulates of Interaction Theory.

This assumption is used to justify the choice of the concepts of knowledge, comprehension, meaning, explanation and consciousness, as defined in Interaction Theory, as a foundation of human thought.

Postulate 8.7.10 (Modes Of Consciousness As Spatialization).
The ontic content of this postulate is that different Symbolic Intuitions of Reality originate from different modes of consciousness. The postulate captures the observation that the workings of the human mind seem to change with the level of consciousness, and is needed in the abstraction process that leads to the definition of how the **Intention** element of the Analytic Structure of Intentionality varies.

The epistemic content of this postulate is that the modes of consciousness that give rise to a Symbolic Intuition of Reality are types of *spatialization*.

The basic idea introduced by this postulate is that it is possible to *identify modes of consciousness with types of spatialization* whenever a Symbolic Intuition of Reality is part of the phenomenology of the mind-world transactions, because modes of consciousness and types of spatialization are coupled in such a way that they consistently vary together. Consequently, the choice to use two distinct terms to talk about to modes of consciousness and types of spatialization is dictated, as a first approximation, by explanatory purposes.

Postulate 8.7.11 (The Identity-based Intellect).
The ontic content of this postulate is the view that, at the level

of consciousness responsible for the mental faculties of problem-solving and mathematical abstraction as we know them today, the fabric of human thought is entirely defined by an archetypical form of the concept of *identity*.

The epistemic content of this postulate is that the mathematical models of the notion of identity based on a construct called **identiton** (Section 15.3)—which are models of equality, isomorphism, equivalence, adjunction, fibration, homotopy, cobordism, etc.—fully describe how the human mind constructs thoughts at the alert, problem solving state of consciousness.

This postulate asserts that in our everyday life we *cannot think* unless our intellect can use some form of identity as a tool to grasp ideas. Note that this assumption does not mention language. This postulate defines the **Intention** element in the Fundamental Correspondence.

The Fundamental Correspondence

The Fundamental Correspondence explains how Interaction Theory models the Abstract Mind, and how the Postulates of Interaction Theory are used in this process.

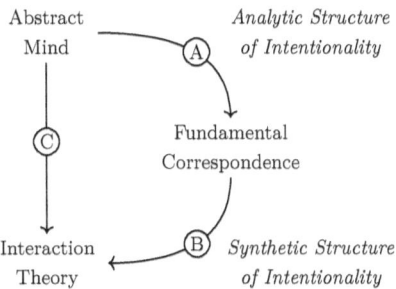

Figure 8.5: The Fundamental Correspondence.

There is a very intuitive way to comprehend how the Fundamental Correspondence works, and is illustrated in the diagram of Figure 8.5, where the Fundamental Correspondence connects the

Introduction

Abstract Mind to Interaction Theory in two steps. Essentially, the construction of the Fundamental Correspondence revolves around the parts of Analytic Structure of Intentionality and Synthetic Structure of Intentionality: some are fixed, some are variable. Of course, the variables are the `concrete` concepts, that define Validation Systems ($\mathcal{M}$-*signature*), and Interaction Models. Everything else is an invariant.

A. The first step is the definition of the Analytic Structure of Intentionality, and consists in the identification of the common patterns through which the Abstract Mind creates concepts, as seen in Chapter 8. This first part of the transition, in which we define the *Intention* and *Action* elements of the Analytic Structure of Intentionality, is based on Postulates 8.7.7, 8.7.10 and 8.7.11. The *Will* and *Object Classifier* elements denote respectively intentionality and the concept created in the mind-world transaction, and therefore do not need to be defined because are the *variables* in this correspondence.

B. The second step is the definition of the Synthetic Structure of Intentionality, which consists of the abstract transformations *Will* → *Intention* (Validation System), *Intention* → *Action* (Interaction Model), and *Action* → *Object Classifier* (Local Identity), and relies on the postulates that define the invariance and universality of the Synthetic Structure of Intentionality, and the structure of knowledge and intelligibility—Postulates 8.7.1, 8.7.3, 8.7.8 respectively.

C. This process of analysis (Analytic Structure of Intentionality) and synthesis (Synthetic Structure of Intentionality) produces the model of Abstract Mind that we call Interaction Theory.

Another, perhaps less intuitive way to think about the Fundamental Correspondence is to think of it as the construction of the Synthetic Structure of Intentionality in reverse: we choose a concept of space, and look for the components of a Synthetic Structure of Intentionality that are consistent with its genesis.

Structure Of Interaction Theory

The structure of Part III reflects the structure of the Symbolic Intuition of Reality introduced at the beginning of the text (Figure 3).

Structure of Interaction Theory

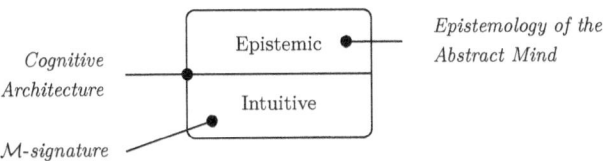

Figure 8.6: The structure of Part III is the structure of the Symbolic Intuition of Reality.

Interaction Theory is presented, step-by-step, as the definition of the *intuitive* and *epistemic* parts of a general Symbolic Intuition of Reality.

Intuitive Part

The intuitive part of Interaction Theory defines intelligibility and a universal type of knowledge.

In "The Common Mathematical Intuition Of Reality", Chapter 9, we introduce the Interaction Model at the basis of the modern conception of mathematics, called OSP Pattern. The OSP Pattern is, perhaps, one of the most important concepts of modern mathematics that allows us to think of any mathematical definition as a particular configuration of three basic patterns.

"$\mathcal{M}$-signatures", Chapter 10, marks the first step toward the definition of Interaction Theory, and tackles the philosophical and mathematical problem of defining a conception of *intelligibility*. This is easily one of the most important and complex chapters of the book. Its most peculiar feature is that, for the first time in the book, we construct a bridge between the philosophical and the mathematical domains by introducing a practical notion that describes the metaphysical structure of thought. This is very exciting and very challenging because it demands us to rewire ourselves with radically new types of abstractions that rewrite the conception of intelligible thought and open up intriguing conceptual landscapes of mathematical and philosophical research. For, while we will explain how $\mathcal{M}$-*signatures* define the metaphysical structure of thought, which is still a very philosophical-sounding idea, the application of the notion of $\mathcal{M}$-*signature* is actually a very practical, very technical enterprise in that it offers a constructive step-by-step description of how these ideas can be mirrored by certain formal constructions, the outcome of which can be described by specific mathematical objects.

"Cognitive Architectures", Chapter 11, takes another step toward the definition of Interaction Theory by defining a conception of *knowledge*. This is another pivotal chapter for its role in Interaction Theory and for the complexity of the ideas presented in it. In the modern conception of mathematics, mathematical knowledge, and hence scientific knowledge, are intrinsically quantitative. This should come as no surprise, given that mathematics is created via a formal system which, by design, is a technology to translate ideas into computational problems. This mindset, condensed in the notion of OSP Pattern, is in this chapter deconstructed and reconstructed in the broader context of intentionality. This is a radical process by means of which we extract the underlying structure of thought that we *use* in the form of OSP Pattern. Once again, the guiding principle we adopt in the quest to reveal the deeper structures at the basis of the human capacity to mathematize, and, consequently, to reveal the workings of the Abstract Mind, are based on the The Sieve. The application of The Sieve to the OSP Pattern, and to the lines of reasoning that lead to its definition, is deconstructed to extract its Observational Structure of Intentionality.

Epistemic Part

The epistemic part of Interaction Theory defines knowledge, comprehension, meaning, explanation and consciousness, and the notions of communication and computation.

In "Epistemology Of The Abstract Mind I", Chapter 12, we use the model of intelligibility introduced in Chapter 10 to begin to construct the foundations of human thought by defining three building blocks of human intellectuality: *knowledge*, *comprehension* and *meaning*.

In "Epistemology Of The Abstract Mind II", Chapter 13, we tackle the problem of how concepts connect us to reality by defining the archetypical structure of an explanatory system. Through the definition of *explanations*, *reality* and *ignorance*, we lay down the foundations for the development of the operational part of Interaction Theory, where practical knowledge is continuously created by resolving the epistemic tension between symbol and meaning through the discovery of explanations.

In "Epistemology Of The Abstract Mind III", Chapter 14, we examine the dynamical features of the mind-world interplay. We use the concepts introduced in Chapters 12 and 13 to define the important notion of *awareness*, and to outline an archetypical structure of *communication* and *computation*.

Key Concepts Of Part III

1. Interaction Theory is a model of Abstract Mind based on the Synthetic Structure of Intentionality.

2. To describe a Symbolic Intuition of Reality, it is sufficient to model a conception of intelligibility and knowledge.

3. Distinct states of consciousness correspond to distinct types of intelligibility, and consist of distinct types of spatialization.

4. An $\mathcal{M}$-*signature* is a model of intelligibility and objectivity that fully describes an Abstract Mind cognitive state, and the emergence of a specific Symbolic Intuition of Reality.

5. A Cognitive Architecture is a model of knowledge in a given $\mathcal{M}$-*signature*.

6. Each $\mathcal{M}$-*signature* gives rise to a local epistemology by defining the meaning of the terms knowledge, comprehension, meaning, explanation and consciousness, and the archetypical structure of communication and computation.

7. The type of knowledge generated by a Cognitive Architecture is neither propositional nor procedural, and characterizes the basic structure of the cognitive technologies definable in a given $\mathcal{M}$-*signature*.

8. There is no privileged frame of reference to define an epistemology of the Abstract Mind, and with it, a mathematical and scientific method of inquiry.

9. A Symbolic Intuition of Reality defines a conception of mathematics as a special relation of epistemic continuity

between symbol and meaning.

Chapter 9

The Common Mathematical Intuition Of Reality

> The question you raise, 'How can such a formulation lead to computations?' doesn't bother me in the least! Throughout my whole life as a mathematician, the possibility of making explicit, elegant computations has always come out by itself, as a byproduct of a thorough conceptual understanding of what was going on. Thus I never bothered about whether what would come out would be suitable for this or that, but just tried to understand - and it always turned out that understanding was all that mattered.
>
> Alexander Grothendieck

What is the structure of our basic mathematical intuition of reality? and How do we define mathematics as we know it today based on this fundamental intuition?

The description of Interaction Theory starts by answering these two questions, because answering these two questions allows us to situate the modern conception of mathematics within the context of the mind-world interplay that produces an array of Symbolic Intuitions of Reality, one of which gives rise to the modern conception of mathematics.

9 The Common Mathematical Intuition Of Reality

9.1 A Strategy To Visit The Mathematical Zoo

A mathematical gadget is any concept used in the process of mathematization. Sets, numbers, abstract relations, geometrical shapes, transformations, formal proofs, computer programs, programming languages and mathematical theories. Anything at all that fuels the thought processes that we observe as our ability to mathematize is a mathematical gadget. So if we want to reason about how any mathematical gadget is defined, we need a strategy to navigate through the humongous zoo of mathematical gadgets that is contemporary mathematics. We need a way to reason about the most basic mathematical concepts and about the most sophisticated tools of modern mathematical technologies seamlessly.

I think that an effective way to reason about a vast array of mathematical gadgets is to try to classify them. Think about a car and a bulldozer. They are both machines but they serve different purposes. They share certain characteristics such as having an engine and doors and a roof but there are elements that only a car possesses. The problem of classifying mathematical gadget is analogous to the problem of classifying mechanical machines. Comparing two cars is straightforward once you know a few key parameters that you need to consider, such as dimensions, weight, engine power, top speed and many more. And once you know how to compare two cars you can venture to compare a car with a bulldozer, because some of the parameters you need to compare cars can be used to characterize a bulldozer. But bulldozers are different from cars because they are designed for a different purpose. Similarly, there are certain families of mathematical gadgets that are like cars, and certain tribes of mathematical gadgets that are like bulldozers. Here is how, in principle, we could classify mathematical gadgets.

Let's say that there are two sets of features $\mathcal{A}$ and $\mathcal{B}$ and that any mathematical gadget O contains a certain amount of features of type $\mathcal{A}$, which we indicate as $\{1, 2, 3 \ldots\}_\mathcal{A}$ and a certain amount of features of type $\mathcal{B}$, which we indicate as $\{1, 2, 3 \ldots\}_\mathcal{B}$. We use the notation $O_{n,m}$ to indicate that O has n features of type $\mathcal{A}$ and m features of type $\mathcal{B}$, and we do not distinguish between features of the same kind. So for example, if we fix the number of features of type $\mathcal{B}$, any two sets of features of type $\mathcal{A}$ that contain the same number of elements, such as for instance $\{73, 55, 39\}_\mathcal{A}$ and

$\{101, 3, 97\}_\mathcal{A}$ are classified as the same mathematical gadget $O_{n,m}$. This simple strategy allows us to compare mathematical gadgets based on the amount of features of each kind they have. Some mathematical gadgets are special, they are the bulldozers, and have also features of type $\mathcal{B}$.

The discussion about the universal patterns by means of which mathematical gadgets are defined follows this car-bulldozer strategy. First, we present a collection of mathematical gadget of increasing complexity, which are ordered according to the pattern we want to reveal, then we proceed to analyze the function of each pattern and the thought processes that it enables.

9.2 The Architecture Of Mathematical Definitions

The following mathematical gadgets, called Set, Function, Magma and Group respectively, are ordered in increasing order of complexity, where complexity is measured as the amount of distinct, irreducible features each gadget codifies in its definition, and where the amount of features is measured as the number of instances of the same feature.

Set

A *Set* is a gathering together as a whole of distinct, definite objects of perception or thought, which are called elements of the set—this is Cantor's definition. Examples of sets are the integers $\{1, 2, 3, \ldots\}$, the four seasons, the episodes of the American TV series Columbo, your dreams, the positions of a Rubik's cube and so on and so forth.

Function

Given two sets A and B, we denote with $A \times B$ the set of all ordered pairs (a, b) where a and b are elements of A and B respectively. A *function* is a set f of ordered pairs of $A \times B$ with the property that for every a in A there is only one b in B such that (a, b) is an element of f. Examples of function are the sets of ordered pairs $(x, x+1)$ or (x, x^2) where x is any number. In contrast, (x^2, x) is not a function: since both $x = 2$ and $x = -2$ satisfy $x^2 = 4$, contrary to the condition of uniqueness.

9 The Common Mathematical Intuition Of Reality

Magma

A *Magma* $(G, \oplus)$ is a set G equipped with a binary operation $\oplus$ that assigns any two elements a and b of G to an element of G denoted by $a \oplus b$. For example, if G is the set all even numbers $2, 4, 6, 8, \ldots$ and $\oplus$ is $+$, the ordinary sum of two integers, then $(G, +)$ is a Magma, since the sum of two even numbers is an even number, and thus an element of G. Note that if G was the set of all odd numbers we would not obtain a Magma by equipping G with the ordinary sum of integers. For, the sum of two odd numbers is an even number, thus not an element of G.

Group

A *Group* is a Magma $(G, \oplus)$ with the following properties:

1. there is an element e of G such that $e \oplus a = a \oplus e$ for every element a of G

2. for every element a of G there is an element b of G such that $a \oplus b = e$

3. $(a \oplus b) \oplus c = a \oplus (b \oplus c)$ for every three elements a, b, c of G

Let's indicate with $a \bmod(5)$ the remainder of the division of a by 5. Thus for example $7 \bmod(5) = 2$. Consider the set $\{0, 1, 2, 3, 4\}$ of the remainders of the division by 5, and equip this set with the operation $a \oplus b = (a + b) \bmod(5)$. So for example $3 \oplus 4 = 2$. It is easy to verify that this is a Group by applying the definitions 1, 2 and 3 above.

If we look carefully at how these four definitions are constructed, we notice that there are three patterns, which we'll call Objects, Structure and Properties.

1. In every definition there are the *Objects* to which the definition applies. In the definition of Set, the objects are the elements of the Set. In the definition of Function, the objects are ordered pairs—of elements of sets. In a Magma, the objects are the elements of a set and so are the objects in a Group. The Objects Pattern is about the distinct features of a given mathematical gadget, and how those features characterize the purpose of that particular definition of that gadget.

9.2 The Architecture Of Mathematical Definitions

2. In every definition, the objects are equipped with a *Structure* which describes how the objects can be manipulated. In the definition of Set, the elements of the set are structureless: we do not require to do anything with them other than stating their existence as definite, distinct entities. In the definition of Function, the elements have the structure of being a certain subset of ordered pairs of elements. In the definition of Magma, the elements of the set G are endowed with a binary operation—the ordinary sum of integers in the example—and so are the elements of a Group.

3. In every definition, the objects equipped with structure satisfy certain *Properties*. In a Set, the elements of the set have no properties. In a Function f, the ordered pairs (a, b) of elements of $A \times B$ have the property that for every a in A there is only one b in B such that (a, b) is an element of f. In a Magma, the element have no properties, and in a Group, there are the three properties that turn a Magma into a Group. The Properties Pattern defines the basic, irreducible structure of our interaction with the context defined by the Objects Pattern.

Any mathematical definition consists of a procedure to specify *Objects*, that we equip with *Structure* in such a way that certain *Properties* are satisfied. The combination of these three patterns is called OSP. For the moment, let's accept that this is the way we define any mathematical gadget that lives in the mathematical zoo, and let's accept also that the three patterns we called Objects, Structure and Properties can be applied only in the following order: first Objects, then Structure followed by Properties.

9 The Common Mathematical Intuition Of Reality

OSP Pattern	Observational Structure of Intentionality
Objects Pattern	*The objects pattern represent the **Data**: what we believe we need to know, to execute a Strategy to achieve a certain Goal.*
Structure Pattern	*The structure pattern represents the **Strategy**: how we want to use the Data to achieve a certain Goal.*
Properties Pattern	*The properties pattern represents the criteria to validate the outcome of the Strategy against the definition of the **Goal**.*

Table 9.1: The Observational Structure of Intentionality of the OSP Pattern.

It is useful, before delving into a detailed description of these patterns, to regard OSP from the point of view of its Observational Structure of Intentionality (Section 3.3). The correspondence we use to interpret OSP from an observational viewpoint is described in Table 9.1, and adds a game-theoretical flavor to this discussion.

Since these three patterns are part of a larger thought process, it is convenient to examine them in two ways: in terms of their *function*, and in terms of the thought process that they *enable*. The next three sections deal with the Objects, Structure and Properties Patterns according to this function-and-enable schema, and explain in detail the Observational Structure of Intentionality underlying OSP.

9.3 The Objects Pattern

The Object pattern defines the data needed to achieve a certain goal according to a certain strategy. Thus, in order to understand how the Objects Pattern operates at a deeper level, we must understand how the definition of an object of thought or perception becomes relevant to the achievement of a certain goal. Definitions are more than just a way to declare the essence of an object or a thought. Every definition is designed to codify the features that capture the reason why a given object of thought or perception is described the way it is, and, as a consequence, every definition reflects in some way the purpose of the observational experience it contributes to create. Very often in mathematics the observation

9.3 The Objects Pattern

that motivates a particular way to define a mathematical gadget is implicit in the definition: it is not declared separately as a backdrop against which the definition is presented. I think it is fair to say that this is primarily a cultural bias, since our current conception of mathematics, to my knowledge, does not incorporate any phenomenological or observational theme. I will use a visual metaphor to bring to light the mechanism by means of which the Objects Pattern encodes the purpose of the observation that motivates a specific definition of mathematical gadget.

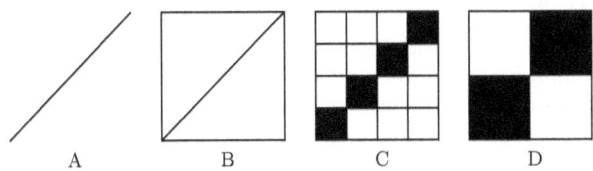

Figure 9.1: Analogy between image resolution and the Objects Pattern.

Let us consider the segment in Figure 9.1-A. We want to examine the relation between the purpose of our observation of that segment and the representation we give of it.

A. A segment is described by means of a straight line. This description provides a visual reference for the concept of segment. In this case only one object is needed to represent a segment, and that object is a graphic representation of a straight line between two points.

B. In this representation, a segment is represented as the diagonal of a square, and the difference between Figure 9.1-A and Figure 9.1-B is that a different geometrical shape is used to represent the same entity.

C. The elements of this representation are the 16 squares obtained by dividing each side of a square into four equal parts. The segment is represented by the four black squares along a diagonal. This representation considers a segment as made of *parts* along with certain spatial relations between them, and uses a particular combination of squares to encode this view. Note that using the four squares along a diagonal is merely a convention, we might decide to use other combinations of black

9 The Common Mathematical Intuition Of Reality

squares to codify a segment, as long as we define how that particular arrangement codifies the purpose of our representation of a segment.

D. This representation is similar to C but uses only four squares. It regards a segment as made of parts, but it interprets it as a set of two points that enclose the segment itself. This representation allows us to reference a segment via its vertices, which are symbolized by the two black squares along the northeast-to-southwest diagonal, and discards the points between them.

These four examples illustrate four alphabets of 1, 1, 16 and 4 letters respectively used to represent the same entity, and we have seen how the purpose of each description is encoded in the number of symbols of each alphabet. This visual metaphor contains the seed of a powerful idea: the idea that *any* number of distinct symbols can be regarded as the representation of some observation, and that the choice of the amount of distinct symbols contains the intrinsic logic supporting the reasoning through which we codify the purpose of the observation into the language of quantity.

The Objects Pattern enables us to regard an object of thought or perception in a way that encodes the primary distinctions that we believe are necessary to characterize the features of our observation. The distinctive feature of the Objects Pattern is quantity: it uses quantity as a way to define the distinctions that are deemed to be necessary to reason about the goal of the underlying observation against a sufficiently variegated backdrop.

9.4 The Structure Pattern

The Structure Pattern characterizes the interaction with the data defined by the Objects Pattern that is consistent with a certain goal. The Structure Pattern is a syntax that dictates the abstract rules by means of which the data of the Objects Pattern can be processed. For instance, in the example introduced in Figure 9.1-C, the Structure Pattern defines that to symbolize the segment of Figure 9.1-A, the 16 white squares must be colored along the northeast-to-southwest diagonal, but it does not define in which order. Thus, the three color configurations of Figure 9.2 are all syntactically correct because they are consistent with the northeast-to-southwest diagonal coloring rule. In this sense, the Structure

9.4 The Structure Pattern

Pattern is to data what inference rules are to formal systems.

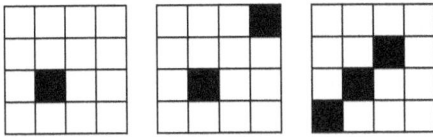

Figure 9.2: Analogy between the spatial relations of squares and the Structure Pattern.

However, if we look carefully to the action of the Structure Pattern on the data defined by the Objects Pattern, we find that there are other, more accurate ways to interpret its nature.

The Structure Pattern As An Independent Pattern.

We have presented the Structure Pattern in the context of the OSP Pattern, and this could lead to believe that a Structure Pattern can exist only if an Objects Pattern is defined. That is not the case. The Structure Pattern is *independent* of the data defined by the Objects Pattern: it is a mere set of abstract rules which are, by design, unrelated to the objects that they combine. We will see this more clearly in the next chapters—the fact that O, S and P are presented as occurring in this order is purely for explanatory purposes.

The Structure Pattern As Identity Through Change.

On the one hand, the Objects Pattern allows us to think about an object of thought or perception in a way that is consistent with the strategy to achieve a certain goal. On the other hand, the Structure Pattern enables us to think about the same goal in terms of the *changes* that are consistent with it. This is why we can interpret the Structure Pattern as a definition of the identity of a certain entity. While the Objects Pattern defines the identity of a problem statically by presenting a collection of objects that symbolizes the data with which we intend to achieve a certain goal, the Structure Pattern defines the identity of a problem via the changes that preserve its definitional integrity.

The Structure Pattern As Objects Pattern.

A third, compelling way to regard the Structure Pattern is the following, which is a direct consequence of the observations we made in the previous paragraph about the concept of identity through change. When a Structure Pattern is applied to the objects defined by an Objects Pattern, the Structure Pattern becomes itself an Objects Pattern that produces *transformations* between data. The Structure Pattern augments the data defined by the Objects Pattern by describing how that data changes. In this sense, the Structure Pattern is a refinement, or a generalization of the Objects Pattern because it introduces in the definition of the data the notion of change.

The Objects Pattern As Structure Pattern.

Similarly, we can regard the Objects Pattern as a Structure Pattern which describes data via transformations that leave the data unaltered. Thus, from an ontological standpoint, there are two identities involved in the use of a Structure Pattern: an *identity by definition* given by the Objects Pattern, and an *identity through change* given by the Structure Pattern itself.

9.5 The Properties Pattern

The Properties Pattern characterizes the validation criteria for a certain goal. It does so by defining the conditions of satisfaction encoded in the Validation System. For example, in the visual metaphor of Figure 9.1-C, the Properties Pattern defines what color combination of the 16 white squares corresponds to the segment of 9.1-A. It dictates that 4 black squares along the northeast-to-southwest diagonal correspond to the conditions of satisfaction encoded by the Validation System of the Fundamental Correspondence.

It is useful to situate the discussion about the Properties Pattern within the broader context of the entire OSP Pattern, and to do so, we can use the ideas presented in Section 8.4 about how formal systems model the Interaction Model. Figure 9.3 recasts the three components of the OSP Pattern with the visual metaphor we used to describe Validation Systems. This representation abstracts the equational properties we used to introduce the Properties Pattern in Section 9.2, and replaces them with the more general

9.5 The Properties Pattern

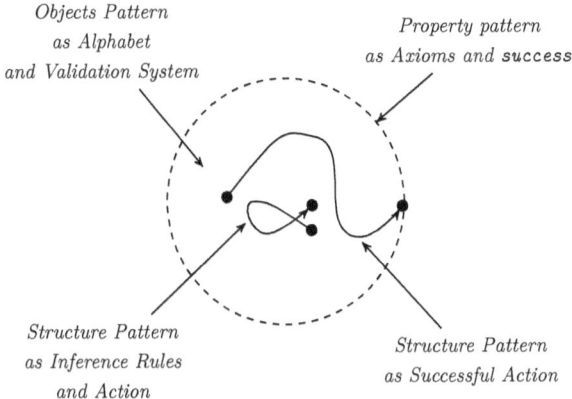

Figure 9.3: Correspondence between OSP and the Interaction Model via the Formal Systems model.

notion of **success** that emerges in the definition of a Validation System. This translation of the concepts underlying the OSP Pattern into the larger context of Interaction Models allows us to draw attention to a broader conceptual framework that seems to underlie the ideas at the basis of the OSP scheme.

Objects Pattern As Alphabet And Validation System

By defining the domain of discourse in which the Observational Structure of Intentionality of the OSP Pattern is embedded, the Objects Pattern becomes effectively a Validation System for the OSP Pattern.

Structure Pattern As Inference Rules And Action

By defining the rules by means of which the data defined by the Objects Pattern is processed, the Structure Pattern models what is, in the parlance of formal system, the notion of inference rules. For, like inference rules, we use the Structure Pattern to *construct or destroy* new data-objects within the domain defined by the Objects Pattern-Alphabet.

9 The Common Mathematical Intuition Of Reality

Properties Pattern As Axioms And success

By defining the validation criteria by means of which the products of the Structure Pattern are to be regarded as consistent with a given goal, the Properties Pattern models the notion of **success** in a Validation Systems. Its action dictates what data configurations become interchangeable within the universe of discourse defined by the Objects Pattern, thus providing a domain-specific notion of identity. In this regard, it must be observed how the equational properties model the more abstract notion of **success** in a Validation System.

9.6 Why We Define Mathematical Gadgets In This Way

In Section 1.1, we saw that the technology of formal systems is essentially a crude interpretation of how we believe the human mind works—at least to the extent in which rational thinking is involved in the mental processes that enable mathematization, or in the thought processes that we interpret as our ability to form mathematical abstractions of the world. So we might be led to think that the reason why we use the OSP Pattern to define mathematical gadgets is that OSP, being a refinement of the formal systems philosophy, is itself a representation of how the mind works. In part, this is true. The three abstract operations captured by the OSP Pattern are indeed the building blocks of what appears to consciousness as the structure of human mind's ability to mathematize. However, this observation about OSP alone, does not do any justice to the metaphysical depth of the background presuppositions involved in this investigation. It doesn't, because formal systems, and therefore OSP, characterize the phenomenology of mathematization, not its nature. If we want to be able to articulate a compelling answer to the question Why do we mathematize the way we do? we are going to need to look at the intrinsic elements at play in the process of mathematization, and we are going to need to interpret OSP not based on what it does, but based on the thought processes that it enables, and on the facts and the assumptions that its very definition subsumes. Another way to think about the question Why do we mathematize the way we do?, relies on the observation that the OSP Pattern marks the conceptual threshold that separates concepts from mathematical concepts: the OSP Pattern encodes in a way that we will discover

9.6 Why We Define Mathematical Gadgets In This Way

in the next chapters when concepts *become* mathematical.

To begin with, I think there are two practical and conceptual reasons why we use the OSP Pattern to define mathematical gadgets: these are what we call the mechanistic and the computational reasons.

Mechanistic

There is a mechanistic reason why we construct mathematical gadgets the way we do. A mathematical gadget constructed with the OSP Pattern provides a concept-specific or problem-specific language with which we are able to formulate questions about a concept or a problem. Mathematical gadgets function like problem-specific languages that allow us to express very accurately how we think about something. I think it is fair to say that the entire edifice of mathematics can be regarded as a very large collection of problem-specific languages to formulate questions about objects and states of affairs in the physical and in the mental world. In this sense, mathematics is a cognitive technology that allows us to generate an unlimited number of thoughts through problem-specific formal languages.

Computational

The effect of equipping a collection of objects with structure and properties, allows us to define and carry out computations on those objects. This is, as we have already observed in Section 1.1, the purpose of formal systems, of which the OSP Pattern is a manifestation. Computation represents the practical side of our current conception of mathematics, and forms the metaphysical structure of any intellectual activity that we recognize as rational knowledge. Not only can we describe certain concepts in the language encoded by a mathematical gadget, we can construct and validate statements about a concept or a problem that that mathematical gadget represents. This is how, at a fundamental level, the notion of computation emerges naturally from the construction of a problem-specific language—in this regard, an analogy that captures these ideas is that of computation as a speech act.

The mechanistic and computational descriptions of OSP are rather intuitive ways in which we can come at the project of studying the deep nature of the OSP Pattern. They simply describe the utilitarian aspects of constructing mathematical abstractions in a certain way so as to be able to systematically generate and manipulate thoughts about objects and ideas.

There is, however, another characterization of the cognitive circuitry at the basis of OSP which is rooted in much deeper structures of the human mind. These structures constitute what we call the *Identity-based mode of thought*.

9.7 The Identity-based Mode Of Thought

The human mind's ability to mathematize seems to be rooted in a very special metaphysical structure of the human mind which I call the *Identity-based mode of thought* or Identity-based intellect (Postulate 8.7.11). A first, approximate description of this metaphysical structure emerges from the following observation: at the alert, problem-solving state of consciousness, we seem to be able to think exclusively by "breaking down and reconstructing things". What this means is that the conceptual circuitry that implements the human mind's ability to mathematize and to articulate abstract thoughts in general, seems to rely on the introduction of some sort of distinction in whatever object of thought it focuses on in order to reason about it. So how can we study this metaphysical structure of thought? I think that what is needed to study this peculiar aspect of the mind is a privileged point of view from which to reframe the entire phenomenology of mathematization within the larger context of the metaphysical structure of thought.

The privileged point of view we are going to take to observe OSP at work, is the notion of *wholeness*. The observation of OSP through the lens of the concept of wholeness consists in an interpretation of the interplay between the basic elements of thought that give rise to OSP Pattern in terms of their relations with the original object of thought or perception that triggers the OSP process. Why wholeness? The reason why the notion of wholeness seems to be a privileged position from which to observe OSP, is that it can be clearly identified before the language-thought interplay takes place. For, when language is involved in the thought processes, OSP is already at work. In contrast, in the fugacious moment before we begin to verbalize any thought, concepts exist in their purest form. They key point is: we need to remember that OSP is a presentation of certain thought processes, and as such, if

9.7 The Identity-based Mode Of Thought

we want to reconstruct the thought processes that we think OSP presents us with, we need to observe the traits that characterize OSP in itself rather than what it produces. This is essentially yet another synthetic process by means of which we attempt to reduce OSP to some basic, axiomatic elements that we can make consistent with the way we become aware of what we think.

If we observe OSP through the lens of the concept of wholeness we can identify the following phases:

The Wholeness Phase

Originally, there is a quiet, languageless thought which is one with the mere awareness of thinking that thought. There is the intention to do something about an object or state of affairs, and that intention defines an abstract goal. It is an undivided, unmediated presentation of an impulse or an intuition that is offered to one's conscious thought without the mediation of language.

The Break Down Phase

By describing a goal with language, we create the Observational Structure of Intentionality for that goal. The Observational Structure of Intentionality for a goal consists of a collection of elements that we believe characterize the purpose of the observation through which we intend to achieve a goal. This is the **Objects Pattern**, and represents the cognitive process of turning a unitary idea into a multitude of objects-data that represents the purpose of the observation of the original unitary idea.

The Reconstruction Phase

We define a way to recompose some of those objects-data in a way that is consistent with the very reason why we have represented the unitary idea with that particular instance of Objects Pattern. This is the **Structure Pattern:** which provides a way to reconstruct what the Objects Pattern has broken down into pieces.

The Identity Phase

Finally, there is the **Properties Pattern**, with which we impose on the objects-data the criteria by means of which we choose to reconcile the distinctions we have introduced with the Objects Pattern by defining a form of Local Identity.

9 The Common Mathematical Intuition Of Reality

An example of the metaphysical structure just described is given by the intuitive notion of set.

9.8 The Deep Structure Of The Concept Of Set

The intuitive idea of set emerges when we think of many as one. The concept of set is used to indicate a certain experience, perceptual or purely intellectual, in which distinct objects or states of affairs acquire a new ontological status as a result of being endowed with a quality of unity, wholeness and cohesiveness.

The definition of set I am going to use to illustrate the metaphysical structure of the OSP Pattern was coined by German mathematician Georg Cantor. Cantor's definition of set is used today in what is known as naïve Set Theory, where naïve signifies that a set theory is not axiomatic and is not based on formal logic—however, all the modern axiomatic set theories still share an implicit form the cognitive machinery outlined by Cantor in his definition of set. In P.E.B. Jourdain's translation of Cantor's "Contributions to the Founding of the Theory of Transfinite Numbers," published in 1915, set is called "aggregate" and is defined as follows:

"By an 'aggregate' (Menge) we are to understand any collection into a whole (Zusammenfassung zu einem Ganzen) M of definite and separate objects m of our intuition or our thought. These objects are called the 'elements' of M."

(Georg Cantor)

Today, the reference to the underlying mental processes is often omitted, as in the following definition:

"A set is a collection of well-defined, distinct objects considered as an object in its own right."

(Contemporary definition)

Cantor describes the idea of set with an extraordinary combination of concepts, by organizing in a single sentence the notions of: *gathering together, whole, definite, distinct, perception*—in some translations from German perception becomes *intuition*—and *thought*. This is an astonishing sequence of concepts that seems to suggests how Cantor regards the mathematical notion of set as part of a larger context in which perceptual experience

9.8 The Deep Structure Of The Concept Of Set

and mathematical abstraction form a continuum. Cantor had the intuition that a mathematical definition of set had to include a reference to the specific thought process that I called many-as-one. His reference to thought and perception is, I think, a crucial indication of the link that he sees between the phenomenology of perception and the genesis of the abstract idea of set. He makes a sharp distinction between perceptual experiences (*definite and separate objects of our intuition*) and experiences that do not involve perception (*...or thought*) as suggested by his use of the disjunction "or".

The Metaphysical Structure Of Cantor's Definition Of Set

Cantor's definition of set orchestrates six cognitive processes, and establishes a clear connection between perception and the abstract notion of set. The tasks carried out by the Abstract Mind to "generate" a set are therefore: acquiring data (gathering together), identifying (whole), identifying (definite), separating (distinct) and presenting to the mind (thought) the objects perceived. Cantor's definition does not explain *how* the human mind—and thus the Abstract Mind—acquires, identifies and separates the objects perceived, and does not involve explicitly any notion of unity and multiplicity. In Cantor's definition, the abstract notion of perception is a deus ex machina that orchestrates the other complex notions involved in the definition, that for this reason, should be all considered indivisibly part of the notion of set.

Ontological Continuity

The definition of set given by Cantor not only creates a new entity with its own ontological status, it also points our intuition to the conceptual roots of an epistemic universe where "set" is consistent with "mind". What makes set consistent with mind, though, lies at a deeper conceptual level. When Cantor constructs his definition of set through a reference to the larger context of the philosophy of mind (*objects m of our intuition or our thought*), he provides the coherence conditions for his definition. For, set is defined by Cantor as a product of the mind by his reference to the fact that the objects being gathered together are objects of perception or thought "before" becoming a set. This point is in common with the contemporary definition, but with an important

9 The Common Mathematical Intuition Of Reality

distinction. Cantor states this principle clearly, without resorting to the linguistic sleight of hand we find in the modern definition where "considered" replaces Cantor's "objects of our intuition or our thought".

From an ontological point of view, both definitions of set, Cantor's and modern, state that:

i) there are objects being perceived

ii) there is an agent perceiving those objects and forming a concept of set as a result of certain thought processes and in full accord to an overarching metaphysical structure

iii) i) and ii) form a continuum of some sort

The property of the concept of set of being ontologically continuous with the cognitive process that generates it from a collection of distinct objects conveys the idea that Cantor's definition of set asserts that the ontology of the entities being perceived and the ontology of the concept of set formed as a result of a certain faculty of the human mind, form a distinct, definite continuum characterized by a quality of immediateness.

Immediateness And Cohesiveness

Cantor's definition of set does not say, for example, that from a collection of well-defined, distinct objects of thought or perception one *eventually* forms the idea of a whole object in its own right. Immediateness means that objects and set coexist, that are presented to conscious thought simultaneously, and that are therefore epistemically accessible and coextensive in distinct but related ways. The quality of immediateness implies also that within the metaphysical context that we call mind, objects and sets are interchangeable notions because the mind itself can conceive sets from objects and objects from sets. A slightly more abstract interpretation of the immediateness codified in Cantor's definition of set, is that objects and sets are a manifestation of the same cognitive process by means of which the mind is able to switch focus from set to object and vice versa. That is to say, the objects *after* the notion of set cease to exist in the form they existed prior to being "settified", yet remain epistemically accessible through their original definition of object (see also Chapter 5 for a discussion about the relation between identity and definitional structure). To

9.8 The Deep Structure Of The Concept Of Set

what extent the notion of set becomes a reference to the objects that participated in the formation of the set is a problem that needs to be investigated.

In summary (Table 9.2), OSP is the manifestation of these three higher metaphysical processes:

OSP Pattern	Synthetic Structure of Intentionality
Objects Pattern	*Validation System.*
Structure Pattern	*Provide a mechanism to construct and destroy objects.*
Properties Pattern	*Equip a set of objects with a domain-specific notion of identity.*

Table 9.2: Functions of the Object, Structure and Property patterns.

Bibliographical Notes

A discussion about the Objects, Structure and Properties pattern appeared for the first time, I believe, in a UseNet discussion in 1998, and was formalized by some of the authors of that UseNet discussion in a series of lectures given during a Quantum Gravity Seminar in 2004 (Baez and Wise 2004), and later by Baez in Baez and M. Shulman 2006. In Baez and Dolan's work, the OSP Pattern is called "stuff, structure and properties", and the discussion is centered around the use of Higher Category Theory to represent the stuff, structure and property concepts in a precise form via properties of functors. I use the term "objects" in place of stuff to relate Baez and Dolan's ideas to an intentional act whose objective is to elicit a specific type of knowledge.

My attempt to situate Baez and Dolan's ideas about the stuff, structure and property pattern in the larger context of the study of the Abstract Mind, leads naturally to a generalization of their approach, as it will appear clear later in the book, where Objects, Structure and Properties become independent entities. Cantor's definition of set is in Cantor 1955. Two papers on the nature and structure of the concept of set influenced more than others some of the ideas I present in this chapter, Lawvere 1994 and Feferman 1989.

Chapter 10

$\mathcal{M}$-signatures

> In the study of ideas, it is necessary to remember
> that insistence on hard-headed clarity issues from
> sentimental feeling, as if it were a mist, cloaking the
> perplexities of fact. Insistence on clarity at all costs
> is based on sheer superstition as to the mode in
> which human intelligence functions. Our reasoning
> grasps at straws for premises and float on gossamer
> for deductions.
>
> Whitehead 1967

> Let those who come after me wonder why I built
> these mental constructions and how they can be
> interpreted in some philosophy; I am content to
> build them in the conviction that in some way they
> will contribute to the clarification of human thought.
>
> L. E. J. Brouwer

In this chapter, we begin to lay down the theoretical foundations for the formulation of Interaction Theory by introducing a notion that codifies a conception of *intelligibility*.

In Interaction Theory, intelligibility corresponds to the fundamental structure of a mind-world transaction, synthesizes what makes the basic, irreducible structure of thought what it is, and does not denote any form of expertise or insight that can be apprehended through experience, training, reasoning or intuition. A way to begin to tackle the problem of defining intelligibility for an Abstract Mind is to focus on what a given conception of intelligibility produces. Intelligibility gives rise to what we call a *metaphysical structure of thought*. By a metaphysical structure of thought, we mean those distinct and irreducible mental functions that enable thought in any understandable or observable sense. The notion we introduce in this chapter to characterize a metaphysical structure of thought is called $\mathcal{M}$-*signature*.

The core idea underlying the necessity to define $\mathcal{M}$-*signatures* may perhaps be expressed intuitively by saying that $\mathcal{M}$-*signatures* characterize a metaphysical structure of thought by defining, at a fundamental level, the things we can *think in*, and the things we can only *think about*. This last sentence might sound counterintuitive: How can we know that we cannot think about something if we don't think about it? And if we think about it, then it means that we can think about it. Well, what I am saying is that we can understand something without having to think about it, by just learning the basic structure of thinking. This is a sort of metathinking, and I'll explain how it works with an analogy.

Imagine a basket full of Lego bricks. That basket full of Lego bricks symbolizes a *mind*. The Lego bricks are designed in such a way that you can connect any two bricks in only two directions: parallel or perpendicular to each other. We can think of a given concept as a set of instructions to build a certain Lego structure, and we can think about the process of thinking, as the process of following those instructions. There can be of course more than one set of instructions to build the exact same Lego structure, exactly like there are many ways to think about the same thing. And this is where the analogy meets reality: it is surely non trivial to try to imagine what we are able to think about, by just looking at the concepts we use when we think, and at the way we are able to arrange those concepts to create complex thoughts. Similarly, it is difficult to imagine what a Lego structure will be, by just looking at the set of instructions to build it. The point is, when we understand the basic rules that define how Lego bricks can be connected to each other, we know whether a certain structure can be built with Lego bricks because we understand that there are certain structures that cannot be built with Lego bricks. The knowledge about what can be built with Lego bricks is not based on the rules to *use* Lego, that is, on sets of instructions, but on the *nature* of Lego itself.

Introduction

Now, to understand what a metaphysical structure of thought is, think about a spherical surface: Can we build one with Lego bricks? Can a "Lego brick-mind" think about a spherical surface? I think it's fair to say that we don't need to try to build a perfectly smooth spherical surface with the Lego bricks to realize that it is not possible to build one. In fact, we don't even need to be able to *think about* a spherical surface at all with a Lego brick-mind. All we need to understand is that there are certain limitations to the shapes that one can build with a certain set of Lego bricks, and that the set of possible shapes is determined by the very structure of Lego bricks. Understanding the structure of Lego bricks, that is, the structure of its geometrical syntax, does not mean or entail being able to conceive complex geometric shapes with our mind's eye, it means being able to determine whether the concept "curved" can be built or not. The Lego brick argument applies to the mind's ability to conceptualize. If we understand the deep structure of our ability to conceptualize, we can determine what classes of concepts are intrinsically non-thinkable. To this end, we are going to explore the deep structure of the mind's ability to articulate thoughts.

This chapter has three primary aims:

1. Motivate the need to introduce $\mathcal{M}$-*signatures* to define intelligibility, and as a fundamental step in the construction of Interaction Theory.

2. Describe the mechanism through which a given $\mathcal{M}$-*signature* gives rise to a metaphysical structure of thought, and characterize any given $\mathcal{M}$-*signature*.

3. Introduce a classification system for $\mathcal{M}$-*signatures*.

As we outlined in the introduction to Part III, the formulation of Interaction Theory consists in the definition of a model of Synthetic Structure of Intentionality. The element of the Synthetic Structure of Intentionality this chapter is concerned with is the Validation System.

Definition 10.0.1. ($\mathcal{M}$-*signature*)
A *metaphysical signature*, or $\mathcal{M}$-*signature* for short, is a finite collection of definite, distinct, irreducible and definitionally closed concepts that enable intelligible thought by defining a Validation System.

10 $\mathcal{M}$-signatures

In Section 8.3 we gave an ontological description of the mechanism by means of which **success** defines a Validation System with these words:

> *"[...] Intention and success are contiguous and mutually exclusive notions from an ontological viewpoint. When there is success, the Intention that propels towards the state of success ceases to exist, and when there is Intention, success is not yet achieved: this is how any given success condition is instrumental to the genesis of the ontological status of a Validation System. Any given success condition provides the conceptual boundary between the Intention to act upon a desire, and the objects and states of affairs after the goal is achieved. The function of a success condition is to signal when the Validation System, intended as the epistemic domain in which any intentional acts takes place, ends. "*

There are several concepts converging in this chapter, and it is perhaps useful to get a bird's-eye view of their interconnections before we delve into the details of the discussion.

1. There are the notions of Validation System, **success** and intelligible thought, which are connected via the definition of $\mathcal{M}$-*signature*.

2. There is the mechanism by means of which **success** defines a Validation System.

3. There is the metaphysical structure of thought, which describes the mental functions that give rise to intelligible and observable thought processes.

These three themes are related in a very precise manner. Recall from Section 8.2 that the notion of Validation System characterizes what qualifies a specific Intention to satisfy the want symbolized by the Will element of the Analytic Structure of Intentionality. Therefore, any given Validation System symbolizes the thoughts that can be articulated within the context set by a given Will. Another way to put this, is to think of any given Validation System as a presentation of a notion of intelligibility, because only the thoughts that are articulated for a specific purpose set by a Will are intelligible. This explains why defining a notion of intelligibility that applies to any given Validation System represents a suitable way to think about

Validation Systems in abstract terms. An example of how this observation applies to a specific instance of Validation System is given by formal systems. There, intelligible means anything whose description can be reduced to, and characterized by, an object that has two distinct states, such as $\{true, false\}$ or $(0, 1)$, because any object with these basic characteristics can be used to symbolize a condition in which a well-formed formula is valid—recall also the example in Section 8.4, and in particular Table 8.3. As a first approximation of intelligibility, we can think of intelligibility in the context of formal systems as a certain conceptual model built on a binary objects such as those we have just described. These observations lead naturally to another observation: given that **success** defines a Validation System, and that a notion of intelligibility characterizes any thought that can be articulated within a given Validation System, there must be a correlation between intelligibility and **success**—this line of reasoning sheds light on the intrinsic interconnections between the first two points. The third point is primarily a descriptive device that highlights the nature of the concepts in the first two points. It is a device that captures the idea that this entire construction is intrinsically metaphysical, in that it defines thought at its most fundamental level and, in the process, attempts to situate the discussion about how intentionality connects the mental and the physical domains.

10.1 Thinking In And Thinking About

An $\mathcal{M}$-signature denotes the threshold intrinsic in the structure of a mind-world transaction, that separates what we comprehend intuitively from everything else.

The need to introduce the notion of $\mathcal{M}$-*signature* is motivated by empirical and theoretical considerations. There is the empirical observation that in the human thought, as it appears to our consciousness in an intelligible unmediated manner, there is a *threshold* which separates the thoughts that we are able to think, from the thoughts that we can entertain without being able to articulate any idea *inside* those thoughts, but only *about* them. And there is the theoretical need to identify and formulate a concept that encodes a coherence condition that subsumes the definitional structure of any given **success condition**.

10 $\mathcal{M}$-signatures

There is a very peculiar feature of human thought that appears in the history of ideas in various guises. It can be found in the Philosophy of Linguistics, in the study of the language-thought interaction, and in the study of the relations between structure (syntax) and interpretation (semantics)—a concept known as syntax-semantics interface—which leads to the various forms of linguistic relativity, a notion that I examined in my early formulations of Interaction Theory. It can be found in the pioneering work of McLuhan, where he examines how the communication medium ($language = syntax$) shapes our thought ($message = semantics$), and in many other fields of research.

The peculiar feature of human thought I am talking about, that we capture with the notion of $\mathcal{M}$-*signature*, is the intuition, vastly corroborated by observation, that the ability to think about something, and the ability to articulate thoughts beyond the grasp of the human intellect as a pure linguistic or symbolic exercise, are two distinct and separate mental faculties.

We have mentioned already in the text the example of a space with 26 dimensions, which is a thought that we can conceive as a linguistic exercise because language is a syntactic engine that allows us to generate arbitrary thoughts. A 26-dimensional space is an example of this feature of the human thought. For, we can *think about* a space with 26 dimensions, but we are not able to *think in* 26 dimensions because our spatial-visual intuition is limited to three dimensions. Similarly, we can think about an imaginary creature that lives simultaneously in the present moment and one minute into the future, but we cannot think in a temporal reference that consists of more than one instant. The presence of this threshold is completely unrelated to the existence of language as a thinking tool, it is intrinsic to the metaphysical structure of thought, and can be detected in pre-conceptual and pre-linguistic contexts. There are many thresholds that define certain specific mental abilities. The examples we have just given illustrate the presence of the thresholds that define our visual-spatial intelligence and our causal and temporal intelligence respectively. But there are more general and more basic thresholds in the human thought, namely: the thresholds that define intelligibility, which is the essence of an $\mathcal{M}$-*signature*. A notion of intelligibility is, by definition, the overarching concept that subsumes any other context in which we are able to articulate thoughts.

10.1 Thinking In And Thinking About

Success And Success Conditions

An $\mathcal{M}$-signature subsumes any **success condition** *for a given Validation System.*

The second reason for introducing $\mathcal{M}$-*signatures* reframes the problem of defining intelligibility we described in the previous section, in the context of the Synthetic Structure of Intentionality.

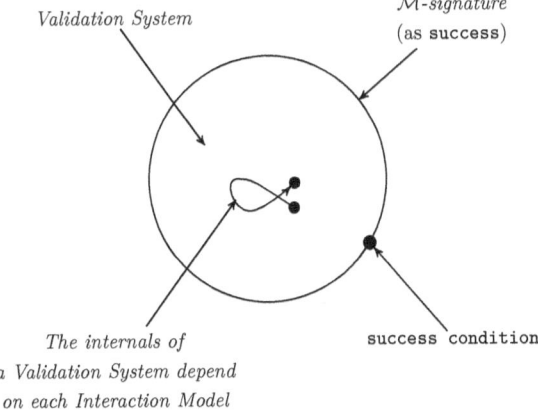

Figure 10.1: Illustration of how $\mathcal{M}$-*signatures* define *intelligible*.

In the reference framework of the Synthetic Structure of Intentionality, a threshold defines *any* intelligible **success condition**, by characterizing what holds together the definitional structure of a Validation System. The threshold-as-**success** is symbolized in Figure 10.1 by the circumference, which is the structure that holds together the Validation System's definitional structure. This is the theoretical motivation for introducing of the notion of $\mathcal{M}$-*signature*.

It is indeed hard to tell the nature of the relation between the threshold I have just described, and the metaphysical structure of human thought. I think that, to the extent to which a definition of rational thought is applicable, the definition of this threshold and the metaphysical structure of human thought are interchangeable notions. This is, so to speak, the practical motivation for introducing the concept of $\mathcal{M}$-*signature*. There might be of course other elements that contribute to the emergence of this threshold, such as physical, chemical or biological causes, and causes related to the complex language-communication-thought interplay that continuously shapes our ability to conceptualize, especially when abstract thought is involved, such as for example the predominance of a symbolic, or an auditory, or visual communication system over other sensory systems. The hypothesis we have put forward in Postulate 8.7.10, is that there is a precise correspondence between modes of thought and states of consciousness—in the form of types of spatialization, as we will see in detail in Part IV—and that distinct states of consciousness—where the distinction might be more fine grained than we can imagine today—corresponds to distinct metaphysical structures of thought in which the very fabric of conceptualization allows for the articulation of radically different thoughts because it is the expression of radically different forms of intelligibility. Identifying and characterizing this threshold is therefore crucially important to comprehend the metaphysical structure of thought, because it provides a definition of intelligibility which is essential to define a Validation System, and, with it, any Interaction Model in which we define the cognitive technologies that relate us to reality.

10.2 How $\mathcal{M}$-*signatures* Define Intelligible Thought

To give an adequate characterization of the notion of $\mathcal{M}$-*signature* as a way to codify a basic intuition of reality, we want to explain how the notion of $\mathcal{M}$-*signature* defines a Validation System. A first step in this direction, is to look for what holds together the definitional structure of a Validation System. This is a project based on a distinct style of reasoning that, at least in principle, holds the promise of giving us access to basic concepts by means of which we can define the fabric of human thought at the level where intentionality steers consciousness.

The questions we ask are:

Z.1 In what way are these—for now hypothetical—basic concepts at the foundation of Validation Systems meant to be basic?

Z.2 What do we need to determine about a concept, for us to be able to say that that concept is basic, in the way it has to be basic to give an account of the definitional structure of a

10.2 How $\mathcal{M}$-signatures Define Intelligible Thought

Validation System?

In this research, we are concerned with the observable structure of human thought as a manifestation of intentionality and consciousness. Therefore, the answer to the first question is that the feature we are concerned with is the human mind's ability to articulate thoughts in terms of Validation Systems—Will *and* Intention—as a result of the very existence of these basic concepts. This characterization of what counts as "basic" in the definition of a concept that codifies the definitional structure of a Validation System, leads us to the answer to the second question. To establish that a certain concept is basic in the way we have just described, we are going to have to be able to say that not only it enables thinking inside a Validation System, but also that it is irreducible with respect to those particular features that make it basic. For, basic and irreducible aren't necessarily synonyms in the context of this discussion, and we must assume that in certain modes of thought, these two features may not coincide. By irreducible, we mean that its effect on intentionality is indistinguishable from the sense and manifestation of intentionality, prior to becoming a distinct, definitionally identifiable concept.

With these methodological premises, we can focus on what concepts enable the minimum form of thought needed to articulate the notion of Validation System (Z.1), and are irreducible with respect to the context in which they are defined the way they are (Z.2). We must note that what determines a basic notion to be basic, in the sense we have given to "basic" in this discussion, might reside in forms of consciousness and thought that, from the point of view of the ordinary alert, problem-solving state of consciousness, we may characterize as prelinguistic or even preconceptual. Whether it is the case that basic irreducible concepts that enable Validation System-thinking are indeed part of the phenomenology of prelinguistic and preconceptual mental states is a vast topic that we will not discuss here. The view on concepts I advocate in this discussion is mildly inspired by Pragmatism, of which I embrace the core tenets in various degrees and nuances. Pragmatism is a philosophical movement spearheaded by American philosopher and mathematician Charles Sanders Peirce in the United Stated around 1870. A core tenet of pragmatism is to define concepts based on the practical effects that they produce.

I regard concepts as empty, meaningless placeholders that become what we use them for, solely and exclusively by virtue of the thought processes that they enable. In Chapter 5 we used a visual metaphor to describe concepts: we depicted concepts as interference patterns of definitional structures. That is to say: our use of the term *concept* is to be intended as a certain system that describes the *interactions between definitional structures*. So for example when we use the term finiteness, we do not mean only finiteness in the ordinary mathematical sense, we mean *any* concept that, under certain well-defined circumstances, *functions* like the ordinary mathematical concept of finiteness by enabling or justifying certain distinct modes of thought, as a result of the problems that we are able to solve with that concept, and of the thoughts we are able to articulate with it.

Our assumption about the concepts we are looking for, are that they exist "definitionally" in the domain of conscious thought, but we do not rule out that they may also exist in an intelligible form in pre- or preterconceptual mental states. In summary, an $\mathcal{M}$-*signature* defines intelligible thought by characterizing the fabric of intentionality at a given state of consciousness. To comprehend how the fabric of intentionality can be characterized, we introduce two central notions of Interaction Theory called **concrete** and **abstract** concepts.

10.3 concrete And abstract Concepts

Based on my observations of a large number mathematical and non mathematical arguments, I have identified a small set of concepts, which I call *substrate concepts*, that stand out for what appears to be their basic role in defining three fundamental Symbolic Intuitions of Reality. The substrate concepts I examine in this book are: *identity*, *finiteness*, and *cohesiveness*.

We classify substrate concepts based on the relations that these concepts have with intentionality. This classification system is inspired by the notion of speech act given in the Philosophy of Language. The resulting intentionality-based concept classification system consists of two categories of concepts, which we call **concrete** and **abstract**. Any intelligible concept belongs to *one and only one* of the following categories at a time.

10.3 concrete And abstract Concepts

Concrete Concepts

The essential traits of a concrete concept are unconscious competence, faculty and action, unintelligibility, intuition and unbiased potential.

Definition 10.3.1. (concrete concept)
A concept is **concrete** if it defines an unintelligible, irreducible, independent, atomic, complete, pre-conceptual, pre-cognitive, deliberate mental ability that characterizes in full the structure of an unconscious act performed freely without intending anything else other than the act itself.

To comprehend the nature of **concrete** concepts, it is essential to recognize that there are concepts that the human mind can grasp only through their *phenomenology*, and never directly through reasoning. This peculiar mode of access to **concrete** concepts extends to concepts a philosophical principle known as *solvitur ambulando*, a Latin expression that means "solved by walking", and that denotes certain types of problems that are solved practically, through direct experience, rather than conceptually, through reasoning. The phenomenology of a concept is the structure of our experience of that concept through its use or formulation. The phenomenology of a concept, therefore, is invariably, solely and exclusively the result of an intentional act: it is necessarily directed at or about or of an object of thought; and the structure of our experience of it rests entirely on this duality. It is not possible to characterize a **concrete** concept through our direct experience of it, because a **concrete** concept is not directed at or about or of an object of thought. It is precisely the phenomenological nature of our epistemic access to **concrete** concepts that sets **concrete** concepts apart from any other concept and that defines them.

Another way to think of **concrete** concepts is by analogy with an archetypical theme common to practically every known philosophical tradition, the dichotomy between *essence* and *substance*. This archetypical principle appears in various guises as *quality* and *quantity*, as *form* and *matter* (Plato), as *act* and *potency* (Aristotle), as *nāma* and *rūpa* in the Hindu tradition.

An intuitive way to convey the nature of a **concrete** concept is through the so-called *centipede effect*, a psychological effect named after the poem "The Centipede's Dilemma", that narrates the story of a centipede that loses its ability to walk and trips over its legs when a toad asks it which leg moves after which. The centipede effect, also known as Humphrey's law, captures a psychological phenomenon everyone is familiar with: consciously thinking of an unconscious activity disrupts that activity. The unconscious ability codified by a **concrete** concept can only be apprehended by applying the **concrete** concept, and not by contemplating it.

We know and understand that **concrete** concepts *exist*, and are not a theoretical abstraction detached from the factual reality of human thought, because we can describe them very accurately by studying how they define the fundamental structure of human thought—as we will do in Chapters 15, 16 and 17. **concrete** concepts are, so to speak, the cosmic principles of human thought as we know it. They are comparable to the laws of Nature because through them it is possible to explain the deep structure of human thought and the genesis of our Symbolic Intuition of Reality.

From a Philosophy of Language viewpoint, **concrete** concepts can be thought of as the cognitive equivalent of a perlocutionary act, in that they do not result in the action that they describe. Any notion of intelligibility is built from, and encoded by, a **concrete** concept.

Abstract Concepts

*The essential traits of an **abstract** concept are purpose, cognition, action, intentionality and effect.*

Definition 10.3.2. (**abstract** concept)
A concept is **abstract** if it describes the intentional application of a **concrete** concept.

A **concrete** concept becomes **abstract** by being intentionally directed to an an end compatible with its nature. The intentional nature of **abstract** concepts does not confer them any teleological feature or intrinsic design. The formation of an **abstract** concept is merely and exclusively intended to be understood as the evolutionary outcome of how **concrete** concepts create our symbolic experience of reality, and in this sense, and for this reason, they represent the phenomenology of **concrete** concepts. From a Philosophy of Language viewpoint, **abstract** concepts can be

10.3 concrete And abstract Concepts

thought of as the cognitive equivalent of an illocutionary act, in that they constitute the action that they describe.

Interpretation Of Substrate Concepts

concrete and abstract concepts are meant to be interpreted in a participatory sense (Section 3.2). A way to visualize the way concrete and abstract concepts should be interpreted is with the metaphor of a layer-cake, in which a layer looks like the graph of Figure 10.2. There, the nodes represent concepts, and the

Figure 10.2: Illustration of the relation between concrete and abstract concepts.

arcs represent relations between concepts. When a set of nodes is tagged as concrete, the remaining nodes become abstract. Upon tagging the nodes of the graph with the two labels, concrete and abstract, the sequences of arcs that form a continuous path from concrete and abstract nodes represent higher concepts. This layer-cake construction evolves in two ways:

- By defining higher concepts as paths, we can construct another graph—another layer—where the nodes are the paths forming higher concepts and the arcs are new relations between higher concepts. This is the upward direction.

- By interpreting each node of a given graph as a higher concept, we can create a new graph which has a distinct path for each node of the original graph, and new nodes and new arcs that represent new relations. This is the downward direction.

We are going to use these definitions to articulate some fundamental questions about $\mathcal{M}$-*signatures* that we need to answer. First and foremost, we need to determine how this intentionality-based concept classification system allows us to classify the substrate concepts and $\mathcal{M}$-*signatures*, then we need to explain how to actually define an $\mathcal{M}$-*signature*, that is, we need to determine what counts as a description of an $\mathcal{M}$-*signature*.

10.4 Classification Of $\mathcal{M}$-*signatures*

The classification system for $\mathcal{M}$-*signatures* we present here is based on the observation that, at the alert, problem-solving state of consciousness, the human mind's mode of thought seems to rely solely and entirely on a notion of identity to carry out its cognitive functions (Postulate 8.7.11). This observation leads to a system for the classification of $\mathcal{M}$-*signatures* that regards identity as the `concrete` concept that gives rise to the default $\mathcal{M}$-*signature*.

Recall the discussion in Section 10.2 about the interpretation of substrate concepts: given a substrate concept, what makes it `concrete` or `abstract` is not an intrinsic feature or quality of that concept, but a manifestation of the level of consciousness in which that concepts forms the basis of intelligible thought—indeed, what makes a concept `concrete` or `abstract` can itself be regarded as a kind of *signature* for that particular level of consciousness in which that concept acquires its quality of concreteness. This observation about how concepts acquire their concreteness, relativizes the act of picking a `concrete` concept as a starting point for the definition of a $\mathcal{M}$-*signature* classification system, in much the same way in which defining the origin of a coordinate system is a mere convention dictated by specific circumstances or goals which has nothing to do with the coordinate system itself, and that does not ascribe to the origin of the coordinate system any intrinsic feature or quality.

Thus, we ask: What substrate concept can we regard as `concrete` and why? Based on the various analyses that have paved the way for the discussion about $\mathcal{M}$-*signatures*, and in particular based on the evidence that led to the definition of the Fundamental Correspondence, there is one substrate concept that stands out for what appears to be its fundamental role in the definition of intelligible thought at the alert, problem-solving state of conscious-

10.4 Classification Of $\mathcal{M}$-signatures

ness: *identity*. Nonetheless, this determination alone does not suffice to designate identity as a **concrete** concept. For, we should prove that the other substrate concepts—finiteness, cohesiveness, etc.—are **abstract** concepts at the same level of consciousness. This is a task that we are not going to undertake here because it does not add anything significant to this analysis.

The classification systems for $\mathcal{M}$-*signatures* that originates from these observations organizes $\mathcal{M}$-*signatures* into two broad categories: *standard* and *non-standard*. Based on the rationale we have just explained, the distinction between standard and non-standard is a pure convention based on which level of consciousness is taken as a reference.

Standard $\mathcal{M}$-*signatures*

The standard $\mathcal{M}$-*signature*, or default $\mathcal{M}$-*signature*, or Identity-based $\mathcal{M}$-*signature*, or sameness-based $\mathcal{M}$-*signature*, defines intelligible thought at the alert, problem-solving level of consciousness. It is the signature of ordinary thought in which we define the modern conception of mathematics. It is the signature upon which we invent formal systems. In the default $\mathcal{M}$-*signature*, intelligibility means anything that the mind can grasp based on a binary reference model that allows the intellect to articulate any thought based on the paradigm "is or isn't", true or false, 0 or 1 and so on. This basic metaphysical structure of thought is what enables the conceptualization of the World we are familiar with. It allows us to construct mental realities in which we comprehend by applying various permutations of this basic binary form of thought that we can articulate in natural language as some kind of dichotomy between being and not being, full and empty, subject and object and so on and so forth. However, the limitations and inadequacies of natural language, allow us to construct descriptions of the default mode of thought that barely scratch the surface of what is a deeply complex realm in which the human intellect is capable of constructing truly majestic hierarchies of concepts. Chapter 15 attempts to offer a more detailed description of the default $\mathcal{M}$-*signature*.

Non-standard $\mathcal{M}$-*signatures*

At the alert, problem-solving state of consciousness, non-standard $\mathcal{M}$-*signatures* are based on substrate concepts other than identity.

10 $\mathcal{M}$-signatures

As we saw in Section 10.3, some of these substrate concept are finiteness and cohesiveness. The conceptual landscape that opens up before us when we begin to explore the metaphysical structure of thought originated by non-standard $\mathcal{M}$-*signature* is imposing and compelling and daunting. The repercussion of the shift from an Identity-based conception of intelligibility to deeply unintuitive forms of intelligibility has a profound impact on our ability to conceptualize experience and, consequently, on our capacity to imagine and define other forms of mathematization. I attempt, with the help of practical examples, to sketch an explanation of non-standard metaphysical structures of thought that originate from finiteness- and Cohesiveness-based notions of intelligibility in Chapters 16 and 17 respectively.

10.5 How concrete Concepts Define Intelligibility

We have already seen how **concrete** concepts define intelligibility, but in a context where the notions explained in this chapter were not known. To explain how **concrete** concepts give rise to forms of intelligibility, we need to revisit with sharper tools some of the ideas introduced in Chapter 8.

To comprehend the nexus between **concrete** concepts and intelligibility, it is necessary to look at how the human mind grasps concept, because concepts, and our experience of them, *coincide*. For example, to *see* coincides with the experience of *seeing*: we do not see the seeing. In Section 8.7, we described how the mental act of grasping an idea is both literal and metaphorical. It is literal, because grasping a concept causes that concept to become part of the internal dialogue through which we tune our consciousness into that concept by changing our perception of it. And it is metaphorical, because there is a strict correspondence between the physical act of *grabbing* an object, and the cognitive act of *grasping* a concept, as described by the Analytic Structure of Intentionality (Section 8.1). There, we introduced the notion of Local Identity (Section 8.7) to describe the precise mechanism through which a notion of identity enables the grasping of a concept—recall that identity is a **concrete** concept at the alert, problem-solving state of consciousness. Intelligibility is precisely the experience enabled by a Local Identity. Any idea, any object, and any state of affairs is turned into a subjective experience by the very grasping of it through the cognitive machinery set in motion by a Local

10.5 How **concrete** Concepts Define Intelligibility

Identity—which we explained in the language of Object Classifiers.

Thus, to give an adequate account of how a **concrete** concept defines intelligibility, it is necessary to explain how that concept is *defined into existence*. The reason why this is so, is that the quest for the definition of a concept is an analytical process of deconstruction of our experience of that concept which, in turn, reveals the very early stages of the formation of the concept in a form that the intellect recognizes, by virtue of having a basic structure compatible with that concept.

But we have already discussed how concepts are defined into existence. In Chapter 8, we introduced the notion of that a concept is a certain device made of Validation System, Interaction Model and Local Identity, and in Chapter 5, we introduced a *metaphysical structure* of concepts via the notion of *definitional interference*. Those two characterizations of concepts, while very abstract, could be used here to describe how **concrete** concepts define intelligibility. But it would make the discussion even more abstract because we'd have to take into account how the definitional structures of **concrete** concepts, **success** and Validation System interact with each other. We want to focus, instead, on giving a very practical account of how the cognitive machinery behind intelligibility works. To this end, we will introduce a very direct, very straightforward, very practical coordinate system.

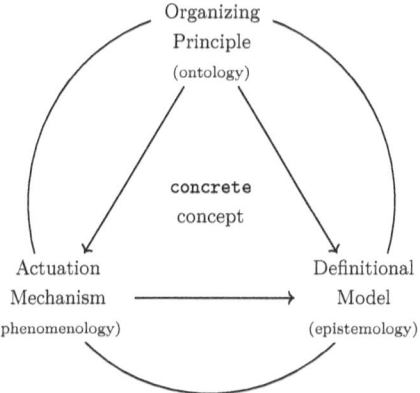

Figure 10.3: The structure of intelligibility defined by a **concrete** concept.

10 $\mathcal{M}$-signatures

This coordinate system to describe how concepts are defined into existence consists of three dimensions: an *Organizing Principle*, an *Actuation Mechanism* and *Definitional Model* (Figure 10.3), which represent respectively the ontology, the phenomenology and the epistemology of a **concrete** concept. This 3-way representation of **concrete** concepts will reemerge in Section 14.1, where we will examine three modes of consciousness that account for the basic structure of our experience of thinking.

Organizing Principle

*An Organizing Principle is an unintelligible metaphysical intuition about the potential of a **concrete** concept.*

An Organizing Principle is known, not understood. The mind has epistemic access to the potential encoded by an Organizing Principle, solely and exclusively via the conscious thought produced by the application of the Organizing Principle to intelligible concepts. The unintelligible nature of Organizing Principles, and the fact that the only epistemic access we have to them is as a metaphysical intuition, is what sets apart **concrete** concepts from ordinary concepts: it is what constitutes the ontological nexus between Organizing Principles and **concrete** concepts. An Organizing Principle is, in a sense, characterized by its liminal nature, by its ability to connect us to **concrete** concepts. The potential of a concept is what a concept is defined to do, without necessarily doing it—recall the distinction between **concrete** and **abstract** concepts given via the analogy with perlocutionary and illocutionary acts. An Organizing Principle can be perceived by the senses of the mind—that's why it is characterized as a metaphysical intuition—without being applied to any thought. For example, the Organizing Principle of identity is *distinctiveness*, because through identity we determine how the objects of thought are distinct from one another. But distinctiveness in itself does not separate the objects of thought from one another, and *is not a functioning concept* as such; it only expresses the *potential* to discern and distinguish. It is an intuition that captures the principle according to which objects of thought are *turned into* distinct and separate entities by a notion of identity within the scope and the horizon of our experience of those objects. Similarly, the Organizing Principle of cohesion is *unity* because through a notion

10.5 How **concrete** Concepts Define Intelligibility

of unity we characterize what holds together objects, but unity in itself does not glue objects together, it acts only as a reference to a certain principle of cohesion through which we ascribe to objects a quality of togetherness. To comprehend an Organizing Principle, we need the notion of Actuation Mechanism.

Actuation Mechanism

An Actuation Mechanism defines the phenomenology of an Organizing Principle.

The Actuation Mechanism describes the features of the conscious thought produced by the application of an Organizing Principle to concepts, and, in doing so, defines the structure of the *spatialization* associated to that Organizing Principle. In Interaction Theory, we *identify* the fundamental structure of the phenomenology of a Organizing Principle with the structure of the spatialization it produces. The last sentence is of crucial importance in this discussion. For example, the Actuation Mechanism of distinctiveness is *division*, because the effect of introducing a principle to distinguish objects of thought results in the—mental experience of the—creation of parts of those objects of thought, namely: the entities that have such and such features and those that don't. In this sense, the Organizing Principle of distinctiveness represents a prototypical form of a mereological frame of reference. We don't see distinctiveness, we see division: distinctiveness becomes intelligible by enabling the mental experiences that we describe as "objects of thought distinct from one another". Division is, therefore, the archetypical structure of spatialization an Identity-based $\mathcal{M}$-*signature*. Similarly, for the **concrete** concept of cohesiveness, the Actuation Mechanism of unity—cohesiveness' Organizing Principle—is *structural continuity*—a generalization of self-similarity—because the application of a notion of unity to concepts results in various forms of correspondences between the part and the whole, which are examples of self-similarity. Unity is an **abstract** concept that we grasp via the concept of self-similarity and its generalizations. Self-similarity is the structure of spatialization in a Cohesiveness-based structure of thought.

10 *M-signatures*

Actuation Mechanisms As Modes Of Consciousness

In Section 14.1 we'll see how the notion of Actuation Mechanism characterizes a mode of consciousness, and how this interpretation is useful to explain the structure of the Epistemology of the Abstract Mind.

Definitional Model

A Definitional Model defines the epistemology of an Organizing Principle by modeling an Actuation Mechanism.

By modeling an Actuation Mechanism, a Definitional Model characterizes *a type of spatialization*. To apply an Organizing Principle to objects of thought or perception, we need to be able to give a representation of that Organizing Principle in a way that is consistent with its intended purpose and phenomenology, as described by its Actuation Mechanism. For example, in the introduction to this chapter, we gave a rudimentary representation of the manifestation of the **concrete** concept of identity via binary objects, such as $\{true, false\}$ or $(0,1)$, because binary objects can be used to separate other objects, thus providing a model of division—the Actuation Mechanism of distinctiveness. These representations of identity encode how distinctiveness works, because through them it is possible to give an account of how a criterion of distinction that is consistent with the purpose of separating objects from one another operates. For example, a mathematical relation that assigns the elements of a set to $(0,1)$ necessarily results in a separation of the elements of the set into two classes, those mapped to 0, and those mapped to 1.

Definitional Models And Interaction Models

By providing a model of an Actuation Mechanism, a Definitional Model becomes a representation of a *prototypical notion of* **success** (Section 8.3). For example, the Definitional Model of distinctiveness is an object that codifies the minimum structure that an Interaction Model needs to encode in its own structure to represent any model of identity. In the Identity-based *M-signature*, a representation of a binary object with certain characteristics that we'll see in Chapter 15, forms the basis for the various definitions of the equational properties through which *any* Interaction

Model reconstructs Identity. Similarly, in the Cohesiveness-based
$\mathcal{M}$-*signature*, a model of self-similarity defines the prototypical
objects through which *any* Interaction Model reconstructs cohesiveness in a way that we'll see in Chapter 17. As we'll see in
detail in Chapter 11, Definitional Models capture the *minimum
structure* that an Interaction Model must possess to codify the
process through which concepts are generated. This fundamental
process is called **reconstruction of the success condition.**

Definitional Models As Models Of Modes Of Consciousness

The notion of Definitional Model will reappear in the form of
model of a mode of consciousness (Sections 14.1), and as the basic,
irreducible structure of the mind-world interaction (Section 11.5).

Definitional Model And Generative Knowledge

As we will see in Section 12.1, in Interaction Theory there is
a special type of knowledge called *generative*, that is neither
procedural nor propositional. Roughly, generative knowledge
defines the shape of a Definitional Model, and with it, how a
Definitional Model functions.

10.6 How Intelligibility Defines Objectivity

A conception of intelligibility gives rise to an archetypical mode
of existence called *objective*, that denotes how the things in the
World participate in a mind-world transaction.

Here we have to deal, once again, with the embarrassing inadequacy of the syntax-laden expressiveness of natural language,
to describe states of affairs that transcend a grammatical structure, and that exist in realities where the dynamic between the
knower and the known is profoundly unintuitive, yet completely
familiar to the human mind. In the discussion that follows, it is
therefore imperative to fully comprehend that the words "object"
and "objectivity" are to be understood as references to what an
object of thought is *about*, as opposite to what it really is in itself, that the word "is" and the dualistic frame of reference of a
"mind" representing the abstract realm, and a "world out there"
representing the material realm, acquire their meaning from an
$\mathcal{M}$-*signature*, and that the epistemic tension between our grasping

10 𝑀-signatures

of an object of thought and what it references "in the world out there" is where this discussion takes place. This discussion is, in other words, naturally situated in the knower-known dynamic *as a whole*.

In the ordinary Identity-based frame of reference, the "object" at the root of the word objectivity, denotes that which exists independently of one's own mind, judgment, disposition and ambition. That "object" is synonym of "things as they really are" as opposite to "things as they exist in our mind", and its very existence signals that, for reasons that will be apparent as we proceed, at some point in the history of human thought, we have felt the need to invent a notion to mark the distinction between the things of the mind and the things in the world. The necessity to define objectivity as the metaphysical quality of reality from which all things *originate*, and to which we strive to *reconcile* through the cultivation of knowledge, emerges precisely from this epistemic tension between a conception of mind and a conception of world. From this perspective, the scientific and philosophical projects are entirely defined by their goal to resolve this tension and reveal the truths underlying the objective reality of human experience. It goes without saying that it is for this very reason that objective claims are like beacons in the human journey to knowledge, and that the common metaphysical theme of the scientific and philosophical projects is a sort of homecoming, a symbolic path to dispel the effects of the separation of Man from Nature.

A core feature of objectivity is, in this sense, its perceived degree of inherent separation from the human experience, its fundamental truthfulness by virtue of not being subjugated by the frame of reference to which human existence is confined: time, space, all kinds of fluctuating forms of psychological, relational, emotional and intellectual identity, and, of course, that ineliminable subjective point of view that creates the epistemic tension between the mind and things as they really are. The dictionary notion of objectivity represents, for these reasons, the origin of all the trajectories of our analytical descriptions of facts and ideas, and is often regarded with the same reverence and appreciation with which we consider absolutes and universals, such as a constant of Nature, or the invariant of a complex process, or anything else that we view as intrinsically veridical, by virtue of displaying properties or qualities that seem to transcend the limitations of human thought and experience.

10.6 How Intelligibility Defines Objectivity

At the basis of objective thought there are two fundamental types of commitment: the commitment to the idea that the world's mode of existence is mind-independent, and the commitment to the idea that a literal interpretation of objective claims constitutes knowledge of the world. This is why, to a certain extent, and with nuances here and there, claims of objectivity are often claims about truths beyond the human domain, and why objective claims constructed according to an objective method of inquiry are believed to give us a privileged epistemic access to the world. There is a sense in which objectivity marks the boundary between the phenomenology of human existence, in the broadest possible sense, and everything else.

By redefining the foundations of human thought around a model of intelligibility, the conception of objectivity based on a mind-world dichotomy changes, and becomes an expression of the epistemic tension defined by the Actuation Mechanism associated to the **concrete** concept at the core of a given $\mathcal{M}$-signature, and is therefore reformulated as a *reference* to the Definitional Model: it becomes the *archetypical way in which the existence of objects of thought or perception is experienced* by an Abstract Mind in the frame of reference of a $\mathcal{M}$-signature—under certain conditions that we will see in detail in Chapter 11. Objectivity is, from the perspective of an $\mathcal{M}$-signature, the archetypical quality that *defines existence* in that $\mathcal{M}$-signature. For example, in the Identity-based mode of thought, where, as we will see in detail in Chapter 15, the Actuation Mechanism is the notion of *division* and the Definitional Model is a **binary object**—anything that has two distinct states—objectivity is a mode of existence that defines the "things out there in the (real) world" as *distinct* from, and as opposite to, the knower and her mental states, and it is in this archetypical form of *separation* that this original and veridical and irreducible quality of the Identity-based mind-world transactions is manifested as consciousness.

We can attempt to condense these ideas in the following definition.

Definition 10.6.1. (Objectivity)
The term *objectivity* denotes how an Abstract Mind in a cognitive state *apprehends* the fundamental mode of existence of the entities it interact with.

10 M-signatures

We are not ready to talk about the "cognitive states" of an Abstract Mind mentioned in this Definition 10.6.1, we will have to wait until Chapter 11 for that (Definition 11.4.1). Nonetheless, we can describe intuitively the condition that gives rise to objectivity as a certain configuration where the **concrete** concept at the core of the *M-signature* where a thought takes place is preserved by the same. For example, in the default *M-signature*, which is based on the **concrete** concept of identity, when the mind-world transaction is such that it preserves (a model of) identity, objectivity denotes the experience of an object of thought or perception that preserves (a model of) identity of that object of thought or perception.

10.7 The Structure Of *M-signatures*

The project of explaining the structure of an *M-signature*, is the project of explaining how the Abstract Mind articulates thoughts when it is in a state defined by a **concrete** concept. It is the project of explaining the essential ways in which the *world appears to, and is understood and explained by an intellect* modeled by the Abstract Mind through the lenses of the rational thought that springs from a **concrete** concept.

There are four topics that a description must cover to qualify as a coherent presentation of an *M-signature*. In Chapters 15, 16 and 17 we'll use this 4-step outline to examine three important *M-signatures*. Given a **concrete** concept, the description of an *M-signature* consists of the following four elements:

- The *phenomenology* of the **concrete** concept at the basis of the given *M-signature*.

- The type of *intelligibility* that originates from the **concrete** concept.

- The Minimal Interaction Model, which defines how thoughts and prototypical forms of knowledge are formed.

- The *epistemological framework* of the given *M-signature*.

Phenomenology Of The concrete Concept

A description of the phenomenology of a **concrete** concept is intended to clarify, at least from a philosophical perspective, the link between the mental and the perceptual manifestations of the

10.7 The Structure Of $\mathcal{M}$-signatures

concrete concept. This requirement is based on the assumption that pervades this study, that the phenomenology of *any* concept—**concrete** or **abstract**—is always both perceptual *and* mental (Postulate 8.7.6).

The outcome of this analysis is a detailed picture of the way the Organizing Principle associated to the **concrete** concept operates via its Actuation Mechanism.

Description Of Intelligibility

The description of an $\mathcal{M}$-*signature* must explain how a **concrete** concept gives rise to a conception of intelligibility. The description of intelligibility, as seen in Section 10.2, consists in the description of the Organizing Principle, the Actuation Mechanism and the Definitional Model. The first two components characterize the structure of intelligibility, and the third defines the conception of knowledge.

This analysis produces a model of the Definitional Model used to construct the process of acquiring knowledge: it literally defines the process of *knowing*, and the meaning of the same. This step is necessary to define the epistemology of a given $\mathcal{M}$-*signature*.

The Minimal Interaction Model

The Minimal Interaction Model is a crucially important element of Interaction Theory that describes the prototypical form of the Interaction Models—i.e. the cognitive technologies definable in a given $\mathcal{M}$-*signature*—and the *use* of Definitional Models in the construction of the same. The Minimal Interaction Model will be defined in Section 11.4. In particular, a Minimal Interaction Model:

- *is* a prototypical model of the Actuation Mechanism of the **concrete** concept at the basis of the given $\mathcal{M}$-*signature*,

- *defines* the fundamental mechanism by means of which thoughts and prototypical forms of knowledge are formed via the type of spatialization that characterizes the given $\mathcal{M}$-*signature*, as described by the Definitional Model,

- *characterizes* the epistemology of the given $\mathcal{M}$-*signature*.

This step is necessary to characterize the epistemological framework defined by a given *M-signature*.

The Epistemological Framework

As we'll see in detail in Chapters 12, 13 and 14, a description of the epistemological framework defined by a given **concrete** concept, consists in the characterization of the following five terms:

- *Knowledge*
- *Understand*
- *Meaning*
- *Explanation*
- *Modes of consciousness*

These are the five dimensions that define the mental space in which thoughts exist, evolve and form our conscious mental experience of the World in a given *M-signature*. These five terms define the horizon in which an Abstract Mind asks questions, and constructs the cosmologies of thought by means of which we relate to the World. These five terms, and their formal implementations in Interaction Theory, define thought at the most fundamental level, and, with it, various notions of computational knowledge and communication.

Bibliographical Notes

The theme of "Thinking In and Thinking About" (Section 10.1) pertains to the fundamental structure of human thought, and its many manifestations can be detected more or less implicitly as an untold assumption in the various studies on the language-thought interaction (McLuhan and Lapham 1964, Chomsky 1986, Chierchia 1995), on the formation of meaning, on the structure of the syntax-semantics interface (Chierchia and McConnell-Ginet 1990, Kripke 1981), on the use of anaphoras (Chierchia 1992), and in the vast literature on the metaphysical interpretation of physical theories and of complex mathematical concepts.

The discussions about "How *M-signatures* Define Intelligible Thought" and "How **concrete** Concepts Define Intelligibility" (Sections 10.2 and 10.5 respectively) has an obvious Platonic influence.

10.7 The Structure Of $\mathcal{M}$-signatures

The analogy between **concrete** and **abstract** concepts and perlocutionary and illocutionary acts is inspired by the works of Searle 1968, Searle 2007, Searle 2009 in the Philosophy of Language. The centipede effect is in Colman 1995, p. 119. It appeared in Pinafore Poems, 1871 Connolly and Martlew 1999, p. 220, and is attributed to English poet Katherine Margaret Craster.

The ternary characterization of **concrete** concepts as Organizing Principle, Actuation Mechanism and Definitional Model is dictated by the practical necessity to explain a distinction between the nature of a concepts, as described for example by the notion of definitional interference, and its phenomenology, still part of the framework of definitional interference. This characterization is, as abundantly illustrated in Chapters 15, 16 and 17 far from philosophical or abstract.

$\mathcal{M}$-signatures create Symbolic Intuitions of Reality, they condense in their structure the workings of a simplified model of human mind, therefore the description of their structure (Section 10.7) is necessarily a multidisciplinary task that spans various topics. For the bibliography that inspired this section I refer the reader to the bibliography in Chapters 12, 13 and 14.

Chapter 11

Cognitive Architectures

> The circumstance that our factual knowledge of the world's arrangements is a process of ongoing interaction with nature has far-reaching implications. It means that as far as we finite knowers are concerned, real things have hidden depths—they are always cognitively opaque to us to some extent because more about them can always come to light.
>
> Rescher 2000

> After silence that which comes nearest to expressing the inexpressible is music.
>
> Huxley 1931

In this Chapter we continue to lay down the theoretical foundations of Interaction Theory by defining a conception of *knowledge*.

Together, the conception of knowledge, and the conception of intelligibility we presented in Chapter 10, define the intuitive part of a Symbolic Intuition of Reality, and constitute the frame of reference to define its epistemic part.

In Interaction Theory we make a sharp distinction between a *conception* of knowledge and a *definition* of knowledge. A conception of knowledge is codified by the notion of Cognitive Architecture, and within a given Cognitive Architecture, each Interaction Model represents a *definition of knowledge*.

The main purpose of this chapter is to explain the relation

11 Cognitive Architectures

between a conception of knowledge and the fabric of intentionality, and the exact mechanism by means of which a given Cognitive Architecture gives rise to the various definitions of knowledge used to construct cognitive technologies, such as for example the OSP Pattern we saw in Chapter 9, and in general any Interaction Model.

11.1 Conceptions And Definitions Of Knowledge

In Interaction Theory, it is necessary to distinguish between the knowledge produced by the cognitive technologies that can be defined inside an $\mathcal{M}$-*signature*, which are encoded by Interaction Models, and the overarching type of knowledge that subsumes *any* cognitive technology definable in the $\mathcal{M}$-*signature*. This distinction is necessary for theoretical and practical reasons. The theoretical reason for making this distinction is that the synthetic description of this peculiar structure of the mind-world transactions requires to define an organizing principle. The practical reason is dictated by the need to identify two types of knowledge that serve intrinsically different purposes: one is a type of knowledge that has no operational requirement, the other is a type of knowledge that defines the terms of our conceptual interaction with the World by shaping the processes by means of which we transform data into information. The former is a type of knowledge called *generative* that we will describe in Section 12.1. The latter consists, for example, of the procedural and propositional knowledge of the Western philosophical tradition, and of many other types of operational knowledge we will discover in Part IV. With these ideas in mind, we introduce two definitions.

Definition 11.1.1. (Definition of knowledge)
An Interaction Model defines a *way to acquire knowledge* based on the structure of thought that originates from the $\mathcal{M}$-*signature* where knowledge is sought, and defines the *type of knowledge* acquired in this way.

The correspondence established by Definition 11.1.1, is between the philosophical content of a certain abstract process to acquire knowledge—which we haven't defined yet—and the Synthetic Structure of Intentionality, via the notion of Interaction Model. It

11.1 Conceptions And Definitions Of Knowledge

tells us that we can *think of* an Interaction Model as a faithful representation of the abstract process of getting epistemic access to a certain reality. It tells us that from an Abstract Mind viewpoint, each Interaction Model encodes a type of knowledge, because it is indistinguishable and inseparable from the experience of knowing an object of thought *through* that Interaction Model. In this sense: each Interaction Model represents *a way of knowing*.

Definition 11.1.2. (Conception of knowledge)
Let $\mathbf{M}$ be an $\mathcal{M}$-*signature*. A *conception of knowledge* is an abstract structure $\mathcal{N}$ defined when the following properties hold simultaneously in $\mathbf{M}$

1. $\mathcal{N}$ is entirely and solely based on the `concrete` concept at the basis of $\mathbf{M}$

2. any set of ontologies that form a spatialization of an object of thought in $\mathbf{M}$ is related via $\mathcal{N}$

A conception of knowledge defines what it means to know something in the structure of thought defined by a given $\mathcal{M}$-*signature*, *irrespective* of the particular Interaction Model employed to interact with an object of thought or perception—point 2 of Definition 11.1.2. A *conception* of knowledge, in this sense, subsumes the structure of *any definition of knowledge* consistent with a given $\mathcal{M}$-*signature*.

A conception of knowledge defines what it means to know something at the most fundamental level, without any reference to a specific *way of knowing it*. On a practical level a conception of knowledge describes the structure of comprehension and meaning in the creative process of acquiring knowledge, as we'll see in detail in Chapters 12 and 13, and in Part IV. On a conceptual level, a conception of knowledge defines the nature of the relations between the objects being known, and the process of acquiring knowledge of those objects. This means that a *conception* of knowledge is necessary to describe the Epistemology of the Abstract Mind, because it defines the fundamental structure of the cognitive processes through which we relate abstract concepts to the objects and states of affairs in the world.

11.2 The Basic Intuition Of Cognitive Architectures

To introduce the basic intuition of Cognitive Architectures, we must examine the notion of Interaction Model from the point of view of its relations with $\mathcal{M}$-*signatures*, rather than, like we did in Section 8.3, from the perspective of its function in the context of the Analytic Structure of Intentionality.

We have already encountered an example of Interaction Model in Chapter 9. There, we described how the construction of any mathematical definition can be mechanized through the OSP Pattern. The OSP Pattern is an Interaction Model in the Validation System defined by the default $\mathcal{M}$-*signature*. We use the OSP Pattern to define mathematical objects, structures and theories through which we acquire mathematical knowledge of the world and of mathematics itself. The process by means of which we create mathematical knowledge via the OSP Pattern has a very distinct structure as depicted in Figure 11.1. We have seen diagrams like the one in Figure 11.1 in Sections 1.1, 2.1 and 2.5. Those diagrams—called ∞-diagram—describe how we use mathematics as a cognitive technology. The fact that we can define any mathematical gadget via the technology of OSP Patterns, and the fact that we can clearly identify and describe the OSP Pattern, allow us to regard the OSP Pattern as a cognitive technology in much the same way in which the fact that we can describe how a lighter works, allows us to regard a lighter as a technology to get a fire started.

Let's examine the use of the OSP Pattern.

A. We use the OSP Pattern to construct a mathematical gadget $\mathcal{T}$ that encodes the questions through which we look at the world **W**. Thus, $\mathcal{T}$ reflects the specific Observational Structure of Intentionality through which we look at **W** (see also Table 9.1).

B. We derive the features and properties of $\mathcal{T}$ through logical reasoning. These are the *data* produced by $\mathcal{T}$.

C. We *interpret* the data produced by $\mathcal{T}$ based on the criteria and belief systems that define the Observational Structure of Intentionality associated to $\mathcal{T}$. These criteria and belief systems are the criteria and the belief systems with which a strategy is validated in a given Observational Structure of

11.2 The Basic Intuition Of Cognitive Architectures

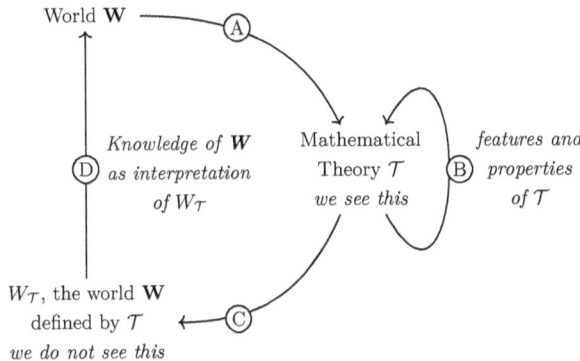

Figure 11.1: Illustration of the modern way to acquire mathematical knowledge.

Intentionality (see Section 3.3).

D. We *acquire knowledge* of **W** by applying a model of truth to the interpretation of the data produced by $\mathcal{T}$. A model of truth is a criterion to ascribe to the data produced by $\mathcal{T}$ a quality that turns that data into information.

For example, consider the problem of studying the motion of a body subject to a force F.

A. We construct a mathematical model to capture the basic law of motion, $F = ma$ (Newton's second law of motion: the acceleration a of an object as produced by a net force F is directly proportional to the magnitude of the net force, in the same direction as the net force, and inversely proportional to the mass m of the object). The Observational Structure of Intentionality associated to this construction is

11 Cognitive Architectures

Goal	Strategy	Data
Understanding the motion of a body subject to a force.	*Look for causal relations between certain data that reflect a given worldview.*	*Force, mass, acceleration, and the background notions of space and time and position and direction.*

The worldview encoded by this Observational Structure of Intentionality is that of an arena called physical world which we choose to describe in terms of space, time, force etc.: these are a choice, though, not a necessity.

B. We derive a description of the motion of a body, such as for example this law $s = \frac{1}{2}at^2$, which relates the space s covered by a body subject to the acceleration a for t time units.

C. We interpret the data produced by (B) based on the Observational Structure of Intentionality associated to this model.

D. We develop the knowledge of the motion of a body by applying a model of truth to the data produced in (C). In this case the model of truth is called "correspondence with measurements": if what we measure corresponds to what the theory predicts then we ascribe to the data of the theory a quality of truth as correspondence with facts.

The mechanization of the process described in this example gives us the modern mathematical picture of the physical world where physical phenomena correspond to mathematical objects—or to their features—and physical laws correspond to the properties of those mathematical objects.

But the same line of reasoning applies also to the more general case in which mathematical theories describe mathematical objects or other mathematical theories. In the general case the model of truth is given by the condition that a formula (theorem) must be well-formed and true. With this observation in mind, we can tackle the general problem of what conception of knowledge subsumes any given Interaction Model.

The basic intuition that motivates the introduction of Cognitive Architectures is that any cognitive technology (Interaction Model) defined in an ambient metaphysical structure of thought (Validation System) must reflect an underlying conception of knowledge, in much the same way in which the OSP Pattern expresses the metaphysical intuition about the orderly nature of the physical world captured by the technology of formal systems.

The notion of Cognitive Architecture captures the underlying conception of knowledge upon which *any* cognitive technology is built, based on a definite and distinct set of answers to the basic questions raised by the project of defining a conception of knowledge for an arbitrary metaphysical structure of thought. These fundamental questions are about the nature, the purpose and meaning of the very act of acquiring knowledge. They are the questions about the nature and meaning of the process of understanding something. They are the questions about the type of knowledge that a Cognitive Architecture gives us access to, and about the interpretation of that knowledge.

11.3 How Cognitive Architectures Define Knowledge

In Section 11.2, we described the basic intuition at the basis of Cognitive Architectures as the idea that the cognitive technologies that can be defined within the arena of a given Validation System, must carry in their structure the seed of an underlying conception of knowledge. This intuition correlates the two entities that give rise to the peculiar mind-world transaction that we refer to as "acquiring knowledge":

- The Interaction Models that encode the cognitive technology to transform data into information.

- The Cognitive Architecture that defines the basic blueprint for all the Interaction Models.

In this section we characterize the exact mechanism by means of which a conception of knowledge is codified in the Interaction Models definable in a Validation System. This fundamental mechanism is called the **reconstruction of the success condition**, and consists in the production of a **success condition** which, as described in Section 8.3, is a model of the notion of **success** that defines a Validation System.

11 Cognitive Architectures

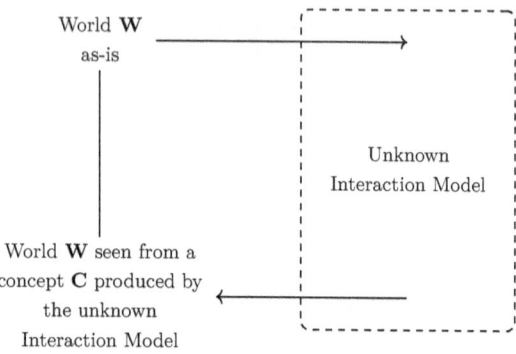

Figure 11.2: Thought experiment to reveal the fundamental structure of Interaction Model defined in a given metaphysical structure of thought.

To reveal the deep structure shared by the Interaction Models defined in a given $\mathcal{M}$-*signature*, we can make the following thought experiment. In Figure 11.2 there is an unknown Interaction Model through which we see the world **W**. What can we say about its structure? To comprehend the basic features of the unknown Interaction Model, we must ask what concepts we have epistemic access to in the $\mathcal{M}$-*signature* where the unknown Interaction Model is defined. In particular, we can ask what is the simplest concept we have access to, and what Interaction Model might produce that concept. By definition, in any $\mathcal{M}$-*signature* the human mind has unmediated epistemic access to the **concrete** concept—this is the definition of $\mathcal{M}$-*signature*. The Interaction Model producing a **concrete** concept, therefore, must be a particularly simple one, since it "does nothing": it produces the pure awareness of an object of thought or perception. This is the meaning of unmediated access. We have postulated that the human mind does not need any cognitive apparatus to interact with a **concrete** concept in a state of consciousness where that concept is **concrete**. Let's call this basic Interaction Model the Minimal Interaction Model. What can we say about a Minimal Interaction Model? In an $\mathcal{M}$-*signature*, by definition, any concept that is not **concrete** is **abstract**, that is, it can be expressed via the **concrete** concept. Thus, any non-Minimal Interaction Model—an Interaction Model that encodes a cognitive technology based on a **concrete** concept—must at least describe a **concrete** concept. This shows that any Interaction Model defined in a given Validation System must reconstruct a model of **success**. This fundamental feature of Interaction Models is how a conception of knowledge is codified in the fabric of Interaction Model, and consequently, in the cognitive technologies we create with them. The **reconstruction of the success condition** captures precisely the relation between the minimality of an Interaction Model and the fundamental definitional structure of a Validation System.

11.4 Classification Of Interaction Models

The classification system for Interaction Models stems from the mechanism of **reconstruction of the success condition** we introduced in Section 11.3.

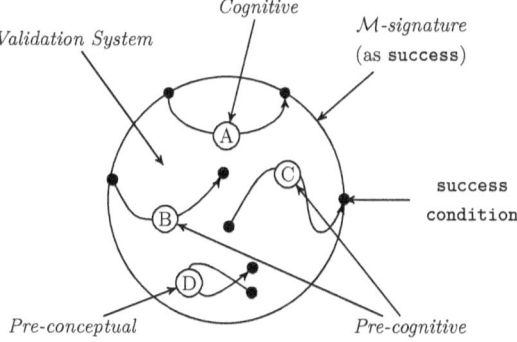

Figure 11.3: Classification system of Interaction Models based on the **reconstruction of the success condition**.

Recall the diagram of Figure 8.2 that describes the relation between Validation System, **success** and **success condition**. We expand that diagram here in Figure 11.3 as follows:

1. The points that form the circumference represent the **success conditions**—as in Figure 8.2.

2. We represent the dimensionless notion of Interaction Model via a one-dimensional object: an oriented path inside the disc—(Figure 11.3-A, B, C and D).

3. We represent the elements of a given Interaction Model as dots •; for example an oriented path • → • may represent a certain set of rules to manipulate sets, such as a morphism $X \to X \times X$ that sends a set X to a cartesian product $X \times X$, and in general any set of rules that we may depict diagrammatically.

4. The notion of continuity, in the form of a continuous line forming a circumference, is used to symbolize that the **success conditions** are *all* models of the *same* notion of **success** that originates from the $\mathcal{M}$-*signature* that defines the Validation System.

We illustrate the **reconstruction of the success condition** as the set of all Interaction Models that begin and end on the circumference (Figure 11.3-A). In particular, we note that two paths inside the disc X, Y can be concatenated when the end of X and the beginning of Y are of the same type—both inside the disc or on the circumference—and that the operation of concatenation of paths, which we'll denote by $\cdot$, is well-defined because the paths it is applied to are *oriented*—they are single-headed arrows. Thus, we can *always* think of a **reconstruction of the success condition** as the concatenation of three paths: one beginning on the circumference and ending inside the disc (Figure 11.3-B), followed by one or more paths beginning and ending inside the disc (Figure 11.3-D), followed by one path beginning inside the disc and ending on the circumference (Figure 11.3-C). Also, we note that path concatenation satisfies the following natural relations:

$$A \cdot A \cong A$$
$$B \cdot C \cong A$$
$$B \cdot D \cong B$$
$$D \cdot C \cong C$$
$$C \cdot B \cong D$$
$$D \cdot D \cong D$$

where the sign $\cong$ indicates a congruence relation—a form of equivalence—between the classes of Interaction Models A, B, C and D defined in Figure 11.3. With this visual metaphor in mind, we can proceed to describe the elements of this classification system.

11.4 Classification Of Interaction Models

Pre-conceptual Interaction Models

A *pre-conceptual* Interaction Model is symbolized by a path beginning and ending in the disc defined by the Validation System (Figure 11.3-D). Its name is intended to suggest that the effect of this Interaction Model on the mind's ability to articulate thoughts—and strictly in the context set by the Validation System defined by a given $\mathcal{M}$-*signature*—does not lead to fully formed concepts. A not fully formed concept is not captured by a definitional structure based on the **concrete** concept from which the $\mathcal{M}$-*signature* originates. Not fully formed concepts are concepts that, for example, in the Identity-based mode of thought cannot be fully captured by sentences such as "This is...", because they do not have enough structure to be grasped by conscious thought (recall Postulate 8.7.10) in an Identity-based fashion. Similar considerations apply to not fully formed concepts in other $\mathcal{M}$-*signatures*.

Examples of pre-conceptual Interaction Models in a given ambient $\mathcal{M}$-*signature* can be represented by diagrams that do not encode, within the syntax defined by the chosen spatialization, diagrammatical models of other types of Interaction Models other than D. The meaning of this last sentence will become apparent when we will explore the structure of three important $\mathcal{M}$-*signatures* in Part IV. For the moment, it is sufficient to think of the various types of Interaction Models described in this classification system as classes of diagrams that symbolize certain sets of rules to manipulate relations ($\rightarrow$) between mathematical objects ($\bullet$), such as the diagrams below.

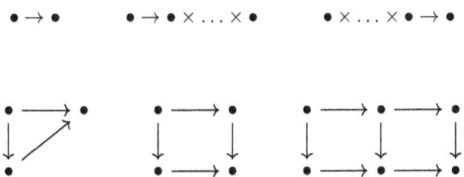

Pre-cognitive Interaction Models

A *pre-cognitive* Interaction Model is symbolized by a path that either begins or ends on the circumference (Figure 11.3-B and 11.3-C)—here the *or* is exclusive. Its name is meant to suggest that these types of Interaction Models are derived from, or conducive to, a **success condition**, but that *do not reconstruct* **success conditions**. Examples of pre-cognitive Interaction Models in an ambient $\mathcal{M}$-*signature* are given by diagrams such as

$$* \to Diag \qquad Diag \to *$$

where $*$ indicates a model of **success condition**, and *Diag* a type-D Interaction Model.

Cognitive Interaction Models

Cognitive Interaction Models, or simply Interaction Models, are the Interaction Models we have seen in the text so far. They represent a cognitive process with the minimum structure and features needed to articulate intelligible concepts, in full accord with a conception of intelligibility defined by the ambient $\mathcal{M}$-*signature* in which they are defined. Interaction Models encode a prototypical form of *computation*, in the steps through which a model of **reconstruction of the success condition** is encoded by the structure of the Interaction Model—The structure of this class of Interaction Models, type-A Interaction Models, is described by the notion of Definitional Model introduced in Section 10.5. We conclude this section with an important definition

Definition 11.4.1. (Cognitive state)
An Abstract Mind is in a *cognitive state* when the **reconstruction of the success condition** takes place.

11.5 Minimal Interaction Models

The Minimal Interaction Model is the core of the Abstract Mind, and models what we have discovered so far about the fundamental structure of the mind-world interplay. The centrality of the notion of Minimal Interaction Model in Interaction Theory is reflected here by the presence of three equivalent definitions of this notion: technical, epistemological and metaphysical. Each definition of Minimal Interaction Model is characterized by a theme.

The *technical* definition revolves around the theme of *minimality*, and characterizes the Minimal Interaction Model via the notion of

11.5 Minimal Interaction Models

reconstruction of the success condition.

Definition 11.5.1. (Minimal Interaction Model – *technical*)
A Minimal Interaction Model is a Type-A Interaction Model.

Recall from Section 11.4 that Type-A Interaction Models are the interaction models that begin and end on the circumference of Figure 11.3. A Type-A Interaction Model is factorable into Type-B, C and D Interaction Models as $B \cdot D^n \cdot C$—as one can verify from the equivalences described in Section 11.4. Definition 11.5.1 expresses the minimality of an Interaction Model via its factorability, and via the correspondence of this property with the reconstruction of the success condition.

The *epistemological* definition characterizes a Minimal Interaction Model via its model-theoretic relation with the Knowledge-Reality Interface (Postulate 8.7.4), its theme is *irreducibility*.

Definition 11.5.2. (Minimal Interaction Model – *epistemological*)
A Minimal Interaction Model is a representation of the basic, irreducible structure of the mind-world interaction at a given state of consciousness.

In Definition 11.5.2, the minimality of an Interaction Model is characterized via the irreducibility of the structure of a mind-world interaction—which we describe in Interaction Theory as the Knowledge-Reality Interface. This is what conveys the epistemological significance of this definition: it is its being centered around an invariant of the mind-world transactions.

The *metaphysical* definition revolves around three fundamental modes of consciousness called spatialization, despatialization and contravariance, that we will introduce in Section 14.1, and depicts a Minimal Interaction Model via the structure of the mental experience that it produces. The theme of this definition of *awareness*.

Definition 11.5.3. (Minimal Interaction Model – *metaphysical*)
A Minimal Interaction Model is an Interaction Model that produces the awareness of an object of thought, which is an experience indistinguishable from contravariance—as described in Section 14.1.

In Definition 11.5.3, the condition of minimality of an Interaction Model is characterized via the irreducibility of the structure of consciousness from which the experience of thought originates. In a slightly more metaphysical tone, Definition 11.5.3 tells us that an Interaction Model is minimal if the experience of thought that it produces coincides with the pure, unmitigated awareness of that thought. This condition accounts precisely for the *only* modes of consciousness we have observed, namely, spatialization, despatialization and contravariance, and translates minimality into the irreducibility of these three modes of consciousness. This definition is called metaphysical because it is based on the most metaphysical, and yet real and practical and familiar of all human experiences: consciousness.

Thus, the adjective "minimal" in Minimal Interaction Model denotes three distinct but related features of the mind-world interaction: the *minimality* of the **reconstruction of the success condition**, the *irreducibility* of the structure of a mind-world interaction, and the *awareness* between the modes of consciousness and the experience of thought. As seen in the description of the structure of consciousness (Section 14.1), any mind-world interaction that alters the way in which the **success condition** is reconstructed, corresponds to a *change* in the mode of consciousness: it is a radically and intrinsically different way of constructing conscious experience. This is, in essence, the reflection crystallized in the definitions just given, which can be further refined—and strengthened—in a so-called 1-out-of-3 fashion, as in the following synthetic definition.

Definition 11.5.4. (Minimal Interaction Model – *synthetic*)
An Interaction Model **T** is *minimal* if whenever one of the below conditions holds, the other two hold:

1. **T** is of Type-A.

2. At any given state of consciousness, the structure of the mind-world interaction that originates from **T** is irreducible.

3. The experience of thought produced by **T** coincides with the basic modes of consciousness defined in Section 14.1.

These definitions of Minimal Interaction Model, complete the theoretical construction of the Abstract Mind by characterizing its *content*, as described in Figure 4.6, in a way that is consistent

11.5 Minimal Interaction Models

with the phenomenology of conscious, rational thought, with the mind's ability to mathematize, and with the way we develop epistemological projects.

Structure Of The Minimal Interaction Model

The structure of a Minimal Interaction Model is the structure of the Knowledge-Reality Interface applied to a Definitional Model, as illustrated in Figure 11.4. It is a model of the fundamental makeup of the mind-world interaction (Definition 11.5.2), it is atomic and irreducible (same definition), it characterizes how the **reconstruction of the success condition** occurs (Definition 11.5.1), and codifies the way in which the three modes of consciousness (Section 14.1) operate and create our mental *experience* of the World.

So far, for explanatory purposes, we have treated $\mathcal{M}$-*signatures* and Cognitive Architectures as distinct, separate and ontologically independent entities, but there is no reason to do so, it is just convenient to think of them in that way. Hence, to describe a Minimal Interaction Model, and, a fortiori, to describe an Interaction Model, it is necessary to acknowledge how the linguistic distinction between $\mathcal{M}$-*signatures* and Cognitive Architectures defines the limits of the interpretation of an Interaction Model.

11 Cognitive Architectures

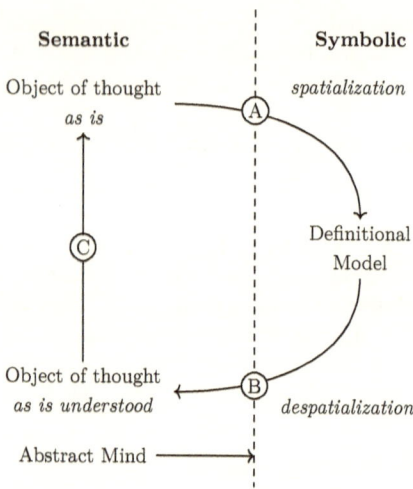

Figure 11.4: The structure of a Minimal Interaction Model

In Figure 11.4, there are no objects of thought or Definitional Models as such. Figure 11.4 is a spatialization of Definitions 11.5.1, 11.5.2 and 11.5.3 that, for explanatory purposes, depicts the mental *experience of an object of thought* in a given $\mathcal{M}$-*signature*, as a three-step process, denoted by the object of thought *as is* and by the object of thought *as is understood*, related via a certain Definitional Model—i.e. via a model of the Actuation Mechanism of a given **concrete** concept. In this sense, a Minimal Interaction Model is an *atomic* process—a notion similar to Whitehead's actual entities—and should not be *identified* with any of the representations given in this book.

The Abstract Mind-As-Sound Analogy

A useful analogy to interpret how the ∞-diagram of Figure 11.4 describes the fundamental structure of the mind-world interplay, is to think of it as the buzzer circuit of Figure 3.4 where the electrical parts correspond to the three steps of Figure 11.4, and the buzz corresponds to the Abstract Mind—denoted by a dashed line in Figure 11.4.

When in Section 3.2 we introduced the participatory mode of thought, we used the example of a buzzer circuit to describe

11.5 Minimal Interaction Models

a simple phenomenon that cannot be captured by the ordinary framework of propositional logic and, hence, by causality. The parts of Figure 11.4 are just like the electric parts of a buzzer circuit, and the sound they produce has an intrinsic unity—what in the philosophical and mystic literature is commonly referred to as the *unity of consciousness*—that we cannot capture by simply expressing the causal relations between some electrical parts. For these reasons, the distinctions we make in Figure 11.4 between the symbolic and semantic parts are purely instrumental to emphasizing the "acoustic characteristics" of a concept that ordinary Identity-based mode of thought and language are not designed to express with the crisp clarity with which they help us grasp the content and meaning of the utterance "there is a hard boiled egg on this table".

Symbolic Part

The symbolic part of Figure 11.4 denotes an inward movement of conscious thought defined by the concerted activity of the modes of consciousness of *spatialization* and *despatialization* codified by a given Definitional Model.

- A given Definitional Model defines how the fundamental structure of spatialization and despatialization operate to give rise to a specific structure and experience of thought.

- The theme of the symbolic part is intentionality, and constitutes the phenomenology of focused thought. For example, at the ordinary alert, problem-solving state of consciousness, focused thought is what we call attention. When the mind is wandering—what some would, erroneously, define as the opposite of attention—focused thought manifests itself as memory, sheer recollection of facts and ideas, reminiscence, analogy, intuition, and free association of ideas without any specific reference to a particular idea. This, too, is a type of attention. When the mind is intensely concentrated, or very stressed, or as a result of prolonged mental fatigue, focused thought, very often, manifests itself as a breakthrough. The very experience of having a breakthrough moment is a type of attention characterized by the collapse of spatialization and despatialization.

- Step A produces a spatialization of an object of thought: the stage in which the experience of the object of thought "takes place".

- In B, the ontologies that form the spatialization of the object of thought are related via models of the `concrete` concept at the basis of the ambient $\mathcal{M}$-*signature*. As seen in Section 10.5, the structure of these models is codified by the Actuation Mechanism, and the models of the Actuation Mechanism are defined by the Definitional Model.

- A and B results in "equations", and the process of constructing these equations is the *acquisition of knowledge* of the object of thought, with respect to a specific spatialization of it, as we will see in Definition 12.3.1. This is the experience of grasping a concept. It is the experience of seeing clearly an idea with your mind's eye. It is the feeling that the mind is pondering over a problem, even if for a split second or less.

Semantic Part

The semantic part of Figure 11.4 indicates an outward movement of conscious thought characterized by the suspension of intentionality, and by the dissolution of any thought into unbridled awareness.

- The role of a given Definitional Model in this phase is to define the provenance of a contravariant mode of consciousness, not its direction or destination. Step C is solely and entirely characterized by the contravariant mode of consciousness, and consists in the complete cessation of any spatialization and despatialization.

- The theme of the semantic part is the suspension of intentionality, and constitutes the phenomenology of abstract thought. This is the experience of climbing up and down an abstraction ladder. It is the experience of self-reflection and self-inspection. It is the mind's ability to observe itself as a whole.

- In this phase the human mind releases the object of thought and empties the stage in which the mental experience of that object was taking place.

- The focus in C is the stage, not the actors or the play. Its purpose is to explore the stage of thoughts rather than do and undo thoughts. The function of the contravariant

11.5 Minimal Interaction Models

mode of consciousness is precisely to enable this type of exploration at a cognitive level.

- Step C defines or redefines the structure of a local epistemology, and marks the process through which we are able to both encompass and challenge the structure of thought ($\mathcal{M}$-signature) and knowledge (Cognitive Architecture). This is the experience of creating meaning. It is the experience of understanding something in a new way. It is the experience of reorienting our thoughts toward a new idea. It is the feeling that the mind is pondering over the way it knows and understands things at a deeper level.

Minimal Interaction Model As Irreducible Object Of Thought

The Minimal Interaction Model, by definition, codifies with its structure the fundamental type of thought an Abstract Mind can grasp. It specifies the line of demarcation between the thoughts that an Abstract Mind can explore from within because they are directly intelligible, and the thoughts that an Abstract Mind can entertain as a fantasy, as a linguistic or symbolic exercise, as a permutation of ideas detached from the senses of the mind. In this sense, we can think of a Minimal Interaction Model as the basic structure of what so far in the text we have been referring to as an *object of thought*. A Minimal Interaction Model is, so to speak, the basic *unit of thought*, the distinctive fingerprint of a particular kind of intelligence.

Bibliographical Notes

It is difficult to tell exactly where my notion of Cognitive Architecture comes from. It is a concept defined primarily by logical necessity rather than by sheer reasoning. What I can tell, is that its current form was heavily influenced by my interpretation of the recent research in n-categories, and by some themes of Process Philosophy, Cognitive Psychology, Cognitive Linguistics, Psycholinguistics, and by various theories about the language-thought interaction. But the introduction of the notion of Cognitive Architecture was essentially a consequence of the distinction between conception and definition of knowledge we introduced in Section 11.1.

But the actual distinction between conception and definition of

11 Cognitive Architectures

knowledge emerged spontaneously as a necessary consequence of the lines of reasoning I was constructing to give an account of a metastructure of human thought that I could make consistent with the phenomenology of mathematics I outlined in The 4 Lenses (Chapter 4).

The very early stages of this research were inspired by some general themes of Philosophy of Process (Whitehead 1979, Rescher 1996, Rescher 2000), by the study of concept formation in Cognitive Psychology (Rosch and Lloyd 1978, N. U. Salmon 2007, Margolis and Laurence 1999, Bruner 1966), by the relation between concepts and language, the priority between language and concepts, by the problem as to whether there can be concepts without language, by linguistic relativity (Whorf 1956, Dummett 1993, Dummett 1993, Carruthers 1996). We find an evolution of some of those early reflections in the ideas underpinning "The Basic Intuition Of Cognitive Architectures" and "How Cognitive Architectures Define Knowledge" (Sections 11.2 and 11.3 respectively).

The "Classification Of Interaction Models" (Section 11.4) is a naïve description of the logic of Interaction Models, and a natural generalization of the OSP Pattern. It is partially inspired by the recent attempts to define n-categories (Leinster 2001a, Cheng 2008). These formulations—strictly speaking not equivalent to each other—imply that the OSP Pattern itself can be thought of in alternative ways, namely: globular, simplicial, opetopic and via path parametrization. Of course the mantra "Any mathematical definition consists of a procedure to specify *Objects*, that we equip with *Structure* in such a way that certain *Properties* are satisfied. " is still valid, but is somewhat relativized by the presence of multiple formulations of n-category.

Chapter 12

Epistemology Of The Abstract Mind I
Knowledge, Comprehension And Meaning

> The sciences do not try to explain, they hardly even
> try to interpret, they mainly make models. By a
> model is meant a mathematical construct which,
> with the addition of certain verbal interpretations,
> describes observed phenomena. The justification of
> such a mathematical construct is solely and precisely
> that it is expected to work.
>
> John Von Neumann (Leary 1955)

> The intentionality of the mind not only creates the
> possibility of meaning, but limits its forms.
>
> Searle 1983

The definition of the Symbolic Intuition of Reality continues in this, and in the next two chapters, with a presentation of the Epistemology of the Abstract Mind.

The project of defining an Epistemology of the Abstract Mind, is the project of defining for any given $\mathcal{M}$-*signature*, the basic cognitive structures required to develop an epistemological project. We identify the cognitive structures necessary to develop an epistemological project with the meaning that the terms *knowledge*, *comprehension*, *meaning*, *explanation* and *consciousness* have in Interaction Theory, and with their mutual relations.

- *Knowledge* is not the justified true belief of the philosophical tradition, because the cognitive apparatuses required to have beliefs, to construct sound arguments, and to define a notion of truth are advanced cognitive functions that emerge only *after* the structures of thought needed to support those functions are formed. The knowledge the Epistemology of the Abstract Mind is concerned with is *archetypical*, it is the basic and irreducible cognitive function from which the advanced cognitive functions needed to develop a theory of knowledge are derived.

- *Understanding* is a generative cognitive process through which the human mind seems to be able to turn knowledge into action according to specific conditions dictated by the structure of thought where knowledge itself is created.

- *Meaning* is the navigation system of the human mind. It is the cognitive function through which the mind evaluates and compares different forms of comprehension, reorients itself towards new goals, and regenerates its own symbolic structure.

- *Explanations* are the arguments that we craft with the tools of a specific system of ideas, and that we send out into the World like sailing vessels: the longer the distance they travel in any direction, the more we believe that Nature is in harmony with our ambitions, and that *that* system of ideas is a privileged frame of reference to understand Nature.

- *Consciousness* is the structure of the experience of any object of thought. It is the most basic syntax that describes how every mind-world transaction creates that symbolic connection with reality we call experience.

The description of the Epistemology of the Abstract Mind offers a conceptual backdrop, and an explanatory framework, to interpret the other fundamental component of an Interaction Theory, a Cognitive Architecture (Chapter 11). The epistemic framework we present in this and in the next two chapters, consists of a correspondence between the deep structure of knowledge, comprehension, meaning, explanation and consciousness, and certain fundamental constructions of Category Theory. To situate the discussion about the structure of knowledge in the Abstract Mind, it is, perhaps, useful to begin by observing that the definition of an Epistemology of the Abstract Mind relies on the answers to a set

Introduction

of fundamental questions that can be categorized into 4 groups:

1. Basic epistemological questions.
2. Questions about the Knowledge-Reality Interface.
3. Questions about the structure and role of consciousness in the Epistemology of the Abstract Mind.
4. Questions about the types of knowledge produced by the Abstract Mind.

Basic Epistemological Questions

The first group of questions is foundational, and pertains to the epistemological scenery where we define the process of acquiring knowledge.

- What is knowledge?
- What does it mean to understand?
- What is meaning?

By answering these questions in the context of the Abstract Mind, we connect two worlds: Epistemology and Intentionality. The answers to these questions reconcile, expand and unify, the creation of meaning as a result of a process of understanding, with the intentional act of acquiring knowledge based on universal cognitive patterns which, as we posited in Postulate 8.7.1, do not change with the levels of consciousness. By answering these questions we characterize, for example, the transition from knowledge without understanding to understanding something fully, the function of meaning in the development of knowledge, the fundamental structure of good explanations, and the structure of higher forms of knowledge.

The Knowledge-Reality Interface

The second group of questions is about the nature and structure of the correspondence between knowledge and reality—what we call in this book the Knowledge-Reality Interface. It is a group of questions that defines how the data produced by cognitive technologies are related to the ideas, the realities, and the objects and states of affairs that those technologies claim to reveal or explain. These questions implicitly address another fundamental

12 Epistemology Of The Abstract Mind I

topic of this discussion: the relation between a conception of *knowledge* and a conception of *computation*.

- What is an explanation?
- What is the nature of the correspondence between a cognitive technology and the reality that that cognitive technology claims to explain?
- Does every cognitive technology give rise to a conception of computation?
- Does knowledge have to be computational to reveal the structure of physical reality?

It is precisely the nature of this correspondence between thought, action and reality that we call knowledge in the broadest and deepest sense, and that we want to integrate in a model of Abstract Mind in a way that we can make consistent with the structure of thought defined by a given $\mathcal{M}$-*signature*. Answering the second group of questions forces us to clarify the exact boundary between knowledge and computability, which is, as we have seen numerous times and from different angles in this book, a manifestation of the philosophy of formal systems. Defining the interface between knowledge and reality, instead, is a radically different problem, because is has a deeply metaphysical nature—based, of course, on a definition of reality—and because is, in many regards, not so important, given that the "real" Knowledge-Reality Interface is, in the schema we are constructing here, and in the traditional conceptions of knowledge, a mere manifestation or reflection of one of the many correspondence theories of truth—the view that truth is correspondence to a fact—a notion that Interaction Theory, as we will see, is generalized via the concept of epistemic continuity.

Modes Of Consciousness

It is difficult to exaggerate the importance of consciousness in every human activity. Consciousness is easily the most pressing and cosmically important feature of human existence, and plays a crucially important role in the formulation of Interaction Theory. In the Epistemology of the Abstract Mind, consciousness is defined and described *via* the structures of thought that it enables. Very broadly speaking, in Interaction Theory consciousness coincides with the specific structure of the spatialization that gives rise of the mental experience of conscious thought. This conception of consciousness, and its relation with spatialization, ties consciousness to the notion of $\mathcal{M}$-*signature*, motivates its presence in this discussion, and situates the study of spatialization in the broader context of the dynamics of the Abstract Mind.

The Types Of Knowledge Produced By An Abstract Mind

The fourth group of questions defines the epistemic boundary of this philosophical analysis by characterizing the type of knowledge defined by a conception of knowledge:

- Is the knowledge needed to define a conception of knowledge, a type of procedural knowledge—the "knowledge how" to do something—or is it a propositional knowledge—the "knowledge that" a fact is true?

- What type of knowledge does a conception of knowledge define?

What we will discover, is that to define a conception of knowledge, we need a new type of knowledge called *generative*, that is neither procedural nor propositional.

12.1 Generative Knowledge

In Epistemology, it is common to make the distinction between *propositional knowledge*—the knowledge that something is true—and *procedural knowledge*—the knowledge how to do something. These two broad categories reflect a classification of knowledge created by the ancient Greek philosophers, who distinguished between *epistêmê*—translated as knowledge—which corresponds to what we call today propositional knowledge, or "knowledge that", and *technê*—translated as skill, craft, art or dexterity—which corresponds to today's procedural knowledge, or "knowledge how". The type of knowledge defined by a conception of knowledge (Definition 11.1.2) is of a third type, which we call *generative*—for reasons that will emerge as we progress in this discussion. Generative knowledge is a superstructure of ordinary knowledge that subsumes procedural and propositional knowledge, but is neither propositional nor procedural. To introduce the notion of generative knowledge, we will dissect the logical structure of two

12 Epistemology Of The Abstract Mind I

examples of knowledge.

Example Of Propositional Knowledge

The knowledge that Roger Federer defeated Andre Agassi in the 2005 US Open Men's Singles final, is an instance of propositional knowledge produced by a distinct cognitive technology that relies on a notion of truth, and on a set of syntactic rules to classify facts into two categories: reality and fiction. For us to be able to truly assert that facts and ideas are either true or false (*epistêmê*), it is necessary to define the structure of our interaction with those facts or ideas. In this example, the structure of the interaction with facts and ideas is a cognitive technology based on an instance of generative knowledge called *causality*, which provides a reference framework to construct models of propositional knowledge of the world based on propositional logic. It is through the technology of causality that in this example we determine the winner of the fourth 2005 Grand Slam in the men's single competition, because the winner is determined by a causally connected sequence of true facts.

Example Of Procedural Knowledge

Baking bread is an example of procedural knowledge. It is a set of skills (*technê*) to achieve a well-defined goal. And the recognition that an object is indeed a loaf of bread, and that it is so as a result of the correct execution of certain preparation steps, is a confirmation that that specific procedural knowledge about making bread has a definitional correspondence with reality and experience. The cognitive technology that enables the validation of the definitional correspondence between a loaf of bread and its definition, is characterized by two sets the features: causal and definitional. It is causal because a loaf of bread is seen in the causal framework defined by the set of steps to making bread. And it is definitional because being a loaf of bread is based on a notion of identity—definitional identity (see Chapter 5).

Generative Knowledge As Metastructure Of Ordinary Knowledge

These two examples suggest that the boundary between propositional and procedural knowledge isn't perfectly defined, and that

12.1 Generative Knowledge

it is often the case that we can regard propositional and procedural knowledge as interconnected and overlapping definitions. We will not elaborate on the distinction between propositional and procedural knowledge any further. What's important about these examples is that they indicate a very distinct feature of generative knowledge which manifests itself through the mechanism by means of which it *enables* propositional and procedural knowledge. There is a use of generative knowledge necessary to define the cognitive technologies based on causality, by means of which a specified interaction with the world gives rise to instances of propositional knowledge, and there is the generative knowledge that goes into defining a definitional correspondence between facts or ideas.

The knowledge created by a Cognitive Architecture (Chapter 11) is of a generative type because it *defines* the conceptual and epistemic framework in which we construct operational models of knowledge such as, for example, propositional and procedural knowledge, which are the types of knowledge produced by Interaction Models.

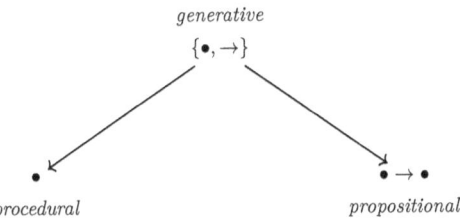

Figure 12.1: Relation between generative, propositional and procedural knowledge.

Figure 12.1 uses the language of diagrams to illustrate how generative knowledge gives rise to propositional and procedural knowledge. The bullets represent instances of propositional knowledge, and the arrows represent instances of procedural knowledge. By drawing a • we mean that "*this* object or state of affairs is true", that is, we can truly say that such and such *thing* has a certain correspondence with reality, and that that correspondence is in full accord with the axiomatic system in which the terms truth and reality and the correspondence between them is defined. By connecting two bullets with an arrow, we mean that "we know how to obtain the target from the source"—where *source* → *target*. This means that we can truly say that source and target are true, in a propositional sense, and that we can define a certain process → that is consistent with the ontological status of source and target, and that yields the target when is fed with the source. The very act of drawing an arrow signifies not only the knowledge how, but also that that particular instance of procedural knowledge preserves a specific type of ontological continuity between source and target: the arrow itself is part of the epistemological framework in which the process it represents takes place. The notion of generative knowledge captures precisely the underlying frame of reference depicted with bullets and arrows, $\{\bullet, \rightarrow\}$, and the style of reasoning, with which propositional and procedural knowledge are defined into existence.

Generative Knowledge In Interaction Theory

As seen in Section 11.1, in Interaction Theory a conception of knowledge is codified by the notion of Cognitive Architecture, and gives rise to a type of generative knowledge. Higher types of knowledge, still of generative type, are also defined in Interaction Theory, and emerge as transitions between $\mathcal{M}$-*signatures*, which correspond to highly specialized forms of consciousness. We'll touch on higher forms of generative knowledge later in this chapter when we'll define consciousness, and in Chapter 18.

Generative Knowledge And Definitional Models

Another way to read the diagram in Figure 12.1 is in terms of its relation with the Definitional Model associated to the `concrete` concept that gives rise to the $\mathcal{M}$-*signature* where the dialectic of thought takes place. As we saw in Section 10.5 "How `concrete` Concepts Define Intelligibility", there is a direct relation between Interaction Models and Definitional Models:

> "... *Definitional Models capture the minimum structure that an Interaction Model must possess to codify the process through which concepts are generated. This fundamental process is called* `reconstruction of the success condition`. "

We will see in detail these *minimal structures* in Sections 15.3, 16.3 and Section 17.3. The fundamental link between the notions of generative knowledge and Definitional Model is that generative knowledge *defines* the structure of a Definitional Model. This statement is not a linguistic device to relate concepts for explanatory purposes, the notion of generative knowledge defines an actual object whose structure we can describe precisely. Each $\mathcal{M}$-*signature*—and the Cognitive Architecture defined in it—has an associated generative knowledge that defines the structure and the terms of the epistemological project that can be defined within that $\mathcal{M}$-*signature*. Much like in the default $\mathcal{M}$-*signature* generative knowledge consists of the archetypical alphabet of symbols $\{\bullet, \rightarrow\}$, in the Finiteness-based $\mathcal{M}$-*signature* generative knowledge consists of a different set of archetypes $\{Diagram, \rightarrow\}$, where $Diagram$ is a graph—bullets connected by arrows—and $\rightarrow$ is a generalization of the ordinary arrow, and in the Cohesiveness-based $\mathcal{M}$-*signature* generative knowledge is the set of archetypes $\{Dots, \rightarrow\}$, where $Dots$ is a graph without arrows, and $\rightarrow$ is a generalization of the ordinary arrow. The various versions of generative knowledge we find in the non standard $\mathcal{M}$-*signatures* give rise to different types of knowledge, much like in the default $\mathcal{M}$-*signature* $\{\bullet, \rightarrow\}$ gives rise to propositional and procedural knowledge.

12.2 The Structure Of Knowledge

The traditional philosophical accounts of knowledge are mostly concerned with the study of the nature of justified true beliefs, and with the relations of these with the entities that they denote. The questions asked in the context of classic epistemology illustrate the spirit and the style of reasoning underlying this philosophical inquiry. What are the sources, the conditions, the structure and the limits of knowledge? What are justifications and beliefs, and what are their sources? What makes justified beliefs justified? What are the relations between knowledge, truth and reality? and so on and so forth. These questions seem to suggest that the traditional philosophical inquiries about the nature and structure of knowledge are, perhaps, based on a background cosmology of thought that sees the epistemological investigation as consisting in the analysis of the relations between various versions of reality, truth and belief, and in the debate about how these acquire their ontological status. This approach to the study of the nature and structure of knowledge is, in many regards, *analytical*, because it reduces the study of knowledge to the study the basic substrate notions of *reality, truth* and *belief.*

12 Epistemology Of The Abstract Mind I

Another approach to the study of the nature and structure of knowledge is conveyed by the questions: What does it mean to know something? Is it possible to characterize knowledge as the difference between the state of an agent before and after knowing something? This is the approach we adopt here, because we regard any concept from (my own interpretation of) a Pragmatist angle. Knowledge, therefore, is regarded in this research as a symbolic placeholder we have invented to denote a certain class of mental processes that enable specific cognitive abilities through which we relate symbolically to the physical and mental realities via our thought processes. This approach to the study of knowledge is *synthetic* and *archetypical*. It is synthetic in the sense that it focuses on the intrinsic structure of knowledge rather than on a set of substrate concepts of reality, truth and belief which could, in principle, be replaced by other concepts. The structures of thought required to have beliefs, and to construct a notion of truth are based on advanced cognitive functions; far too advanced to be taken seriously as candidate fundamental mental processes. Thus, from a synthetic perspective, the study of knowledge is the *study of the cognitive metastructures activated by the state of knowing something*. This approach to the study of knowledge is also archetypical, because it is concerned with what knowledge is at a fundamental level and in any state of consciousness. We humans have different ways of knowing based on how we feel, and the experience of knowing without understanding, without having any coherent form of justified true belief, is far too common and far too present and effective in our lives to be ignored. I think that to fully appreciate the deep nature of knowledge, it is indispensable to understand what the phenomenology of knowledge tells us about the deep cognitive structures at work when we create knowledge, when our perception of reality transitions to a state where we *feel* we know. We know very well, both intuitively and practically, what it means and what it feels to know something. It means being able to make decisions, to solve problems, to discover and understand relations between data, ideas and answers. Knowing is distinctively associated to a precise intellectual feeling, often reflected at a physiological level.

The characterization of knowledge we present here is a schema to reframe the traditional philosophical accounts of knowledge within the framework of $\mathcal{M}$-*signatures*. It is a method to do away with the questions raised by specific *definitions* of knowledge, and zero in on

12.2 The Structure Of Knowledge

an overarching *conception* of knowledge where those questions can be asked the way we do, but can also be imagined and formulated in other ways, depending on the underlying metaphysical structure of thought in which they are articulated.

At the core of a definition of knowledge, there are the cognitive processes through which knowledge is pursued. And cognition is the polymorphic manifestation of various forms of spatialization, through which the mind creates cosmologies of axioms about the nature of reality—reality intended as a mind-world interconnected whole rather than in a dualist fashion—and where the dialectic of thought evolves from *metaphysical intuitions* to *formal definitions*. These spaces of thought can be regarded as conceptual incubators where the human mind is or becomes able to think in any intelligible way within the boundaries set by the $\mathcal{M}$-*signature* where the thinking occurs. These mental spaces are where the knowledge we are interested in takes place, because it is in its purest form. We have countless examples of what happens in these spaces of the mind, but because we rarely define a conception of knowledge—or, perhaps, because we do not make the sharp distinction between a conception of knowledge and a definition of knowledge—we seldom recognize the nature of the events that take place in these mental spaces. In these spaces of thought, knowledge is the evolution of the relations between the basic ontologies that form the space via the **concrete** concepts that define the $\mathcal{M}$-*signature* in which those relations exist.

From the Abstract Mind viewpoint, and from an Interaction Theory viewpoint, the traditional accounts of knowledge are not satisfactory because they do not relate the genesis of justified true beliefs, and other traditional descriptions of knowledge, to the metaphysical structure of thought where those descriptions take place. There is nothing axiomatic about the fundamental structure of human thought, and thinking that this is not the case should be either rigorously justified by a philosophical argument that employs this assumption, or incorporated explicitly in a theory that claims to explain the nature and structure of knowledge.

A way to begin to comprehend the phenomenology of knowledge, is to look at the structure of the spatialization where any intelligible acts of *knowing* takes place. We will illustrate briefly the phenomenology of our physical knowledge of the world, and use this as a stepping stone to derive the cognitive superstructures with which we will introduce the conception of knowledge we use in this book.

12.3 Knowledge In Physics And Mathematics

In the structure of thought defined by the default $\mathcal{M}$-*signature*, we can define a space of thought that represents physical reality, as the collection of the ontologies of energy, mass, spacetime, forces etc. that form the arena where we develop our physical knowledge of the World. The *structure of knowledge* in the space of thought defined by the ontologies of energy, mass, spacetime, forces etc. is the process through which we correlate those ontologies *via* the **concrete** concept at the basis of the default $\mathcal{M}$-*signature*, that is, through various forms of mathematical identity—i.e. equality, isomorphism, equivalence, adjunction, fibration, homotopy, cobordism, etc. What this means, is that in the epistemic domain defined by *this specific* spatialization of physical reality, what we call *knowledge* is the process of establishing logically inferred, Identity-based correspondences between energy, mass, spacetime, forces etc. Thus, the structure of the modern physical knowledge of the world in the mental workspace defined by the default $\mathcal{M}$-*signature*, has *necessarily* the form of equations between these ontologies, as illustrated by the following famous equations, which relate the ontologies of mass, energy, force, etc. via various forms of *mathematical sameness*:

- force *equals* mass subject to acceleration, as in Newton's second law $F = ma$

- changes in the electric field **E** *correspond* to changes in the magnetic field **B**, as described by Maxwell's equations

$$\nabla \cdot \mathbf{E} = \frac{\rho}{\epsilon_0}, \nabla \cdot \mathbf{B} = 0, \nabla \times \mathbf{E} = -\frac{\partial \mathbf{B}}{\partial t}, \nabla \times \mathbf{B} = \mu_0 \left(\mathbf{J} + \epsilon_0 \frac{\partial \mathbf{E}}{\partial t} \right)$$

- energy *equals* mass, as in Einstein's $E = mc^2$

We don't read these equations mathematically, we read these equations metaphysically; and what they reveal when we read them in this way, is the common pattern captured by the conception of knowledge (Definition 11.1.2). The crucial point is that there is no necessity to spatialize the physical world in this way: *our representation of the World via mass, energy, forces, spacetime*

12.3 Knowledge In Physics And Mathematics

etc. is a convention dictated by the way we ascribe, consciously and unconsciously, the sensory perception-based structure of our thoughts to the world. We see and grab objects *therefore* the world is *made of* stuff. We move in space, we get old, and all things appear to undergo certain processes of atomical and biological decay *therefore* the world is *made of* space and time, and so on and so forth.

We can summarize the observations made in the previous paragraph as follows. Let $\mathbf{M}(\mathcal{C})$ be an $\mathcal{M}$-*signature* based on the `concrete` concept $\mathcal{C}$, and let $\{a, b, c, \ldots\}$ be a collection of ontologies that form the spatialization $\mathbf{S}$ of an object of thought or perception in $\mathbf{M}(\mathcal{C})$. In the dialectic of thought of $\mathbf{M}(\mathcal{C})$, by *knowledge of* $\mathbf{S}$, or $K_{\mathbf{M}}(\mathbf{S})$ for short, we mean the cognitive process of constructing all the valid relations between the ontologies of $\mathbf{S}$ via models of $\mathcal{C}$.

For example, let $\mathcal{C}$ be the philosophical notion of *identity*, and in $\mathbf{M}(\mathcal{C})$, let us consider a certain spatialization $\mathbf{S} = \{a, b, c, \ldots\}$ of an object of thought or perception. The *knowledge* $K_{\mathbf{M}}(\mathbf{S})$ of the object of thought spatialized as $\mathbf{S}$, is the construction of all of the valid generalizations of identity relations as in Figure 12.2.

	spatializations →		
	$a = b$	$b = c$	$\ldots$
models of	$a \cong b$	$b \cong c$	$\ldots$
identity	$a \Leftrightarrow b$	$b \Leftrightarrow c$	$\ldots$
↓	$a \dashv b$	$b \dashv c$	$\ldots$
	$\ldots$	$\ldots$	

Figure 12.2: Example of Identity-based knowledge

where $=, \cong, \Leftrightarrow, \dashv, \ldots$, denote various *mathematical forms* of the abstract notion of identity, namely: equality, isomorphism, equivalence, adjunction, fibration, homotopy, cobordism, etc.

This schema can be generalized naturally in an arbitrary $\mathcal{M}$-*signature* as in the following

Definition 12.3.1. (Acquiring Knowledge)
Acquiring knowledge is the process of relating a specific set of

objects of thought via models of the **concrete** concept that defines the $\mathcal{M}$-*signature* where the dialectic about those objects of thought takes place.

When the process of Definition 12.3.1 is applied iteratively to itself, the equations become the ontologies of a higher spatialization, and the Definition 12.3.1 gives rise to higher types of knowledge in the form of equations between equations, equations between equations between equations, and so on. This leads to

Definition 12.3.2. (Knowledge)
Knowledge is the totality of the relations based on the **concrete** concept that define a given $\mathcal{M}$-*signature*.

Definition 12.3.2 creates what we call a *cognitive superstructure*. It characterizes the fundamental structure of a conception of knowledge in a given metaphysical structure of thought. Its implementations are all the instances of the **reconstruction of the success condition**, which we introduced in Section 11.3.

12.4 Why Knowledge Is Locally Superadditive

Superadditivity captures the idea that the whole is more than the sum of its parts. In Interaction Theory, and often in everyday experience, knowledge displays the property of superadditivity with respect to the context defined by the Interaction Model where the dialectic of thought takes place (see Sections 8.2 and 8.3), thus the characterization of *locality*. From Definition 12.3.1, it follows that the acquisition of knowledge consists in the determination of relations between objects of thought—the constituents of an Interaction Model—via models of the **concrete** concept where the Interaction Model is defined.

Intuitively, if an Interaction Model $\mathcal{S}$ consists of the following objects of thought $\{a, b, c, \ldots\}$, and if the $\mathcal{M}$-*signature* where $\mathcal{S}$ is defined is based on the **concrete** concept of *identity*, *knowledge* in $\mathcal{S}$ is a measure of the identities in $\mathcal{S}$. The most basic knowledge in $\mathcal{S}$ is the knowledge of the *objects of thought* themselves via equations of the form $a = a, b = b, \ldots$. The term "equation" and the symbol $=$ denote various mathematical forms of the abstract notion of identity, such as those illustrated in Figure 12.2. Higher forms of knowledge in $\mathcal{S}$ involve higher forms of identity between

12.4 Why Knowledge Is Locally Superadditive

higher objects of thought, such as $a = b, b = c, \ldots$ but also $a = b+c, c = a-b, b = a \oplus c, \ldots$ where $+, -$ and $\oplus$ denote certain operations in $\mathcal{S}$.

We can explain how superadditivity is codified in Definition 12.3.1 with the following ad-hoc mathematical analogy. Let the objects of thought of $\mathcal{S} = \{a, b, c, \ldots\}$ represent *sets*. In $\mathcal{S}$, we construct the collection $K(\mathcal{S})$ of all mathematical forms of identity. The elements of $K(\mathcal{S})$ are, therefore, the elements in Figure 12.2. In particular, each element a, b, c,... of $\mathcal{S}$ corresponds to can be projected on the following list

$$K(\mathcal{S}) = \left\{ \begin{array}{cccc} a = a & a \cong a & a \Leftrightarrow a & \ldots \\ b = b & b \cong b & b \Leftrightarrow b & \ldots \\ c = c & c \cong c & c \Leftrightarrow c & \ldots \\ \ldots & \ldots & \ldots & \ldots \end{array} \right\}$$

which represents all possible equational identities.

In $K(\mathcal{S})$ we define a positive function f as follows. Let X and Y be two elements of $K(\mathcal{S})$, then $0 \leq f(X) \leq f(Y)$ if the number of identities in X is smaller than the number of identities in Y. From this definition, it follows that $f(a = a) = f(b = b)$, since both equations involve just one identity each, namely a and b. And it follows also that $f(a = a) < f(a = b)$, since in $a = b$ there are three distinct identities involved: there is a, which symbolizes $a = a$, there is b, which symbolizes $b = b$, and there is $a = b$. Note that, therefore, $a = b$ is an equation of equations: $(a = a) = (b = b)$!

In mathematics a function g is superadditive when

$$g(X + Y) \geq g(X) + g(Y)$$

where $X + Y$ denotes an object that contains X and Y. If g is our knowledge-detecting function f, then the inequality above reads: the knowledge of the whole $(f(X + Y))$ is greater than the knowledge of the parts $(f(X) + f(Y))$. If X and Y are two elements of $K(\mathcal{S})$, $X + Y$ is a set of two equations. For example, if $X = (a = a)$ and $Y = (a = b)$, $X + Y = \{(a = a), (a = b)\}$. We can extend this rationale to the case when X and Y are sets of equations in $K(\mathcal{S})$, and define $X + Y$ as the (set-theoretic) union of X and Y. In general, given two sets X and Y, it is possible to construct a form of generalized identity—an identity like the ones depicted in Figure 12.2—between X and a subset of $X + Y$, which we denote by $sub(X+Y)$. The most trivial is of course between X and itself as member of $X+Y$, but in general there can be infinitely many other ways to find an image of X inside $X+Y$. For example, another way consists in swapping in $X+Y$ an element of X with an element of Y. Thus, in general, the following generalized identity holds $X \cong sub(X+Y)$, and therefore

$$f(X \cong sub(X+Y)) \geq f(X)$$

by definition of f. Here, again, $\cong$ is a mathematical model of the abstract notion of identity. It follows that $X+Y$ has at least two identities besides the trivial $X+Y = X+Y$, namely, $X \cong sub(X+Y)$ and $Y \cong sub(X+Y)$, thus

$$f(X+Y) \geq f(X \cong sub(X+Y)) + f(Y \cong sub(X+Y))$$

and

$$f(X \cong sub(X+Y)) + f(Y \cong sub(X+Y)) \geq f(X) + f(Y)$$

which proves that f is superadditive.

The Paradox Of Locally Superadditive Knowledge

The notion that knowledge grows with the accumulation of explanations, like a shopping bag where we throw more and more potatoes until it's full, is indeed an enticing perspective, because we identify the achievement of total knowledge with the cessation of every form of suffering, and with the establishment of a blissful, lasting experience of human existence for every human being. The problem with this additive notion of knowledge is that it is true until it's false, and we know by experience that the history of human thought, and, with it, the history of human knowledge, advances through intellectual and scientific revolutions, not with the number of books published. Perhaps, as we'll see later in this chapter in the discussion about the relations between rational explanations and reality, the path suggested by the mystic or revelatory types of epistemic access to the world, is not there as an alternative, systematic method of seeking explanations, but to remind us that we should look carefully at the reasons why rational human knowledge evolves when we rip that shopping bag open.

Knowledge In Interaction Theory

In Interaction Theory, knowledge is the construction of relations between a certain collection of ontologies that represent an object of thought, based on mathematical models of the Definitional Model (Section 10.5).

12.5 The Structure Of Comprehension

When we don't understand something, we often ask "What does it mean?" because we have the intuitive feeling that to understand something we need to relate what we don't comprehend to the things that we already understand. Questions about meaning are essentially the *only* type of questions we are able to ask about what we don't understand, and indicate a first peculiar feature of understanding: understanding is an integral part of the subjective experience and, as such, it is intrinsically different from knowledge, which has a non-experiential objective nature. In contrast, when we *do* understand something, we are able to articulate questions that penetrate the nature of what we know, and that create or suggest new meanings and new ways to understand what we already know and understand. Why is it so? and Why do we want to know the meaning of something that we don't understand? Shouldn't we first understand something in order to ask what it means?

These observations indicate that whatever comprehension is, and whatever cognitive processes are connected to it, its phenomenology is, at least superficially, related to the subjective phenomenology of knowing and meaning, which explains why knowledge can be easily communicated and meaning cannot. Another observation that can be made along this line of thought, is that it is possible to know something and its meaning, and still not understand it. For these reasons, to reveal the structure of comprehension, we will first separate comprehension from knowledge and meaning, so that this concept can be studied in isolation and at a higher level of abstraction, and then we will dissect it with some of the tools we introduced in Chapter 3.

Knowledge Without Comprehension

A first observation that may, perhaps, help reveal the nature of the distinction between knowledge with and without comprehension, is that we seem to understand things always and only in relation to a specific context in which we conceive a strategy to achieve a goal. There is no such thing as understanding a concept or a fact in an abstract and purposeless sense: without a specific purpose, we simply *acknowledge* a concept or a fact, in much the same way in which notice a stone lying on the side of the road. Without a purpose, our knowledge of the world is solely and entirely *propositional*, it is the knowledge that a certain thing exists, that it is in such and such way by virtue of giving rise to a certain mental or perceptual experience: and nothing else. This observation, inspired by the intentionality-laden worldview that permeates this book, seems to suggest that the Observational Structure of Intentionality may be taken as a model of the process of understanding, because understanding is an intentional act, being fully characterized by a specific goal.

Meaning Without Comprehension

The impulse to ask What does it mean? when we don't understand something, is motivated by the intuitive knowledge, or mental habit, or cognitive bias, that when we know the meaning of something we are able to understand when and where that specific knowledge is applicable. As we'll see in the next section, meaning has precisely this function: it acts as a compass to navigate through ways of understanding. Yet, the knowledge of when and where a certain knowledge applies to a problem or an idea does not say anything about the structure of that particular knowledge. The reason why we ask What does it mean? is that in most cases we have the hope or the illusion that we can extrapolate the intrinsic structure of something that we know, by simply applying it successfully to a sufficiently large number of cases. Thus, the notion of meaning is connected to a *procedural* knowledge of the world by virtue of characterizing a specific purpose, and to the Observational Structure of Intentionality model, via the intentionality-laden worldview of this book.

These observation are summarized in the following hypotheses

- Knowledge without comprehension is a form of propositional knowledge.

- Meaning without comprehension is a form of procedural knowledge.

12.5 The Structure Of Comprehension

- The Observational Structure of Intentionality codifies the philosophical notion of comprehension.

The first two points of the above hypotheses are entirely observational; they are part of the phenomenology of knowledge and meaning. To verify the validity of the third point, we must give an account of the full correspondence between comprehension, and the act of performing a deliberate observation. Note how by separating comprehension from knowledge and meaning, we have merely emphasized the structure of the mutual relations between knowledge and comprehension and meaning and comprehension, but the actual structure of comprehension is still unknown. In other words: so far, comprehension has been a structureless conceptual placeholder with no purpose other than that of connecting knowledge to meaning.

The Structure Of Comprehension

In the parlance of the Observational Structure of Intentionality, to *comprehend* something is always in relation to a strategy to achieve a goal. Thus, to comprehend necessarily means that—or is a reference to the fact that—certain data are consistent and coherent with a strategy that is believed to lead to a well-defined goal in an intelligible way. In other words, it is *purpose* that defines a strategy that turns the data we acquire through a cognitive technology into information. This last observation raises two questions: What is it that makes data consistent and coherent with a strategy? and What is it that characterizes data, based on their capacity to enable and activate a specific strategy to achieve a goal?

By answering the first question within the frame of reference of the Observational Structure of Intentionality, we reveal the connection between the data, and what makes a strategy successful; we relate the knowledge of something, that is, *any data*, to the intentional act that transforms that knowledge into action (strategy). By answering the second question, we explain what differentiates knowledge from comprehension: the data needed to act upon a specific want. To put this more prosaically: by explaining what characterizes the data that activate a strategy, we explain what turns certain data into information.

First, we observe that the existence of multiple sets of data characterized by the property of activating a specific strategy is consistent with the Observational Structure of Intentionality. For there exist strategies that admit different sets of data to function, in much the same way in which an event (goal) can have different causes (strategies). This observation indicates that to answer the first question, we need to focus only on the role played by a given data with respect to a strategy, rather than on the data itself, because the uniqueness of the data is irrelevant in this characterization.

Next, for a given data to be consistent with a given strategy, it must activate the strategy, and for this statement to count as an adequate answer to the first question, we must define what it means to activate a strategy. In the context of the Observational Structure of Intentionality (Section 3.3), a strategy is activated if the agent is able to interact with the environment as a result of the application of the strategy, and in the way specified by the strategy. To answer the second question, we observe that for a dataset to be consistent with the purpose defined by a given strategy, the data must be structured in such a way that every part of it is necessary and sufficient to enable the strategy. In other words, the data must be a *minimal* set of data. A given dataset is minimal with respect to a given strategy, if any other data that activates the same strategy is not properly contained in that data. To summarize, we can define comprehension in a given $\mathcal{M}$-*signature* as follows

Definition 12.5.1. (Comprehension)
Comprehension is the determination of a minimal set of data that activates the strategy of an Observational Structure of Intentionality where the purpose of that apprehension is defined.

Thus, the process of understanding is *always* the process of characterizing what makes certain data minimal with respect to a strategy to achieve a certain goal. This conclusion conveys a truth anyone who has tried to define a concept is very familiar with: the project of constructing a good definition is essentially the project of comprehending what concepts are necessary to characterize the idea that one wants to define, and what concepts are unnecessary. This is why I compare the construction of definitions, and thus, comprehension, to sculpting. To really understand something you have to *define* it.

12.5 The Structure Of Comprehension

A Mathematical Characterization Of Comprehension

Here we use the language of diagrams to crystallize informally the characterization of understanding given in Definition 12.5.1.

Let the bullets and arrows of the diagram of Figure 12.3 represent data and strategies respectively. An arrow between two bullets denotes a known strategy that is *activated* by the data at the source of the arrow and that *leads* to the data at the tip of the arrow. Two bullets can be connected by multiple arrows to signify that we know multiple strategies the lead to the same goal. Strategies can be combined by juxtaposing adjacent arrows in the natural way, by connecting a tip to a source.

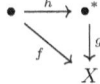

Figure 12.3: A diagrammatic representation of the process of understanding

The condition of minimality that encodes the process of understanding as defined above, is expressed by saying that for a certain data $\bullet^*$ to be minimal, it must have the property that *any other strategy conducive* to the goal X, such as f in the diagram above, can be expressed *via* $\bullet^*$ by juxtaposing h, the strategy that leads to $\bullet^*$, to g, the strategy that leads from $\bullet^*$ to X. In symbols, we write $f = g \circ h$. In Category Theory this condition is called *universality*. We say that $(\bullet^*, g)$ are universal because any other arrow that points to X factors via g.

Clearly, this definition of understanding is tightly coupled with the particular type of spatialization where comprehension is sought, since each element of the space in which an object of thought is spatialized, represented by the specific set of bullets and arrows in which we seek a universal conditions, defines the possible types of comprehensions in that space of thought. This is in line with the ideas we presented in Chapter 3.3, where we examined how the data that activates an Observational Structure of Intentionality is itself part of a process, each state of which is codified by a goal-strategy-data structure.

12 Epistemology Of The Abstract Mind I

The Cognitive Superstructure Of Comprehension

The universal property we have seen in the previous section is a special case of a general superstructure that generalizes and extends in many directions the intuitive notion of "factoring arrows". In Category Theory, this generalization takes many forms, such as, for example *Kan extensions* and *functor factorization systems*. Nonetheless, for the purpose of this discussion about the deep structure of comprehension, the way in which comprehension can be represented in mathematics with higher structures is not so essential. What matters is that we are gradually constructing an exact correspondence between the basic elements of the Epistemology of the Abstract Mind, and precise mathematical structures that, given the level of abstraction of this topic, live in the rarefied zoo of Category Theory.

Comprehension In Interaction Theory

In its most general form, the cognitive process of understanding is formulated in Interaction Theory via types of factorizations of the entities that codify the cognitive strategies—the Observational Structure of Intentionality, in the parlance of the Synthetic Structure of Intentionality. The core of the cognitive machinery of understanding, is that the factorization itself is expressed via the superstructure of knowledge that characterizes the ambient $\mathcal{M}$-*signature* where knowledge is sought. For example, if A factors via B, as in $A = B \times C$, the solutions of this equation are to be interpreted according to the superstructure of knowledge codified by the equational properties that define knowledge itself, as seen in Section 12.2. We will see concrete examples of how understanding works when we'll examine three important $\mathcal{M}$-*signatures* in Part IV.

12.6 Meaning And Semantogenesis

We have invented the notion of meaning to explain how ideas, objects and states of affairs are or become relevant to a specific context, and to express the intuitive and experiential belief that, by doing so, what we understand can be made consistent with the things we understand, and with the explanations we formulate about those things (see Sections 12.5 and 13.2).

Our aim in this section is to introduce two important cognitive

12.6 Meaning And Semantogenesis

processes that, together, explain in the framework of the Abstract Mind, and in the language of Interaction Theory, what is commonly and philosophically referred to as *meaning*. We call these thought processes *protosemantic* and *semantogenetic*. As usual in this book, our analysis of a concept is synthetic, and focuses entirely on the structure of the definitional interference of the concept at hand. To begin to comprehend the synthetic structure of meaning via its definitional interference, it is useful to start with a few observations about the phenomenology of meaning, because they help us situate meaning in the broader context of the Epistemology of the Abstract Mind.

Phenomenology Of Meaning

When do we use the word "meaning" in a question? and How do we get a sense of meaning in our transactions with the World, even when it's not apparent, even when meaning itself is not spelled out? These two questions are useful to unearth the basic structure of the phenomenology of meaning, because they invite us to recognize meaning in our transactions with the World in two ways: when we comprehend our experience, and when we create knowledge by articulating answers and explanations.

Meaning In Our Transactions With The World

We ask "What does it mean?" with the instinctive expectation to confirm, one more time, the practical and abstract knowledge that any coherent answer to that question has the potential to increase the quality and the degree of our understanding of the World, our judgment, and our ability to make effective decisions.

The use of meaning in every transaction with the World improves the quality of our experiences. We have the intuitive knowledge that the very grasping of meaning, without necessarily knowing what meaning is and how it functions at a cognitive and metaphysical level, is sufficient to set in motion certain thought processes that improve and, often, amplify our ability to penetrate the structure of our thought processes, and to increase and sharpen our capacity to relate our ideas to the World. In this broad sense, meaning acts as a prototypical conduit for the development of our understanding of the World. Any change in our knowledge occurs through the use of meaning, and, conversely, we search for changes in our knowledge to find or create meaning.

12 Epistemology Of The Abstract Mind I

By relating meaning to decisions, judgment and the development of knowledge, these observations on the occurrence of meaning in our transactions with the World relate the phenomenology of *meaning* to the phenomenology of *information*. This is a crucially important feature of meaning that, as we will see in this section and in detail in Chapter 20, has profound epistemological, physical and mathematical implications. But for now, let's focus on the phenomenology of meaning, and on what it reveals about this concept.

Meaning And Knowledge

Another feature of the phenomenology of meaning is that meaning can be easily communicated. We do not *explain* meaning, we *communicate* it. This observation indicates the subjective and normative nature of meaning, and corroborates the necessity to make a distinction between the origin of meaning, and our subjective and interpersonal experiences of it. We communicate meaning by describing its use and implications. We say: "This means that you can do A, B, C to such and such things, in such a way that C, D, E, and by doing so, you can F, G, H in such and such way." In other words, meaning is communicated by relating instances of knowledge via the same *equational structures* by means of which we construct knowledge (see Section 12.2).

The Dual Nature Of Meaning

The last observation on the relation between meaning and knowledge justifies and explains the organization of the most influential and interesting foundational theories of meaning: the so-called mentalist and non-mentalist theories. We won't describe these theories because they are already eloquently described in the abundant literature on the subject. We will briefly present a few remarks on the mentalist and non-mentalist theories of meaning in light of these observations on the phenomenology of meaning. The leitmotif that seems to permeate both the mentalist and non-mentalist foundational theories of meaning, and that is indeed present in the characterization of meaning given in Interaction Theory, is that meaning has a dual nature: it is ontologically objective by virtue of being communicable—ontologically objective means that its existence is agent-independent—and is epistemically subjective by virtue of originating from ways of understanding—epistemically subjective means that its function is agent-relative.

12.6 Meaning And Semantogenesis

The epistemically subjective nature of meaning dominates the mentalist foundational theories of meaning, which are centered around the general conviction that the nature of meaning can be explained in terms of the mental states of language users. One of the most influential analyses of meaning, due to P. Grice, characterizes meaning via the explanatory priority of speaker-meaning over expression-meaning, and this approach relates this view, in part, to the dual nature of meaning that emerges from the analysis of its phenomenology. Meaning has a subjective nature by virtue of its direct relation to ways of understanding and to the genesis of information, and this subjective nature conditions its objective phenomenology.

The ontologically objective nature of meaning permeates the various non-mentalist foundational theories of meaning, which attempt to explain meaning via its use rather than via its mental content. These theories are focused on the mechanism by means of which the use of an expression determines the meaning of the same. Thus, from a non-mentalist perspective, the characterizations of meaning is ontologically objective by virtue of being centered around the agent-independent features that make agent-relative expressions meaningful.

In light of this brief examination of the phenomenology of meaning, we can make, as a first approximation, the following claims.

- Meaning is strictly related to the notion of information, because any process that ascribes meaning to data, turns that data into information. The opposite is also true: the loss of information is a direct cause of a loss of meaning. Note, however, that the loss of information is itself an informative event, and, as such, it *does* convey meaning, but of a different kind of the meaning lost with the loss of information (more about this in Chapter 20 where we use Interaction Theory to define information). For the purpose of this discussion, we can ignore the subtleties related to a definition of information.

- The phenomenology of meaning and information, indicates that meaning and information always *coexist*.

- Meaning shares with knowledge the property of being easily communicable and ontologically objective.

- Meaning is communicated, not explained. It is communicated as hierarchical forms of knowledge dictated by the equational structures that define a knowledge system, as defined in Section 12.2.

These introductory observations on the phenomenology of meaning have unearthed, at least at an archetypical level, the connections of meaning with language and reasoning. With these ideas in mind, we have enough information to examine the function of meaning.

Function Of Meaning

The problem of explaining the function of meaning, is the problem of explaining the necessity, the structure, and the mode operation of meaning in the Abstract Mind framework. By explaining the necessity of meaning, we justify why meaning exists by virtue of operating on the ways of understanding the way it does. This is, so to speak, a definitional necessity. By explaining its structure, we justify what makes the action of meaning what it is, and the principles that define meaning beyond a specific realization of this concept. By explaining its mode of operation, we justify how the function of meaning fulfills the purpose of meaning. Once again, to explain the function of meaning, we use The Sieve.

From now on, meaning is an empty, meaningless—no pun intended—conceptual placeholder, defined for the sole purpose of denoting a certain cognitive process whose phenomenology corresponds to what we ordinarily identify as the significance of an object of thought or perception. Note how in this analysis we are neither concerned with the phenomenology of meaning, nor are we interested in characterizing meaning according to a particular foundational theory of meaning.

Meaning is a solution to the problem of identifying what matters to a context or an idea. This observation indicates that meaning is intrinsically an agent-relative notion: it requires a counterpart to *function* as meaning. It also indicates, in more concrete terms, that meaning is related to a goal by virtue of signifying what matters to the achievement of that goal, or by virtue of explaining how the achievement of that goal matters for other goals. This is the peculiar feature of meaning that intrinsically relates this concept to the organized contexts in which certain goals are defined. Even the

12.6 Meaning And Semantogenesis

most abstract and elusive contexts in which we use meaning adhere to this schema. For example, the meaning of life characterizes what matters to human existence, and why it matters to human existence to be clear about what's really important in life. These reflections help us introduce the second step of The Sieve: how meaning solves the problem it was invented to solve. The analysis of this point requires an abstraction leap.

The realization that the notion of meaning is agent-relative, together with the fact that meaning denotes what matters to a goal, suggests that meaning is directly related to the way we understand. Nowhere does this observation become more apparent than in the taxonomy of intentionality given in the Observational Structure of Intentionality. The fundamental difference between understanding and meaning is that meaning is a reference to a *goal*, and understanding is a reference to the *data* that activate a *strategy* to achieve that goal. By referencing a goal, meaning ascribes to the strategy that leads to that goal the quality of being the conduit to that goal. Meaning is, therefore, a reference to the goal of a given Observational Structure of Intentionality in the broader context of other goals and other Observational Structures of Intentionality.

And this is precisely the mechanism through which meaning is described in Interaction Theory: as a transformation between ways of understanding. The conception of meaning as transformation between ways of understanding explains the necessity and structure of meaning. It explains its necessity, by virtue of codifying the agent-relative nature of meaning in the form of an abstract relation between multiple ways of understanding. It explains its structure, by dictating that the agent-relative nature of meaning must emerge from coherent ways of understanding. It is precisely the coherence condition that defines the types of meanings that can be constructed between ways of understanding. To crystallize these observations diagrammatically, if we denote with A and B two ways of understanding—as described in Section 12.5—we can depict meaning as an arrow $A \to B$, where the very *existence* of an arrow *is* the meaning that connects A to B. With these premises, meaning is *equivalent* to an existential condition of a certain kind consistent with the definitional structure of A and B, and its function is conveyed by its existence or definability as an arrow between certain objects that represent ways of understanding. Another way to phrase this concept is to say that meaning is represented

12 Epistemology Of The Abstract Mind I

by an existential condition—mathematically, though, this can be conveyed in less strict forms where the notion of existence is less absolute, and certain mathematical objects have various degrees or modes of existence; recall, for instance, the various version of the mathematical representations of the philosophical notion of identity. This last point answers implicitly the last question of The Sieve: the knowledge needed to create meaning is the knowledge needed to define or prove the existence of a coherent transformation between ways of understanding.

We are finally ready to introduce the two thought processes that define meaning.

Protosemantic Transformations

Protosemantic transformations are transformations between Interaction Models. This broad concept is used here to introduce semantogenesis, and in Chapter 20 to define information in Interaction Theory. Here we focus on meaning only.

Definition 12.6.1. (Protosemantic transformation)
A structure-preserving morphism between Interaction Model is called *protosemantic*.

By defining a transformation between Interaction Models we define how to gauge the extensive properties of understanding. This is, perhaps not so surprisingly, why in the theoretical study of the nature of information we are often led to think of information as a framework to measure the extensive properties of knowledge. The parallel between knowledge and meaning that these observations about the nature of meaning are revealing is exactly the following: knowledge as equational system of objects of thought (Section 12.2), and meaning as equational system of ways of comprehension those objects of thought.

A transformation between Interaction Models is protosemantic in that it defines the archetypical structure of the transformations between the ways of understanding definable within those Interaction Models. This is the content of the next thought process.

12.6 Meaning And Semantogenesis

Semantogenesis

Semantogenesis, from Greek σημαντικός, (*semantikos*, significant), and γένεσις, (*genesis*, origin), is a transformation between ways of understanding, as described in Section 12.5. But why does a transformation between ways of understanding create meaning? and What exactly *is* meaning, beyond its definitions and functions?

The way meaning is defined in Interaction Theory conveys what I think is, perhaps, the most intuitive way of explaining the deep nature of meaning. Meaning is to knowledge what a compass is to a map. The notion of meaning acts as an *internal compass* with which the human mind navigates through various Observational Structures of Intentionality, by reacting to the changes in the goals and in the strategies to achieve those goals that are continuously set by the stream of thoughts and observations triggered by mental and sensory perceptions. This activity of reorienting thoughts toward new goals along a trajectory of minimal datasets is the experience of *understanding through meaning* that we all have when we learn. Meaning allows the Abstract Mind to direct its focus by virtue of informing the intentional process that stirs the perception of the objects of thought. These observations, and the definition of comprehension (Definition 12.5.1), lead to the following:

Definition 12.6.2. (Semantogenesis – *philosophical*)
Semantogenesis is a framework to characterize the extensional properties of comprehension.

The parallel between knowledge and meaning as equational systems we introduced informally, can be further refined, still informally, as follows: *information is to knowledge what meaning is to understanding*. The notion introduced by this comparison is that information and meaning measure the extensive properties of knowledge and understanding respectively. This intuitive idea is captured by the following:

Definition 12.6.3. (Semantogenesis – *technical*)
Semantogenesis is a mathematical object that classifies types of understanding.

12 Epistemology Of The Abstract Mind I

Roughly, an object that classifies ways of understanding is a prototypical process to characterize how meaning guides our mind when we think. A practical example of this insight is the notion of mass. The notion of mass allows us to *classify* physical objects, and a *measure* of mass is one of the basic ingredients that allow us to define systems of ideas we call physical theories to construct inferences about the interactions between physical objects that have mass.

The Cognitive Superstructure Of Meaning

Obviously, the definitions given above are heavily theory-laden. They are, because we are stretching certain analogies we borrow from the iconography of mathematics and physics, to convey ideas those very iconographies are based on— explain a metaphor with another metaphor at your own risk! But we are not completely in the dark. The limitations of certain metaphors and intuitions become more and more visible as we get closer and closer to the rock bottom of the structure of human thought. But the process of getting to this depth, and the increasing difficulty to describe what we discover, are themselves as informative as the ideas we are trying to convey. The idea we want to convey is that the nature of meaning, its superstructure, intended as a certain conceptual gadget to capture the extensive properties of understanding, is not that different from the function that certain concepts acquire by virtue of explaining other concepts. Explanations are measured by their capacity to improve our ability to interact with the World symbolically, in much the same way in which meaning is measured by its capacity to improve our ability to comprehend those interactions. In this sense, the apparent difficulty to capture meaning rests on the instinct to give a quantitative interpretation of an equational framework to describe meaning that is intrinsically qualitative.

Qualitative means *local*, which is the practical and conceptual opposite of *extensional*. It means that whatever outcome we derive from a quantitative approach to meaning, its validity and consistency and coherence are to be sought and weighed outside of the context in which the results are found. Therefore, *locality is the cognitive superstructure of meaning*. It is the set of conditions, relations and transformations through which a specific equational representation of meaning connects instances of knowledge and

12.6 Meaning And Semantogenesis

understanding to other instances of knowledge and understanding.

Meaning And Semantogenesis In Interaction Theory

From the observations on the cognitive superstructure of meaning, it follows that a formal description of meaning is, necessarily, a description of the qualitative, subjective and local nature of certain extensive properties of understanding. Here we want to convey with four intuitive examples how the abstract ideas about the superstructure of meaning are translated in the framework of Interaction Theory. Meaning is not created in just one way. Yet it is possible to recognize an underlying theme or strategy in the way meaning is created, as the discussions that follow illustrate.

Intuitive Example Of Semantogenesis

To begin to get a feel for how the superstructure of meaning is translated in Interaction Theory, we need to comprehend how and why Definition 12.6.3 is useful to capture the qualitative nature of meaning. Thus, we ask: Why is a mathematical gadget that classifies ways of understanding useful to characterize meaning? For starters, we can think of an object that classifies other objects, or a *classifier* for short, as a morphism that assigns real numbers to a given set of mathematical objects. For example, consider the set of integers $\mathcal{N} = \{1, 2, 3, 4, \ldots \infty\}$ and the morphism Γ that assigns each element n of $\mathcal{N}$ to an element of the set $\{0, 1\}$ according to the rule $\Gamma(n) = 0$ if n is even, and $\Gamma(n) = 1$ otherwise. Γ symbolizes a range of meanings. This is a basic way to quantify something. It is a way to give a quantitative representation of a qualitative concept. We can think of Γ as meaning because it "discerns" natural numbers based on their divisibility by 2. This deceptively simple mechanism codifies the basic idea of meaning when we *interpret* the symbols 0 and 1 in $\{0, 1\}$ as meaningful information about $\mathcal{N}$. What this basic construction reveals, is that, as a first approximation, the mode of existence of meaning is the mode of existence of the properties that define a classifier. Meaning is not a symbol or an entity: meaning is codified by a hierarchy of existential properties and, for this reason, coincides with the existence of an object or with the satisfaction of a certain property. This idea, translated and adapted to more sophisticated constructs such the the objects that codify the modes of understanding in a given $\mathcal{M}$-*signature*, is a basic semantogenetic framework. For this reason, the *definition* of a semantogenesis is strictly related to a specific *purpose*, since it is possible, in principle, to define multiple classifiers for the same object. A way to visualize how and why semantogenesis is always connected to a specific purpose, as one would expect thinking about meaning, is to picture meaning as a path between two villages. There can be many ways to go from one village to the other, each defined by the specific characteristics of the terrain between the two villages.

Protosemantic Transformations And Semantogenesis

With this intuitive example in mind, we ask again: What is a semantogenesis? To give an intuitive answer to this question, we observe that, since an Interaction Model defines a *way to acquire knowledge* (Definition 11.1.1) that is used to define the equational properties with which we define comprehension, and since in the definition of comprehension the only thing that matters is how the equation $f = g \circ h$ is constructed, what we are really interested in to define meaning, is not so much what an individual Interaction Model *is* per se, but how it acquires its definitional identity *through change*. For this reason, in what follows, we think of morphisms between Interaction Models, that is, protosemantic transformations.

Let $\mathcal{U}$ be a certain object made of all transformations between Interaction Models within a given $\mathcal{M}$-*signature*. We can think of $\mathcal{U}$ as the collection of all protosemantic transformations (Definition 12.6.1) definable in a given $\mathcal{M}$-*signature*. $\mathcal{U}$ represents the set of all conceivable *modes of comprehension* within a given $\mathcal{M}$-*signature*. And let **P** symbolize a set of *values*. A semantogenesis is a certain class of morphisms

$$\Gamma : \mathcal{U} \to \mathbf{P}$$

Recall the definition of comprehension given in Section 12.5: $\mathcal{U}$ is where a certain type of knowledge takes place, it is where comprehension is defined as "the knowledge that f is $g \circ h$" according to the equational properties that define a given $\mathcal{M}$-*signature*. The role of **P** is precisely that of quantifying comprehension. Γ *creates* meaning by quantifying a class of modes of comprehension.

Semantogenesis As A Measure Of Ways Of Understanding

Another way to interpret the general idea of a classifier of ways of understanding is to give a slightly more sophisticated formulation

12.6 Meaning And Semantogenesis

of the notion of semantogenesis that further illustrates a mechanism through which meaning is formed by virtue of being codified by a certain mathematical object. To this end, we need the mathematical notion of *measure*. By a mathematical measure we mean a certain set of rules to assign a non-negative real number to each subsets of a set. The rules that make up a measure are reasonably intuitive, and reflect to a certain extent, the way we commonly interact with quantities in the physical world. For example, if we denote with $\mu(A)$ and $\mu(B)$ the measures or two sets A and B respectively, and if A is a subset of B, we want $0 \leq \mu(A) \leq \mu(B)$, and if A and B have some elements in common, we want $\mu(A \cup B) \leq \mu(A) + \mu(B)$ where the strict equality holds only when A and B have no elements in common. By defining a measure on a set, we implicitly codify the notions of "more" and "less" for that set. A measure on a collection of ways of understanding does exactly this: it weighs the ways of understanding by virtue of defining what understanding is more than other ways of understanding, and by doing so it confers to each way of understanding a meaning.

Semantogenesis As Partial Order On Ways Of Understanding

A collection of ways of understanding is not just a representation of certain types of implicit knowledge: it is also a *space*. And on that space, we can define an Interaction Model, that is, a certain way to relate the elements of that space to each other. We can, in other words, enrich a collection of ways of understanding with a certain structure that codifies *how* we form meaning. Much like we did in the previous example, here we want to be able to create meaning by constructing utterances such as "this way of understanding comes before that way of understanding". The "before" and "after" are, so to speak, a bit like the "more" and "less" we saw in the previous example. Roughly, when we equip a collection of objects with a set of rules to construct statements for *some* pairs of elements about which element of the pair comes before the other, we are defining what in mathematics is called a *partial order*. Note that not all pairs of elements need to be comparable in a set equipped with a partial order. In this case, meaning is codified by the rules that define a partial order on a set of ways of understanding, and is created each time we compare two ways of understanding according to those rules.

12 Epistemology Of The Abstract Mind I

Bibliographical Notes

For a general reference on the traditional views on epistemology, evolutionary epistemology and the process of acquisition of knowledge I'd recommend Audi 1993a, Audi 1993b, Goldman 1986, Nagel 1986, Dretske 1980, Campbell 1956, Bradie 1994.

The distinction I introduce between conceptions and definitions of knowledge, and the existence of an archetypical form of knowledge underpinning the ordinary philosophical notions of knowledge (propositional and procedural)—what we call generative knowledge (Section 12.1)—stem from my interpretations and reworking of various category-theoretic constructions, such as, but not limited to, categorial logic (Jacobs 2001, MacLane and Moerdijk 1994, Goldblatt 2006), hyperdoctrines (Lawvere 1969a, Lawvere 1970, Seely 1983) and internal language (à la Mitchell-Bénabou), Kripke-Joyal semantics (MacLane and Moerdijk 1994), doctrines (Lawvere 1969c, Kock and Reyes 1977), and Lawvere theories (Lawvere 1963, Adámek, Rosický, and Vitale 2010, Peter T Johnstone and Wraith 1978, Hyland and Power 2007).

The theme of generative knowledge is in Lo Vetere 2016a. I begun to formulate the idea of a deep structure of knowledge (Section 12.2) and comprehension (Section 12.5) back in the 90's, when I was comparing certain constructions used in Topos Theory—the category-theoretical notion of pre-sheaf and the use of split fibrations (à la Bénabou, Peter T Johnstone 2002a, Jacobs 2001, Streicher 1999, Bénabou 1985, Gray 1966) to reconstruct identity—to certain communication patterns used in Neurolinguistic Programming. The other candidates to become the structures of knowledge and comprehension where various types of (functorial) factorization systems, such as in category-theoretic generalizations of Galois Theory (Borceux and Janelidze 2001).

The category-theoretical interpretations of knowledge as equational system based on models of **concrete** concept, of understanding, as (generalized) limits (including Kan extensions), of meaning, as measure of the extensive properties of forms of understanding, and its correspondence with information as extensive property of types of interaction-theoretic knowledge appear in a number of unpublished manuscripts of mine, and in Lo Vetere 2016c.

The traditional philosophical discussions about meaning, which I use here to introduce the interaction-theoretical version of meaning,

12.6 Meaning And Semantogenesis

were inspired, among others, by Carnap 1947, Kripke 1981, Eco 1994. My views on the meaning of concepts are strictly non-operationalist, that is, I disagree with Bridgman's view *"we mean by any concept nothing more than a set of operations; the concept is synonymous with the corresponding set of operations"* (Bridgman 1927).

Physics Without Space

The discussion about what drives the choice of space, time, mass and energy as the foundations of Physics, and of reality, has become increasingly pressing in the past century. Those elements are a choice, as are other physical quantities that could equally be taken as fundamental. I think that an eloquent example of this trend is a theory that reinterprets physics itself in the language of Topos Theory called Topos Quantum Theory. Topos Quantum Theory is a theory of physics proposed by C. Isham, A. Döring, J. Butterfield and others (for example in Isham 1997, Flori 2013b and Flori 2013a) in which every element of the ordinary way of thinking about the basic structure of a theory of physics—or, in other words, what a theory of physics should be—is constructed inside a mathematical object called Topos. A Topos is a category (recall Definition 4.3.1) with some extra structure that makes it a sort of generalized space where it is possible to define a theory of sets—see also Section 4.1 "The Generative Lens".

Inside a Topos it is possible to define the basic ingredients needed to construct mathematics, in much the same way in which mathematics is commonly defined from a theory of sets. "The key idea in this [Topos Quantum Theory] approach is that constructing a theory of physics involves finding, in a topos, a representation of a certain formal language, that is attached to the system under investigation [...]. Thus the topos approach consists in first understanding at a fundamental level what a theory of physics and associated conceptual framework should look like and, then, applying these insights to quantum gravity." (Flori 2013b)

Roughly, the structure identified by the authors of Topos Quantum Theory is the following:

1. A states space S.
2. A field that represents physical quantities Q. A field is a function that maps states to real numbers, $f : S \to \mathbb{R}$.

3. Any proposition of the form "The quantity Q is in a subset T of the real numbers, $Q \subseteq T \in \mathbb{R}$, which corresponds via f to a subspace of S. The subsets T form a Boolean algebra denoted by $Sub(S)$.

4. The states of S are identified via a morphism that sends $Sub(S)$ to $\{0,1\}$, which codifies the state of S that are in $Sub(S)$, when the value is 1, and those that aren't.

From a metaphysical viewpoint, I think this transition could be best described as a move toward a "physics without space", where the role of the arena where things exist is itself parametrized as part of the modeling process. I have no doubt that this approach can and will work, the question is: where is the mind?

Chapter 13

Epistemology Of The Abstract Mind II
Explanations, Reality And Ignorance

> ... [T]he mathematics of the future, like that of the
> past, will include developments which are relevant to
> the philosophy of mathematics [...] They may occur
> in the theory of categories where we see, once again,
> a largely successful attempt to reduce all of pure
> mathematics to a single discipline.
>
> Abraham Robinson

> In the future, as in the past, the great ideas [of
> mathematics] must be simplifying ideas, the creator
> must always be one who clarifies, for himself, and
> for others, the most complicated issues of formulas
> and concepts.
>
> A. Weil 1950

One of the main goals of the philosophical and scientific projects is the production of explanations, and we, as a species, invest a huge amount of mental energy to improve our ability to seek and construct increasingly sophisticated explanations. But what is an explanation? And why do we want to explain things? In this chapter we present an answer to the first question from an Abstract Mind's perspective, but before we get to the core of the argument, we want to briefly answer the second question.

I think there are three main answers to the second question, one is *transcendental*, one is *practical*, and one I'll let the reader decide;

13 Epistemology Of The Abstract Mind II

but this classification is, perhaps, illusory. There is the transcendental impulse toward the unknown that is deeply embedded in the human nature, that defines us as a species, and that propels us through our existence. From this transcendental perspective, we want to explain how everything works to satisfy our burning need to understand our own nature. There is the practical impulse to understand enough of how everything works, propelled by the hope or intuition that with enough knowledge, we can transform ourselves into eternal, godlike creatures that design their own spiritual, intellectual and biological evolution. In the Western culture, this intuition manifests itself first as a purely intellectual titillation, and evolves effortlessly into the grandiose project of living forever, conquering the Solar System and, with enough time, the entire Universe. And there is the third reason why we want to explain things: it is for the pure spiritual pleasure that comes with dissolving yourself into a journey where the categories of the mind do not apply. Knowledge is its own reward.

If we look at the problem of seeking and constructing explanations in its entirety, we notice that this problem is actually twofold. There is the problem of defining an optimal epistemological and methodological framework to characterize explanations. And there is the problem of characterizing the cognitive act of seeking and constructing explanations *in itself*, regardless of the specific epistemological and methodological framework used to produce explanations. The argument we present here is concerned with the second problem, and offers a way to define what it means, at a cognitive level, to seek and construct explanations, and what it means that a certain statement *explains* something. The most interesting and historically important solutions to the first problem—the problem of characterizing an optimal epistemological and methodological framework to characterize explanations—are, among others, the Deductive-Nomological model, the Statistical-Relevance model, and the Causal-Mechanical model. We do not discuss these models here, because they are covered already by the vast literature on the subject, and because this omission does not affect the content of this book. The first problem is also the problem of the nature of mathematical explanations. It is the problem of understanding what mathematical explanations tell us about physical reality, which we covered in Chapter 2, and that we used to situate and motivate the Abstract Mind model and the basic ideas of Interaction Theory.

We explain explanations and explanatory systems in three steps: we begin by describing the elements of an explanatory process, and then define an abstract explanatory process, its interpretation, and the cognitive superstructures that define explanations. Roughly:

Definition 13.0.1. (Explanation)
An explanation is a statement that links the explanandum to a set of accepted truths where it is believed that there is an epistemically objective correspondence between the explanans and those accepted truths.

We will see that the actual conceptual machinery that produces explanations is a complex device with many moving parts, most of which lie at the intersection of philosophy and mathematics, and that for this device to produce *good* explanations—the only explanations everybody should be interested in—a lot of things have to fall into place. But before we get to the main theme of this chapter, there is an important point that requires an adequate introduction, because it underpins this entire analysis and is central in this research.

13.1 The Principle Of Epistemic Relativity

Ordinary explanations, the explanations we construct in the frame of reference of the Identity-based $\mathcal{M}$-*signature* (Chapter 15), are hierarchies of statements of the form "Y follows from X, because A, B, C,..." constructed according to the syntax of a language defined in a certain mathematical gadget that represents the explanandum from the point of view of a set of accepted truths. But what makes an explanation an explanation isn't merely a matter of syntactic structure and correctness, it is far more that that: it is how profoundly an explanation validates or disproves the basic belief system from which it originates. Recall how in 1905, when Einstein postulated that the speed of light is independent of the frame of reference—a notion popularized as "the speed of light is constant"—his theoretical discovery forced us to profoundly revisit a host of worldviews about the fabric of reality and about our place in the Universe that had been the bedrock of over two centuries of scientific hypotheses and explanations since the advent of Newtonian physics.

The exploration of explanatory systems beyond the ordinary

13 Epistemology Of The Abstract Mind II

Identity-based $\mathcal{M}$-*signature* is based on the observation that in Nature there seems to be no indication that the **concrete** concept of identity is a privileged frame of reference to define mathematical thought and a scientific method. Identity, and its many manifestations, are anthropomorphic projections that we transfer more or less consciously in our efforts to come at the world with a demand that it endorse a worldview in harmony with what we think is the only safe, reliable, rational structure of human thought. Our efforts to connect symbolically with the fabric of reality through a systematic method should be measured primarily by how much they deepen our own self-conception as creatures of a living Universe. Nature has no requirement to be in harmony with human thought and ambition and when its elegant design seems to speak the language of human thought, our first impulse should be to cherish that moment, and immediately reclaim and comprehend the awe and the naïveté that led us to that epiphany: there's always time for pride. We condense these reflections in the following principle:

Definition 13.1.1. (Principle of Epistemic Relativity)
There is no privileged frame of reference to define an epistemology of the Abstract Mind, and with it, a mathematical and scientific method of inquiry.

The principle of epistemic relativity is really the principle of epistemic unity, in that the purpose of the relativization it introduces, is instrumental to the definition of a unifying framework for the definition and the development of a holistic conception of human thought. Thus, epistemic relativity must be understood not as a weak paradigm of knowledge, but, on the contrary, as a principle of interdependence that deepens and amplifies the meaning of knowledge along trajectories that are tightly intertwined with the deep structure of human thought, and with the symbolic significance of human work and ambition.

13.2 Explanations And Explanatory Systems

To explain what makes a coherent explanation a coherent explanation, we need to familiarize ourselves with the elements that form an explanatory process, with the way those elements are used in the explanatory process to construct an explanation and why, with

13.2 Explanations And Explanatory Systems

what defines the limits of our interpretations of an explanation, and with the exact mechanism through which explanations are used to advance our knowledge.

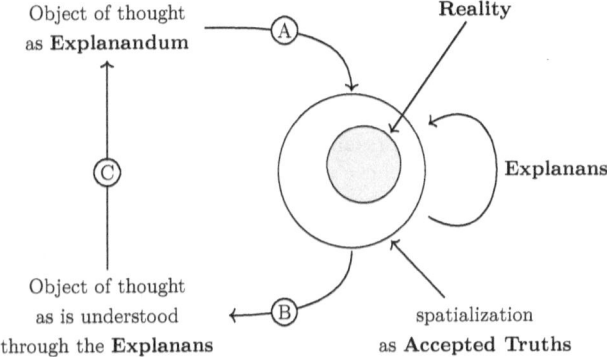

Figure 13.1: The elements of a rational explanation in the ∞-diagram.

The conception of what an explanation is or should be, is based on the structure of the Knowledge-Reality Interface (Section 13.8), and is, therefore, represented by an ∞-diagram. This view is reflected in the elements of an explanatory process.

Two elements define at the most fundamental level an explanatory process in Interaction Theory: a conception of knowledge, and a conception of the relation between rational explanations and reality. These two elements form the backdrop against which the explanatory process is designed by answering the question: What should an explanation be? In particular, in the framework of Interaction Theory, these two basic elements are parameters that describe the state of an Abstract Mind. As a result, in Interaction Theory explanations have always a relative character because they are intimately related to the specific $\mathcal{M}$-*signature* in which they are sought and constructed.

Explanandum And Explanans

The terms *explanandum* and *explanans* (Figure 13.1) are distinct entities that indicate respectively a fact or object of thought being explained, and the statement we construct to explain it. In this description of explanations, therefore, the explanation of a hypothetical fact or idea that encodes its own explanation would be problematic—there is of course in this observation a clear Gödelian echo of the discussion about the Limits of Formal Systems we saw in Section 8.4.

Accepted Truths And Self-explanatory Ontologies

The *accepted truths* represent a certain spatialization of an explanandum where we define a *reality* (Figure 13.1-A). The accepted truths are facts or ideas about a reality in which the explanandum exists. We agree, by definition, to not understand or explain the accepted truths, but only to acknowledge or define them in a way that is or can be made consistent with the type of epistemic access we want to have of that reality through our explanations. The accepted truths are *self-explanatory ontologies* by virtue of not requiring to be explained or understood by *definition*.

For example, in Newtonian Physics, a notion of physical reality (explanandum), which symbolizes a certain form of material existence, is spatialized in a way that reflects a certain philosophical position, according to which everything that exists in that specific form of material existence is made of space, time, forces, mass and energy. Thus, in Newtonian Physics, space, time, forces, mass and energy are *accepted truths* about the physical reality Newtonian Physics is concerned with, and are *assumed* to be consistent with the type of physical inquiries (explanans) we conduct about that physical reality. In contrast, in Particle Physics, the *same* notion of physical reality Newtonian Physics is concerned with, is spatialized in a different way, and the accepted truths resulting from this spatialization are energy and spacetime, whereas mass and forces are carried by particles (energy), that is: they are no longer assumed, they are *explained*.

In summary

Definition 13.2.1. (Accepted Truth)
An *accepted truth* is a spatialization of an explanandum, in which a reality is defined.

13.3 Epistemic Continuity, Truth And Reality

An explanatory system **P** must contain a *belief system*, or a set of *background presuppositions*, that defines a form of underlying cohesion between an explanandum and an explanans produced by **P**. The nature of the cohesion between explanandum and explanans transcends the explanatory system, and is therefore postulated. Cohesion between explanandum and explanans means that **P** must be equipped with a framework—a criterion—to decide or stipulate that if A explains B in **P**, then B must have a certain degree of correspondence with an accepted definition of truths, which may or may not extend to a correspondence with reality.

Another way to describe the principle of cohesion between explanandum and explanans that defines **P**, is what we call *epistemic continuity*: the conviction that a well constructed chain of good explanations leads to good explanations. In this sense, and within the horizon of a given explanatory framework, explanations are construed as entities that can be added to a corpus of knowledge like we throw potatoes in a shopping bag. This is the conviction that—the ordinary definition of—rational thought applies to reality to a degree sufficient to grant us a certain privileged epistemic access to the structure of the things that exist.

Definition 13.3.1. (Epistemic continuity)
Epistemic continuity is the belief that an explanandum has an intrinsically protosemantic cohesiveness that corresponds to the symbolic cohesion inherent in the structure of an explanatory system **P** to a degree consistent with the explanans produced by **P**.

To get a feel for what epistemic continuity means practically, let's consider the case of the explanation of a physical phenomenon. The notion of epistemic continuity stipulates that there is a correspondence that we trust, between the syntactic and semantic cohesion of the explanations that we construct with our symbol systems, and a hidden organizing principle underlying the things we explain. The syntactical correctness that holds together a mathematical theory becomes a reliable reference to the kind of truth that holds together the things that we explain with that mathematical theory. When a physical experiment corroborates the prediction of a physical theory, we *interpret* this result as

13 Epistemology Of The Abstract Mind II

confirmation of the correspondence between the syntactic and semantic cohesion of that particular physical theory, and the physical phenomenon deciphered by that theory. This interpretation is based, so to speak, on a form of semantic induction: when a physical theory is experimentally confirmed, what holds together that theory globally, becomes a global reference to what holds together all the physical phenomena explained by that physical theory, it becomes a law of Nature. For example, we do not check every pair of stars in the Universe to satisfy ourselves that our theories about nucleosynthesis are correct: we imply the validity of our theories by continuing to interpret a range of physical phenomena with them, until we find a phenomenon that we cannot explain with those theories.

The catch with epistemic continuity applied to mathematical predictions, is that mathematics as we know it today is a non-evolutionary system of thought—mathematical truths are eternal—but Nature, as far as we know, is a highly evolutionary living system. We have invented the laws of Nature because we believe that the regularities of physical phenomena that we predict mathematically are eternal, and that the fabric of reality is either not involved in the evolutionary process of the Universe, or is an invariant of that process.

Truth

The term *truth* denotes a certain class of facts or ideas for which the correspondence with reality encoded in the explanans has the highest known degree of correspondence with reality. The present framework to explain explanations, therefore, regards truth as having a dual nature, *factual* and *metaphysical*. Truth shows its factual nature when is used to mark a correspondence with a reality in a purely mechanical, computational fashion. Truth shows its metaphysical nature when is used to signify that a factual correspondence between a theoretical fact and the elements of a reality, corroborates the system of ideas in which epistemic continuity is needed to justify the coherence conditions used to *construct* the explanations of that reality. The accepted truths are true in the sense just described. Reality is *true*. Awareness is *true* (Definition 14.2.1).

Definition 13.3.2. (Truth)

13.3 Epistemic Continuity, Truth And Reality

An object of thought $\mathcal{O}$ is *true* if it corresponds to a self-explanatory ontology via a Minimal Interaction Model (see Section 11.5).

Truths are stipulated or derived. They are stipulated when they are the content of a coherent *definition*. They are derived when they are deduced from the rules of an Interaction Model.

Reality

The term *reality* denotes an arena derived from a subset of accepted truths where the explanandum is represented. Explanations are about the facts that occur in an ontologically objective construct, such as physical reality, or about ontologically subjective constructs, such as perception. Reality is the arena where the explanandum exists, and is in general distinct from the arena where the explanations about it are constructed and studied. The dichotomy between the ontological domain of an explanandum, and the ontological domain of the explanans, is captured by the ∞-diagram.

Reality is an accepted truth which exists in an self-explanatory manner, and is intrinsically *true*—although, it should be clear that the category of veridicality does not need to apply to a reality for it to be real. This observation suggests that *the way in which reality is true*, is a way of being true that is assumed to be, to a certain degree, compatible with the type of details our explanatory methods of that reality are designed to reveal, and that the nature of the relation between the veridicality of reality and the truths we determine with our explanations are themselves true, or conducive to an intelligible degree of truth. In this sense, the most basic reality is the experience created by a Minimal Interaction Model (Section 11.5), which coincides with the unmediated, intuitive experience of an object of thought. The mantra is "Reality is the Interaction Model".

A notion of *reality* is part of the underlying belief system or of the background presuppositions that define an explanatory process **P**: its role in **P** is to signal the ultimate truth—ultimate with respect to the definition of **P**.

Definition 13.3.3. (Reality)
A *reality* is the experience created by an Interaction Model.

From these observations, it follows that the description of *a* reality is necessarily the description of the epistemology that originates from a definition of knowledge created by a given Interaction Model, and that therefore the most basic form of reality is the experience created by a Minimal Interaction Model.

Real means: my conscious experience of what I see of the world through this Interaction Model is intrinsically true regardless of its implicit mode of existence, *therefore*, my experience has an epistemically objective significance that transcends its ontologically subjective nature, because I *believe* that the Symbolic Intuition of Reality from which this experience originates is in a relation of epistemic continuity with the world that causes this experience. The feature of *being real*, which is commonly referred to as *a* reality is, therefore, a reference to an instance of epistemic continuity that through its intrinsic nature connects and reconciles the subjective and objective domains of knowledge, existence and meaning. These observations imply that the description of a reality based on the notion of epistemic continuity is a *prototypical correspondence theory of truth*.

13.4 A Canon Of Good Explanations

Good explanations relate the element of the explanans to the explanandum via a broad body of knowledge rather than via the explanation itself.

Good explanations are constructed according to a specific *structural canon* that defines what makes good explanations good. The structural canon of good explanations stipulates the degree to which a given statement that claims to explain a certain fact or idea is to be trusted by virtue of having an adequate correspondence with the structure of an accepted reality. The core idea of the structural canon of good explanations is that good explanations are constructed in such a way that the details of the explanation are *not* connected to the explanation via the explanation itself, but via a broader body of accepted background knowledge that provides the epistemological arena where the explanation is sought and constructed.

For example, in Physics, good explanations are explanations difficult to vary without questioning or contradicting a substantial

13.4 A Canon Of Good Explanations

body of accepted, reliable, experimentally confirmed, or theoretically sound knowledge. We could explain the movement of the Earth around the Sun along an elliptic orbit as the effect of an invisible unicorn pushing our planet. However, it might as well be a flying blue pig pushing our planet around the Sun—it's blue, because it lives in the sidereal space; by the way, this is an explanation of why pigs flying in the interstellar space are blue. And why not justifying the orbit of the Earth around the Sun as the effect of the collective belief that the Earth orbits around the Sun? The problem with these explanations is that the unicorn, the flying blue pig and the collective belief are related to the orbit of the Earth around the Sun only *via the explanation* itself. This is what makes these explanations *bad* explanations: it is the *method* used to construct the causal link between the various elements of the explanation. The good explanations of the orbits of the planets in the Solar System are based on Newton's Law of Gravitation, and on Einstein's General Relativity, which provide a much wider framework to explain and relate many other physical phenomena. As a result, if a hypothetical physical experiment, carried out by multiple independent teams around the world, refuted Newton's (or Einstein's) theory of gravitation, the entire edifice of modern physics would collapse or would be indeed profoundly shaken at its foundation. In contrast, if we switch the unicorn with a blue pig, or if we stop thinking that the Earth orbits around the Sun, nothing happens to the system of ideas we have constructed to explain other physical phenomena, because no explanatory systems capable of producing predictive knowledge of the world is based on the physics of unicorns and blue pigs.

Good explanations are constructed by relating the details of the explanandum to a wider body of verified beliefs, actions and facts outside of the scope of the explanation. The validation of the background beliefs, actions and facts is carried out to the maximum extent the reduction to root causes is possible, relevant, meaningful and consistent with the background presuppositions upon which the causal frame of reference is applicable to the beliefs, actions and facts being explained.

The structural canon of good explanations, therefore, consists of these two features:

- They are connected to a wider body of knowledge in full accord with the frame of reference where the *questions* answered by the explanations are formulated.

- They do not break the definitional structure that justifies why a certain frame of reference is the technology chosen to *formulate* the questions answered by those explanations.

For example, a physical explanation that assumes, states or implies an illogical conclusion can never be a good explanation because it breaks the very presuppositions upon which physics uses logic to gain a sort of privileged epistemic access to the structure of physical reality, and in which the questions about physical reality are formulated according to logical rules.

It is precisely this structure that separates good explanations from bad explanations, not testability. For, there are testable bad explanations. It is possible to successfully test bad explanations with experiments that verify the existence of the effects or the facts described by the explanation, but that fail to challenge whether the details of the explanation are related to the explanation via the explanation itself, or via a body of theoretically or empirically confirmed knowledge.

A Conception Of Knowledge Is A Self-explanatory Statement

A conception of knowledge, as introduced in Section 11.1 and in Definition 11.1.2, captures the cognitive process through which the Abstract Mind constructs knowledge at the most fundamental level, and constitutes an elementary and irreducible explanation. To understand why a conception of knowledge is also an elementary explanation, or, more correctly, a self-explanatory statement, we need to recall that in Definitions 11.1.1 and 11.1.2 knowledge is defined equationally with respect to the structure of an ambient $\mathcal{M}$-*signature*. Self-explanatory statements *are* knowledge because they are in a one-to-one correspondence with the entities they explain by virtue of being irreducible in the $\mathcal{M}$-*signature* where they are defined. This concept can be expressed symbolically as follows: if we denote self-explanatory truths with bullets and explanations with arrows between bullets, we can *identify* bullets with those arrows ↻ in which the source coincides with the target. This observation allows us to dispense with bullets and use only arrows to describe the Epistemology of the Abstract Mind. In Category Theory, the process of defining an object only via arrows

13.4 A Canon Of Good Explanations

is known as a 1-sorted representation.

Relation Between Rational Explanations And Reality

In the Western intellectual tradition, we construct rational explanations because we believe that this is the best way to acquire deeper knowledge of the things that exist: the direction of this transaction is from the mind to the World. What makes the knowledge produced via rational explanations deeper than the unmediated knowledge we get from our senses, are its predictive character, and the symbolic structures through which that knowledge is related to broader contexts, both actual and imaginary.

Other philosophical and intellectual traditions have different approaches to epistemology, such as the traditions that cultivate mystic and revelatory types of knowledge: the direction of those transaction are from the World to the mind, and often through the human body, or through other material objects.

The connection between explanations and reality we are concerned with in this book is of the first kind. As we saw in the introduction to Chapter 1, the Western intellectual tradition is based on the belief that we called the *intelligible universe hypothesis*, according to which the realities we are aware of—both the ontologically subjective and the ontologically objective realities—are to a certain extent intelligible. The role of rational explanations is to relate our ideas to the nature and structure of these realities, and in this sense, it is a sort of epistemological homecoming through the mind to an original mode of being. It is precisely the theme of epistemological homecoming that separates the Western tradition from the mystic and revelatory epistemologies: the latter do not see any separation between the mind and the World, and thus no need to connect to the World rationally; their view is that knowledge is a dialogue between Man and Nature that occurs naturally when we let it happen by shutting up the rational chatter, and let Nature do the talking.

The classification of the relations between rational explanations and reality we made in Section 3.2, consists of two fundamental belief systems: the belief that reality is intrinsically related to the human explanations by virtue of its nature—what we called the *strong math-reality connection*—and the belief that the structure of reality is to some extent permeable to the human explanations, but is not "written" in the language of the human explanations—what we called the *weak math-reality connection*.

Explanations Are Locally Superadditive

Explanations are superadditive because the more we explain an idea, the more we uncover of that idea. Superadditivity is deeply embedded in the ordinary conception of explanations, and reflects, to a certain extent, the philosophical position according to which explanations are intrinsically concerned with the extensive properties of the facts and ideas.

13.5 The Explanatory Process

An explanatory process is a process that produces explanations according to the *structural canon* for good explanations. The production of a good explanation consists in the formulation of a *theory* in a certain system of thought that codifies the ontologies of the abstract explanatory process introduced at the beginning of this chapter. To fully appreciate the meaning of the term "theory" in Interaction Theory, and to comprehend how theories explain things within the horizon defined by the Epistemology of the Abstract Mind, we must first make an important distinction between *explanations* and *demonstrations*, and we need to clarify the exact mechanism which sets theories apart form other, milder types of explanations. This discussion reflects this rationale.

Explanations And Demonstrations

The distinction between an explanation and a demonstration depends on the context and purpose of the scientific or mathematical inquiry in which the argument supporting the explanation is sought and developed. When mathematics is used in natural sciences, its role is primarily descriptive: mathematics is required to describe *how* a physical phenomenon occurs, which is often sufficient to make predictions about it, but it rarely explains *why* it is, necessarily, so—sometimes because it can't, sometimes because it is not required. The type of whys given by mathematical theories in natural sciences are merely a manifestation of higher structural relations codified in or emerging from the definition of those theories, because non-explanatory arguments are often enough to advance knowledge.

13.5 The Explanatory Process

For example, Newton's Law of Gravitation

$$F = G\frac{m_1 m_2}{r^2}$$

is a so-called inverse square law, that is, it varies with the inverse of the square of a distance r. Newton's mathematical description is sufficient to make predictions about the position of celestial bodies. But the reason why it is an inverse square law and not something else, is not part of that mathematical description, or of the theoretical framework used by Newton to derive his law. This means that, from a Newton's Law perspective, the argument needed to justify why Newton's Law of Gravitation is an inverse square law is an *explanation*. In this sense, the mathematical framework of Newtonian gravitation is demonstrative, not explanatory. In Interaction Theory, we are interested in the basic structure that an argument constructed within a given $\mathcal{M}$-*signature* must have to count as an explanation, and *not* on the technologies used to construct demonstrations, such for example the technologies used in applied mathematics.

To understand what turns a demonstration into an explanation, we need to recall the discussion of Section 3.2 about the two ways of grasping concepts: anthropic and participatory. In the framework of formal systems, and in Interaction Theory, theories are developed in an anthropic mode of thought, and may be interpreted in a participatory way. What distinguishes explanations from demonstrations is precisely the point in which we decide to switch from an anthropic mode of thought to a participatory mode of thought. We call this decision the *setting of a reductionist signpost*.

Setting The Reductionist Signpost

We determine what turns a demonstration or a description into an explanation, by deciding when to stop the reductionist process of seeking increasingly basic causes, or by realizing that we have reached a theoretical or technical impasse that is either ineliminable with the tools available in the framework where it emerges, or is insurmountable with the theoretical knowledge available. It is at that point that we begin to switch from an anthropic mode of thought to a participatory mode of thought, and start seeing demonstration as models of broader systems of ideas that we have

13 Epistemology Of The Abstract Mind II

to humbly explore in many directions to fully comprehend their philosophical, theoretical and technical implications.

There are two scenarios in which this change of perspective occurs. When we stop asking *why*, we stipulate that we have reached the point in which the elements of thought are taken as self-explanatory—or it is satisfactory and coherent to do so—being those elements part of the spatialization of an object of thought. Alternatively, we decide to revisit the background presuppositions upon which we have constructed the spatialization where we ask why, and change the perspective of our inquiry. The change of perspective is often associated with various nuances of the problem of challenging the basic definitions upon which we have attempted to develop a specialized type of knowledge up to that point. These questions tend to be quite radical, uncomfortable, unforgiving and brutal, such as: What should mathematics be? What should a theory of physics look like? Why are we explaining things in this way? and How do we define the coherence conditions for a scientific project? What are the limits of our explanations? How can we justify the methods of our intellectual explorations without having to ask these questions all over again? What are the things we cannot explain with our current theoretical and explanatory framework, and what does it mean that we cannot explain them? and so on and so forth.

For example, in Newton's Law of Gravitation the accepted truths are the elements of a spatialization of the notion of physical reality where the world is made of space, time, forces, mass and energy, and in which the fabric of space allows us to measure the distance between two points with a real non negative number. This description changes radically as soon as Particle Physics attempts to describes the same world, because the spatialization of that world changes quite radically. Suddenly, the fundamental elements—the accepted truths of the Newtonian reference model of physical explanations—are particles: there are force carriers—particles that give rise to the forces between particles we observe and measure—and the fabric of space changes. When we compare Newtonian Physics to Quantum Mechanics we are forced to face the cumbersome and uncomfortable questions such as What theory of physics is right? What should a physical theory of space and time and forces look like? It gets even worse when we try to reconcile General Relativity with Quantum Mechanics. The same questions about what a theory of physics should be, come back even stronger than they do in Newtonian Physics. For example, in Loop Quantum Gravity—a modern theory that attempts to unify General Relativity with Quantum Mechanics—space is quantized.

Explanations And Theories

In Interaction Theory, a *theory* is a generalization of the notion of theory we encounter in the context of formal systems, based on a *definition* of knowledge codified by an Interaction Model. A definition of knowledge provides the arena where it is possible to formulate the relations between certain mental constructions and a reference reality. This general schema is a direct application of the paradigm illustrated in the ∞-diagram. For example, in the default, Identity-based $\mathcal{M}$-*signature*, the definitions of knowledge are mathematical models of the philosophical notion of identity, therefore *any* theory *necessarily* consists in the construction of *equational relations* between the elements of a certain spatialization. In Interaction Theory this paradigm is generalized in two directions: $\mathcal{M}$-*signatures* and Cognitive Architectures.

There is the framework of the Cognitive Architectures and Interaction Models used to systematize and organize the *definitions* of knowledge, and, as a result, the construction of "equational" systems that describe a certain problem and which define forms of computation. These notions of computation are the language in which theories are formalized in Interaction Theory.

There is the framework of $\mathcal{M}$-*signatures* used to characterize the fundamental structure and the meaning of the equational systems defined by the Cognitive Architectures. This framework characterizes knowledge at an archetypical level, and provides a way to systematically construct systems of thought in which the fundamental notion of identity is replaced by other concepts that reveal new types of rational thought.

13.6 Explanations, Meaning And Ignorance

The interpretation of an explanatory process characterizes the nature and content of a correspondence between explanans and explanandum produced according to the *structural canon* of good explanations. By characterizing a correspondence between explanans and explanandum, an explanatory process defines also the types of subsequent spatializations that an object of thought can be subject to. The problem of characterizing the relation between explanans and explanandum is captured by questions like What does an explanation reveal of a reality? and What does it mean that a certain explanation *explains* something about a reality? This is a basic problem at the foundation of what we think the philosophical and scientific projects *should be* and *should do*, and how we *should go* about them.

Another crucially important requirement an interpretation of an explanatory process must meet to qualify as a coherent characterization of a correspondence between explanans and explanandum, is that it must contain an implicit definition of *ignorance*. This is a fundamental feature of the interpretative framework of explanations that is, I think, not sufficiently recognized. The interpretative framework of an explanation, not only should guide the development of higher types knowledge and meaning within the horizon of the theory from which the explanation originates, it should also inform the very rational and intuitive processes of formation of that theory. And this is only possible if an interpretation contains the metalanguage to define the structural limits of a knowledge that can be constructed upon the frame of reference offered by an explanation. The ultimate significance of the interpretation of an explanation, is not so much in what it describes of a correspondence between explanans and explanandum, but in how well it defines their mutual distance, and the nature and structure of their differences, which are broadly referred to as ignorance. The interpretation of an explanation that fails to augment the quality and depth of our ignorance, turns essentially an explanation into a dignified demonstration.

Here we will outline the aspects of the interpretation of an explanatory process that are more interesting from the point of view of Interaction Theory. These are:

- The meaning of truth
- Quantitative reality-explanation relation
- The paradox of fully explainable realities
- Qualitative reality-explanation relation

13.6 Explanations, Meaning And Ignorance

The Meaning Of Truth

Being true indicates the existence of a certain correspondence between a fact and other axiomatically true facts, and marks the success condition of any interaction between a cognitive technology and Nature. Not so surprisingly, one the many effects of this conception of knowledge is that the increase of knowledge often comes in the form of an increase of ignorance. It is in the *logic of knowledge* that the negation of knowledge is itself a new order of knowledge. The meaning of truth in relation to this explanatory framework is therefore not an extensive property of reality because it is not additive.

Quantitative Reality-explanation Relation

In the quantitative interpretation of the relation between reality and human explanations, reality has a discrete nature, or can be made equivalent to a representation of its own nature that can be fully explained by a *finite number* of explanations, after which: *nothing else can be known about that reality because there is nothing else to know about it.* The quantitative interpretation, being a reflection of the superadditive conception of knowledge (Section 12.4), leaves us with the questions: What is the relation between that reality and our explanations if that reality can be fully explained? What does *fully* mean, by the way? We are left with rather dramatic questions such as What's the difference between reality and our explanations, if we can fully explain reality? Have we explained what we thought we were explaining? What's the point—or what's wrong—with the philosophical and scientific projects if we can reach a set of ultimate explanations?

The Paradox Of Fully Explainable Realities

We have already discussed in this chapter the foundational problems raised by the locally superadditive nature of knowledge. I think the existence of fully explainable realities is problematic, because it forces us to challenge the ontological status of a reality that we can fully explain. If fully explainable realities exist, then we need to define what "real" means. What's really real, our explanations, the reality they explain, or both? and if both are real, In what way are they both real but different? and we have to ask, once again, the uncomfortable question of why we think that the very project of seeking explanations the way we do makes sense. And if both our explanations, and fully explainable realities are real—have the same mode of existence—then we need to rethink another foundational notion at the core of the scientific project, the relation between ontological continuity and epistemic continuity, because by claiming that a reality coincides with its explanation we collapse two entities, reality and explanation that we assume to have distinct modes of existence. If fully explainable realities don't exist, then we have one more reason to not ignore the problems posed by the superadditivity of human knowledge.

Qualitative Reality-explanation Relation

When the reality-explanation relation is qualitative, reality forms an ontological and epistemological continuum, and it can never be fully explained. This inexplicability and indefinitude sets an intrinsic limit to the philosophical and scientific projects. A qualitative relation between reality and explanation poses new questions every time we are able to explain parts of it. This is the reality we are familiar with: it is a reality where we content ourself to keep playing the game of knowledge, because the relation between explanations and reality is never complete, and it is in this incompleteness that we find the deepest knowledge—dramatic, but true. A reality fully explained is, in a sense, *defined* by those explanations: those explanations and that reality are, so to speak, equivalent. In contrast, a reality that cannot be fully explained is not definable in the language in which we construct the explanations of that reality, and *transcends* that language along dimensions that, when uncovered, constitute radical knowledge. The parallel between explanations and theorems in a formal language is inevitable—a reality fully explainable is compared to a formal system that is complete and consistent, and a reality not fully explainable is compared to a non trivial formal system where certain theorems cannot be proven—see also Section 8.6.

13.7 The Metastructure Of Explanations

As seen in the description of the explanatory process, in Interaction Theory explanations are theories in a given $\mathcal{M}$-*signature*. The superstructures we present here translate and generalize the terms *theory* and *metatheory* in the language of $\mathcal{M}$-*signatures* and Interaction Models.

13.7 The Metastructure Of Explanations

Endoexplanations As Theories

Endoexplanations are theories defined in a given $\mathcal{M}$-*signature*, and take, therefore, the form of statements constructed according to the structure of a local epistemology—the epistemology of an Interaction Model. Much like in the default $\mathcal{M}$-*signature* a theory is a collection of logically connected statements that define higher order equational relations between certain ontologies, in any given $\mathcal{M}$-*signature* a theory is a statement in which certain ontologies of an Interaction Model are related via higher equational relations based on the **concrete** that gives rise to that $\mathcal{M}$-*signature*. The prefix *endo-* indicates that a theory is formulated within the horizon of a given $\mathcal{M}$-*signature*, and that its limits are, therefore, set by the types of self-referential constructs definable within that $\mathcal{M}$-*signature*. The purpose of endoexplanations is to construct true accounts of a reality consistent with the ability to articulate rational thoughts created by an $\mathcal{M}$-*signature*.

Exoexplanations As Metatheories

Exoexplanations are theories defined on transformations between $\mathcal{M}$-*signatures*. As we'll see in Section 14.1, this definition implies that we can think of exoexplanation as true statements about modes of consciousness, or, more intuitively, as statements about the (subjective) experience of thought. Much like endoexplanations are tools to explain specific *applications* of a specific type of rational thought to objects and states of affairs in the World, exoexplanations are tools to explain our *experience* of those endoexplanations. Clearly, this description implies that the most elementary form of exoexplanation is a statement about the transformations that do not alter a given $\mathcal{M}$-*signature*: what is generally referred to as *awareness* (see Definition 14.2.1).

Explanations In Interaction Theory

In Interaction Theory, endo- and exoexplanations are defined on the equational systems defined by a given $\mathcal{M}$-*signature*. The way the equational system at the basis of a given $\mathcal{M}$-*signature* defines an explanatory framework is twofold:

- An equational system defines the symbolic structure of knowledge (Section 12.2) upon which the syntax of explanations is *constructed*.

13 Epistemology Of The Abstract Mind II

- An equational system defines the symbolic structure of inference that codifies the form of epistemic continuity (Definition 13.3.1) upon which the explanations are *interpreted*.

To comprehend the fundamental structure of explanations in Interaction Theory, and to establish the limits of the interpretations of what we see and comprehend when we look at the World through the conceptual lens of a coherent explanation, we need to explore what connects an explanation to the mechanism through which a given **concrete** concept defines intelligible thought, and understand how this nexus defines the conceptual structures that differentiate an explanation from any other utterance or thought. To this end, we need to recall two of the elements we introduced in the analysis "How **concrete** Concepts Define Intelligibility" in Section 10.5, namely, Organizing Principle and Actuation Mechanism, and, with that analysis in mind, we need to examine how utterances interact with the definitional structure of the elements that they reference. The reason why we need to look at the fundamental nexus between an explanation and the cognitive machinery that defines intelligible thought will become apparent as we progress in this argument.

The Deep Structure Of Explanations

In Section 10.5 we saw that for any given **concrete** concept, the Actuation Mechanism defines the phenomenology of the Organizing Principle, and that this mechanism dictates how an Abstract Mind grasps reality symbolically at the most fundamental level via the types of spatialization defined by the Definitional Model. Consequently, given a **concrete** concept C, *any* thought or utterance formulated by an Abstract Mind in the state defined by C consists necessarily of various manifestations of the Actuation Mechanism associated to C. It consists necessarily of various presentations of the constituents of the Definitional Model arranged according to certain rules dictated by a given Interaction Model. This is, abstractly, how a **concrete** concept shapes the construction of thoughts and utterances and, a fortiori, of explanations. To comprehend *practically* how a **concrete** concept shapes the construction of explanations, we need to take a closer look at what differentiates the various types of utterances.

Utterances (and thoughts) range from statements, to questions, commands, and exclamations. Definitions and explanations are

13.7 The Metastructure Of Explanations

statements. A way to comprehend what differentiates an explanation from any other type of utterance is to look at the relation between an utterance and the definitional structures of the entities referenced by that utterance.

First off, we can ignore questions, commands and exclamations because, by definition, they do not affect the definitional structure of the entities they reference. What differentiates an explanation from any other expression, is that an explanation is concerned with *how* a certain entity referenced in the explanans acquires or changes its definitional structure, whereas definitions stipulate *that* certain entities constitute the spatialization where that Actuation Mechanism exists as phenomenology of the Organizing Principle. It follows that, at the most fundamental level, explanations must necessarily be concerned with how an instance of Actuation Mechanism acquires or changes its definitional structure, or, in more intuitive terms, must be concerned with the *genesis of the definitional structure of an Actuation Mechanism*. To get a feel for what this observation means practically, let's show the intrinsic difference between definitions and explanations in Physics: the *how* vs. *that*. We stipulate *that* mass and energy exist, and explain how the physical transformation of mass into energy via Einstein's equation $E = mc^2$ sheds light on the genesis of the definitional structure of mass and energy, by showing *how* they can turn into each other. One can of course define new entities *through* an explanation. For example, one could define mass and energy directly via Einstein's equation, as those fundamental constituents of Nature, say, X for energy and Y for mass, defined by the relation $X = Yc^2$, but that would be a *new type of spatialization*, not an explanation!

Lenses And Images

Explanations are *necessarily* concerned with how the definitional structure of an instance of the Actuation Mechanism *comes to be*, and the concept that captures the definitional genesis of an Actuation Mechanism becomes the lens through which the World is seen and understood through an explanation based on that Actuation Mechanism. These observations lead to the following two definitions, which we'll see in action in Part IV.

Definition 13.7.1. (Lens)
A *lens* is a concept that faithfully describes the definitional genesis

of an Actuation Mechanism.

In Interaction Theory, the concept through which we *understand* the World we see through a lens is called an *image*.

Definition 13.7.2. (Image)
An *image* is a concept that faithfully describes what we see through a lens.

For example, in the Identity-based $\mathcal{M}$-*signature*, the definitional genesis of *division*—division is the Actuation Mechanism of identity (see Chapter 15)—is conveyed by the term *change*, which evokes the experience of a plurality of identities that participate in a process of becoming. Change is the *lens* of identity because it describes at the most fundamental level the genesis of the definitional structure of division by answering the question: What is the origin of division?. By virtue of characterizing the genesis of the definitional structure of division, change becomes, at an archetypical level, a way to grasp and use the abstruse idea codified by division: the idea of something that comes *before* the mental act of discerning the objects of thought. Through change we see *process*. Process is the *image* of identity. Process is how change is manifested in the World. It is what the World is made of when we look at it through the explanations we formulate in the metaphysical structure of thought defined by the Identity-based $\mathcal{M}$-*signature*. Through the notion of process we *understand* the World in terms of entities that acquire and lose their definitional identity in a symphony of changes.

The application of these ideas will become apparent when we will derive the fundamental structure of the explanations that an Abstract Mind formulates when it is in the Identity-based, Finiteness-based and Cohesiveness-based states.

13.8 The Structure Of The Knowledge-Reality Interface

By a Knowledge-Reality Interface we mean a distinct cognitive apparatus through which the human mind is able to relate thoughts to objects and states of affairs in the world at a metaphysical and epistemological level. The phenomenology of a Knowledge-Reality Interface is the phenomenology of those mental states that we

13.8 The Structure Of The Knowledge-Reality Interface

might associate, as a first approximation, to the mental experience of the objects and states of affairs in the world, such as the seeing, the hearing etc. The approximation of this description consists in reducing this phenomenology to the mental states that are not opaque to consciousness. In the context of the Abstract Mind, the structure of the Knowledge-Reality Interface denotes the paradigm (Figure 13.2) according to which the human mind is able to establish a symbolic, metaphysical and semantic correspondence between the features and properties of abstract constructs, and what we see of the World through them.

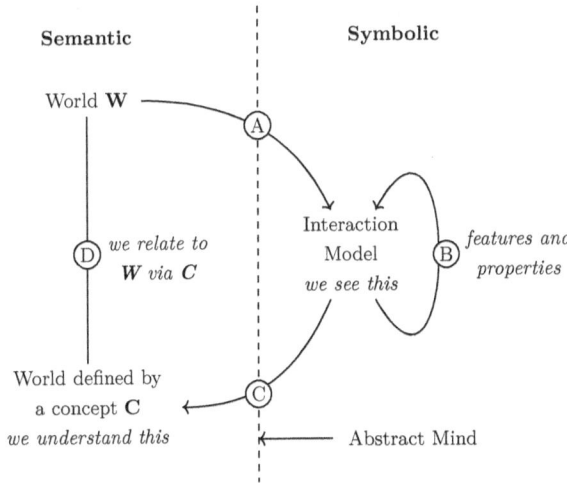

Figure 13.2: The Knowledge-Reality Interface.

We have already seen the ∞-diagram in Figure 13.2 in Sections 1.1 and 11.2, and in Chapter 2. It is the general schema through which we create mental maps of reality via Interaction Models. The distinction between mental states and experience of the world created by the human mind through its ability to conceptualize, is purely functional to the purpose of this description, which is the separation of the cognitive act of reasoning (Figure 13.2-Symbolic), from the act of looking at the world through the lens of concepts (Figure 13.2-Semantic). We have deduced this structure by observing how the human mind seems to articulate rational thoughts,

13 Epistemology Of The Abstract Mind II

and by observing how we have been creating mathematical abstractions of the world since we have historical evidence of them. One of the conclusions suggested by this research, is that the cognitive apparatus underlying the Knowledge-Reality Interface, namely, the Interaction Model, has a specific structure described by the **reconstruction of the success condition** (Section 11.3), and that the knowledge of this cognitive mechanism allows us to replace an Interaction Model with another, as long as the new Interaction Model is built according to the same mechanism. The invariance of the Knowledge-Reality Interface, is captured by Postulate 8.7.4.

Definition 13.8.1. (Structure of the Knowledge-Reality Interface)
A Knowledge-Reality Interface is a cognitive apparatus through which the human mind is able to relate thoughts to objects and states of affairs in the world at a metaphysical and epistemological level. The structure of this fundamental cognitive apparatus, is an invariant of Interaction Theory (Postulate 8.7.4), and is described by an ∞-diagram.

In Interaction Theory

Postulate 8.7.4 dictates how the cognitive processes that represent the primary features of the Abstract Mind are defined in Interaction Theory. It defines the architecture of the fundamental mind-world interactions by means of which the Abstract Mind creates knowledge and meaning.

13.9 Knowledge, Computation And Reality

One of the questions raised by the problem of defining a conception of knowledge, concerns the directions in which the ordinary definitions of knowledge and computation, and their mutual relations, can or should be generalized in order to remain consistent with the overarching conception of knowledge we are developing in this book. For example, if we discovered that an Interaction Model other than the OSP Pattern successfully decodes parts of physical reality, would that discovery automatically provide us with new definitions of knowledge and computation?

The structure of the epistemological problem raised by the definition of a conception of knowledge can be summarized as

13.9 Knowledge, Computation And Reality

follows:

1. Does knowledge have to have a predictive character in order to qualify as knowledge? This question demands us to focus on the relation between computation—which forms the basis of the predictive character of mathematical and scientific knowledge—and the structure of our epistemic access to the world—which currently revolves entirely around the formal systems philosophy.

2. Is a correspondence theory of truth necessary to define knowledge? That is to say, Do we have to have a mechanism to validate information against an axiomatically true portion of reality to characterize that information as knowledge?

3. What is the relation between computation and predictability? Currently, the latter is a consequence of the former, and there is no notion of predictive knowledge that does not have a computational foundation.

I think there are two sensible approaches to the problem of characterizing how definitions of knowledge (Definition 11.1.1), computation and their mutual relations can be generalized within the broader context of the conception of knowledge introduced in Definition 11.1.2. There is the view that a definition of knowledge and a definition of computation are two sides of the same coin, and that the distinction we make between them is purely instrumental to the use we make of these two terms: to emphasize the predictive nature of knowledge we talk about computation, to emphasize its semantic value we talk about knowledge. And there is the view that knowledge and computation are distinct and separate concepts that are related *via the structure* of the Interaction Model in which they are defined. I am inclined to believe that the latter approach offers a more balanced way to deal with this problem.

These two perspectives give rise to the following three views on knowledge and computation.

Conservative

This view is based on the hypothesis that the OSP Pattern is the *only* instance—or one of many equivalent instances—of an archetypical cognitive apparatus of the human mind that gives rise to an intelligible notion of computation, and, consequently, to the

13 Epistemology Of The Abstract Mind II

conception of predictive knowledge we are familiar with through the technology of formal systems. This idea, which is essentially a manifestation of the strong math-reality connection (Section 1.2), leads to the questions: How can we interpret cognitive patterns other than the OSP Pattern? Are cognitive patterns different from the OSP Pattern still forms of knowledge? And if so, what knowledge do they represent? And what alternative conceptions of computation could we construct without the technology of formal systems? What are these alternative notions of computation, and in what way would they relate us to physical reality? I think this position, although plausible in principle, is based on a rather extreme slant on what defines knowledge. It is imperative in this type of analyses, to maintain an intellectual rigor and a cold detachment from the themes that are suggested by an anthropic mode of thought. The view that the OSP Pattern is the only possible model of computational knowledge, is an example of an easy-to-close case in which one can quickly define the boundaries of what is humanly knowable, by invoking the notion that what is already known defines what we can't even imagine because we don't have the intellectual maturity and sophistication to ask the right questions.

Pragmatic

This view regards Interaction Models as the entities that codify definitions of knowledge and the corresponding definitions of computation. In this sense, the pragmatic view is a generalization of the strong math-reality connection. According to this view, *every* Interaction Model gives rise to a definition of computation, or to a definition of knowledge, or both, and characterizes what connects them. In the case of the OSP Pattern, knowledge and computation are related via the syntactic nature of the ordinary definition of knowledge. This is a radical view that, when embraced, demands us to rethink computation and knowledge not in terms of the quantitative and qualitative correspondences that these concepts allows us to establish between our ideas and the objects and states of affairs in the world, but on the structure of an Interaction Model, no matter how exotic that might be. To see how this would work, let's consider an Interaction Model consisting only of Objects Patterns. The Objects Pattern is an example of pre-conceptual Interaction Models—recall Section 11.4—and can be thought of, for example, as a sequence of sets $\mathbf{S} = \{S_0, S_1, S_2, \ldots\}$. According to the prag-

13.9 Knowledge, Computation And Reality

matic view, **S** models a type of knowledge and computation. If we assume, for example, that the sets in **S** are also ordered, and have only a finite number of elements, for example from a master set like an alphabet, then **S** can be turned into, or thought of as a formal language whose syntax is unknown. If we do so, then the problem of defining the knowledge encoded by **S** becomes the problem of interpreting **S**: it is the problem of finding the minimal set of inference rules that generate **S**. Why? Because syntax is the basic frame of reference we use to relate **S** to what we think it describes (recall the ∞-diagrams). The absence of a technology to relate **S** to reality—in the example the technology of formal systems—leaves us without a way to *use* **S**. This is the main reason why the pragmatic view, while theoretically intriguing, seems too abstract and impractical. The problem is, when the **reconstruction of the success condition** is missing or incomplete, our awareness of an object of thought is incomplete—we will discuss this further in "How Data Becomes Message" in Section 14.6.

Definitional

This is the view that *some* Interaction Models define intelligible notions of computation and knowledge, and is the view that we adopt in this book. The characterization of what counts as knowledge and computation, according to this view, is based on the mechanism by way of which an Interaction Model is related to the **success condition** (Section 8.3). The discovery of this mechanism leads to the identification of three types of knowledge that originate from an Interaction Model, called *cognitive*, *pre-conceptual* and *pre-cognitive*, that we introduced in Section 11.4.

Bibliographical Notes

The traditional accounts on the nature and structure of physical and mathematical explanations that inspired this research are, among others, in Lewis 1973, W. C. Salmon 1998, Popper 1959. The traditional accounts on the Deductive-Nomological model are in Popper 1959, Braithwaite 1953, Beales and Gardiner 1953, Hempel 1942, Hempel 1966, Hempel and Oppenheim 1948. For the Statistical Relevance model I'd recommend W. C. Salmon 1971, Greeno 1970, Jeffrey 1969, and for the Causal Mechanical model W. C. Salmon 1984, W. C. Salmon 1998, Woodward 2003, Woodward 1989. A traditional view of the philosophical dimensions of

13 Epistemology Of The Abstract Mind II

the problem of mathematical explanations and of its relations to reality is in Baker 2005, Baker 2009, Baker and Colyvan 2011, Cellucci 2008, Steiner 1978a, Steiner 1978b. The reinterpretation and rework of a theory of explanation at the basis of Interaction Theory—especially the theories of explanations that originate from the non-standard *M-signatures*—are heavily influenced by the style of reasoning of Higher Category Theory, and by the ways in which it unifies and transcends mathematical thought in ways we are only beginning to understand. Among the many works of distinguished philosophers, physicists and mathematicians that shaped my understanding of the deep structure of explanations, I would like to mention Lawvere 1969a, Lawvere 1979, Lawvere 1986, Penrose 1989, Feynman 1965, Barwise and Moss 1996, Deutsch and Marletto 2014.

Of particular importance in this research is the notion of epistemic continuity, which emerged in my work spontaneously as a natural consequence of the fundamental structure of ideas underlying Interaction Theory. I didn't have to introduce epistemic continuity, it was already there, written all over this research. All I had to do was crystallize it in a definition and explain its conceptual necessity (Lo Vetere 2016b). Similarly, the structure of the Knowledge-Reality Interface is another fundamental theme of this research that I did not have to explicitly introduce because it was already present in most of the arguments I was developing. All I had to do (Lo Vetere 2016d) was polish up the underlying idea and articulate it in a definition and a diagram—the ∞-diagram. The theme of the relations between knowledge, computation and reality, which is another leitmotif of this research, is a slightly more sophisticated formulation of what is known as the syntax-semantic interface problem (see also Chapter 20): how to explain the emergence of meaning from syntactic rules. I do not embrace any computational theory of mind. This belief is reflected throughout this book in the participatory structure of the *M-signatures* and Cognitive Architectures interplay, where local definitions of computation—local with respect to an Interaction Model—provide a conceptual substrate to develop knowledge, but never in an absolute way. This approach is one of the main themes of the categoric-theoretic forms of *internalization* (MacLane and Moerdijk 1994, Jacobs 2001, Goldblatt 2006, Peter T Johnstone 2002b), together with categorification, which I have briefly described in the Motivic Lens (Section 4.3).

Chapter 14

Epistemology Of The Abstract Mind III
Consciousness, Communication And Computation

> We will need to use some very simple notions of
> category theory, an esoteric subject noted for its
> difficulty and irrelevance.
>
> Moore and Seiberg 1989

In Interaction Theory, consciousness is defined as a *change in the perception* of an objects of thought, and is *identified* with its effects on a metaphysical structures of thought, as described by the $\mathcal{M}$-*signature* model. Consequently, in Interaction Theory the study of consciousness is the study of the changes occurring within a given $\mathcal{M}$-*signature*, and between $\mathcal{M}$-*signatures*, and of the conditions that enable those changes, and is therefore part of the study of the *dynamic properties* of the Abstract Mind. This is why we talk about consciousness in the Epistemology of the Abstract Mind, because this conception of consciousness ties the phenomenology of consciousness to the dynamics of the exact structures of thought by means of which an Abstract Mind develops abstract knowledge of the World. The other aspects of the dynamic behavior of the Abstract Mind we review here is the mechanism conducive to the changes in the cognitive states of an Abstract Mind, which in Interaction Theory constitutes the archetypical structure of communication, which gives rise to the metaphysics of language and computation. We conclude the chapter, the introduction to the Epistemology of the Abstract Mind, and Part III, with the formulation of a conception of mathematics in an arbitrary $\mathcal{M}$-*signature*.

14.1 Modes Of Consciousness

By defining consciousness via the changes it causes in the way the human mind perceives its own thoughts, we provide a conceptual framework to answer questions like: What are the conditions that enable or trigger new ways of understanding? What are the conditions to develop a new type of knowledge? and, in general, What are the dynamical and evolutionary structures of an epistemological project? My use of the words "enable" and "trigger" might suggest that in this description we use a causal framework to conceptualize the fluctuations in the modes of consciousness. It might suggest that I think about the changes in the modes of consciousness, like I think of billiard balls smashing into each other. This is not so. In this description, causality is an *analogy*, not a reference to the dynamics of the modes of consciousness. It's just another manifestation of the inadequacy of ordinary language to describe anything that does not adhere to the subject-object dichotomy. When we describe the changes in the modes of consciousness, we are bound to use a language that is constructed to convey only two ways of grasping concepts. There are *principles*, which are eternal and abstract, in the sense that they do not participate to the reality we describe with our intellect and are causally inefficacious, and there are *ordinary concepts*, which are mostly defined by their causal efficacy, and form the raw material that the human mind seems to be designed to grasp. The former can only be grasped as metaphysical intuitions; the latter are continuously constructed and destroyed like we erect and knock down buildings. The definitions of $\mathcal{M}$-*signature* and of the Epistemology of the Abstract Mind shatter this simplistic system of thought, and show how to reframe the very structure of human knowledge by modeling the mind-world interplay. It is from this expanded perspective that this description of the dynamics of the modes of consciousness should be understood.

To begin to describe consciousness via its effects on the metaphysical structures of thought, we need to ground consciousness to the Epistemology of the Abstract Mind. This is the content the following fundamental

14.1 Modes Of Consciousness

Observation 14.1.1.
The conditions for change in $\mathcal{M}$-*signatures* and Interaction Models are characterized by specific properties of *understanding* and *meaning*—as defined in Sections 12.5 and 12.6 respectively.

The properties of understanding and meaning characterize *the structure of change*, are empirical in nature, and derive from the observation of several forms of mathematical, linguistic, and creative types of reasoning I carried out in my research. Observation 14.1.1 captures the idea that there are certain *configurations* of the elements of a local epistemology that *enable*, without necessarily triggering, specific transitions between $\mathcal{M}$-*signatures*, and between Interaction Models within a given $\mathcal{M}$-*signature*. Epistemologies are always *local* because they are defined in a given $\mathcal{M}$-*signature*.

It should be clear that these conditions on the structure of understanding and meaning must occur at two levels, cognitive and conceptual, which correspond to the structure and phenomenology of thought respectively.

At a *cognitive* level, these conditions capture the structure of the **concrete** concept at the basis of a given $\mathcal{M}$-*signature*—as described by the Organizing Principle, Actuation Mechanism and Definitional Model in Section 10.5—and denote specific irreducible *modes of consciousness* that manifest themselves via the *abstract structure* of the cognitive apparatus in which comprehension and meaning operate.

At a *conceptual* level, the conditions that at a cognitive level denote the modes of consciousness, manifest themselves as *understanding* and *meaning*—as Defined in 12.6—in full accord with the structure of thought enabled by the ambient $\mathcal{M}$-*signature*.

From these observations, it follows that the characterization of consciousness we are defining is necessarily of the form "When one or more elements of a local epistemology have such and such properties, the Abstract Mind *can* change its state from a certain $\mathcal{M}$-*signature A* to another $\mathcal{M}$-*signature B*." Note the use of the italic typeface in "can". My use of the terms "state" and "change" are not meant to be interpreted in a temporal, deterministic fashion, and the description of consciousness we are giving here isn't syntactic, it's *synthetic*. The characterization of the conditions that enable certain changes of $\mathcal{M}$-*signature* doesn't turn the Abstract Mind into an automaton. This characterization simply defines the basic structure of the phenomenology of consciousness, and ratifies the intuitive knowledge that the phenomenology of consciousness is indistinguishable from consciousness itself.

To introduce the structure of consciousness, we need to introduce three concepts that we will use as an intuitive, visual analogy. The language we use to describe these concepts is, once again, borrowed from Category Theory, but is free from technicalities.

Initial And Terminal Objects

To explain the conditions that give rise to the three modes of consciousness we introduce in "Consciousness And Spatialization" (Section 14.3), we need to introduce the category-theoretical notions of initial and terminal object. Recall the informal definition of category (Definition 4.3.1). Given a category **D**, an object of **D** is *initial*, and is denoted by 0, if for every object c of **D**, there is a unique arrow $0 \to c$. Similarly, a object of **D** is *terminal*, and is denoted by 1, if for every object c of **D**, there is a unique arrow $c \to 1$—the uniqueness of arrows, in Category Theory, is always up to isomorphism, or other equivalence relations.

Objectification Of Understanding

We have seen the objectification of understanding in the description of semantogenesis in Section 12.6. I use the verb "objectify" as synonym of bracketing. To objectify an abstract structure means to regard that structure as a whole, and to use that whole to define higher order concepts.

14.2 The Structure Of Awareness

In Interaction Theory, the notion of *awareness* is instrumental to the definition of the three fundamental modes of consciousness, and captures the observation that in the human mind, the most elementary form of awareness corresponds to a mental state in which an object of thought is self-evident in the context in which it manifests itself to consciousness. This observation leads to the intuitive notion that to characterize awareness it is sufficient to require that the *phenomenology* and the *epistemology*—as defined in Part III—coincide.

14.2 The Structure Of Awareness

Before we introduce the definition of awareness used in Interaction Theory, there is a question that, I think, Observation 14.1.1 raises, and that can be easily answered by giving a definition of awareness. The observation that the conditions for change in $\mathcal{M}$-*signatures* can be expressed via certain properties of understanding and meaning, might seem to pose the problem as to whether this characterization is a chicken-and-egg problem, in which the causal relation between changes in $\mathcal{M}$-*signature* and properties of understanding and meaning is undefined. What comes first? A change in the $\mathcal{M}$-*signature*, or the emergence of certain properties of the structure of comprehension and meaning? To solve this puzzle, we need to remember that the structure of the Epistemology of the Abstract Mind is based on a *conception* of knowledge codified by the definition of $\mathcal{M}$-*signature*, and that each *definition* of knowledge, which corresponds to an Interaction Model, is used to construct the models of understanding and meaning described in this chapter. Hence, to determine what comes first, a configuration of understanding and meaning, or a change of $\mathcal{M}$-*signature*, we must look at the hierarchical relations between these concepts. The definition of $\mathcal{M}$-*signature* determines the definition of Interaction Model, and this in turn defines its own local epistemology, and, with it, the structural properties of comprehension and meaning.

With this clarification in mind, the observation that the most elementary form of awareness corresponds to a mental state in which an object of thought is self-evident, is translated in the parlance of Interaction Theory by the following definition

Definition 14.2.1. (Awareness)
Given a **concrete** concept **C**, and an object of thought $\mathcal{O}$, we say that the Abstract Mind is *aware* of $\mathcal{O}$ if the epistemology of $\mathcal{O}$ corresponds to the phenomenology of the Actuation Mechanism associated to **C**.

An effective way to visualize the notion captured by Definition 14.2.1 is to draw or imagine a dot

•

This dot represents $\mathcal{O}$ as the Abstract Mind sees and understands it based on the **concrete** concept that currently shapes what's intelligible: it is a representation of the full correspondence between the experience of $\mathcal{O}$, and the object of thought $\mathcal{O}$ experienced through the **concrete** concept that defines the structure of thought with which I am writing this paragraph, namely, identity. Being aware of $\mathcal{O}$ means seeing $\mathcal{O}$ as it appears to the human mind in the irreducible, intelligible form defined by the **concrete** concept of identity.

For example, in the default, Identity-based $\mathcal{M}$-*signature*, the correspondence between the epistemology of $\mathcal{O}$, and the phenomenology of the Actuation Mechanism associated to the **concrete** concept of identity consists in the following *interpretation* of •

- *knowledge* corresponds to a special form of Interaction Model called Minimal Interaction Model (Section 11.5) characterized by the property of being irreducible. The structure of the Minimal Interaction Model dictates the structure of understanding, meaning and explanation, and the interpretational framework in which they are meant to be understood.

- *understanding* is constructed and interpreted based on the Minimal Interaction Model, and can be described visually by a diagram

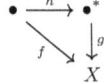

where • and •* *coincide* with X.

- *meaning* is constructed and interpreted based on the Minimal Interaction Model, and has the structure of the semantogenetic process applied to the form of understanding defined above.

- $\mathcal{O}$ is self-evident by virtue of having only an irreducible *explanation*. It is an accepted truth by virtue of coinciding with our experience of it.

14.3 Consciousness And Spatialization

In Interaction Theory the term consciousness is a mass noun that denotes the dynamic nature of $\mathcal{M}$-*signatures*, and is not

14.3 Consciousness And Spatialization

defined or regarded as an individual concept on its own right. Consciousness has a *local appearance* and a *universal nature and presence*, as crystallized by Postulate 8.7.2. The reason for not defining consciousness directly is, therefore, twofold, and can perhaps be explained as follows.

- Because Interaction Theory is a participatory system of ideas in which, by design, the question "What is consciousness?" can be asked only with respect to a context defined by specific transformations between $\mathcal{M}$-*signatures*. Consciousness, in other words, is a localized reference to certain states of the Abstract Mind that has, nonetheless, the universal nature of the metaphysical structures of thoughts described by Interaction Theory.

- Because in Interaction Theory, as seen in Postulate 8.7.2, consciousness is both "the stage and the play": it is the *stage* when regarded abstractly as the arena of thought, and the *play* when it becomes the thinking.

These are the reasons why we focus on the structure of consciousness rather than on defining consciousness directly, because although my knowledge of consciousness and $\mathcal{M}$-*signatures* is incomplete, there are distinctive and very recognizable patterns that seem to indicate that the changes in $\mathcal{M}$-*signature* obey certain laws. Those laws are the closest thing to what we could safely call a definition of consciousness, which is, for this reason, more accurate to characterize as metastructure of the human experience. Thus, we ask: Can we define consciousness via its structure? Pragmatically, yes. Practically, I have to acknowledge that the deeper I study the structure of consciousness through the mathematical language of Interaction Theory, the more I am persuaded that consciousness is a much bigger mystery than my minuscule mind can comprehend.

Postulate 8.7.2 is used here to introduce the three irreducible modes of consciousness that seem to constitute the fundamental *structure of consciousness*, and that are used in Interaction Theory to define a conception of consciousness for the Abstract Mind. Figure 14.1 is a representation of the structure of consciousness via the three fundamental modes of consciousness of Interaction Theory, namely: *spatialization*, *despatialization* and *contravariance*. In Figure 14.1, the bullet represents an object of thought, $\mathbf{C}$ is a `concrete` concept, and $\mathbf{C}(\bullet)$ denotes the unmediated experience of the object of thought, or, as discussed in Definition 14.2.1 the *awareness* of that object of thought. **C(•)** is, so to speak, the "sleight of mind" through which the human mind seems to be able to become aware of its own thoughts at a pre-cognitive, pre-conceptual, and pre-representational level. The structure of thought that hinges on the three arrows of Figure 14.1 forms the conceptual framework for the definitions of the Minimal Interaction Models that we'll see in Part IV.

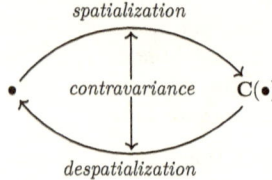

Figure 14.1: A representation of consciousness via the three basic modes of consciousness of Interaction Theory

Spatialization, despatialization and contravariance are the three basic and irreducible *modes of consciousness* that, together, seem to form the fundamental structure of conscious thought. These modes of consciousness enable and define the human experience of thought at any given point in time—once again, the category of time, just like causality, is used in this book purely for explanatory purposes, and is just another example of how embarrassingly inadequate ordinary language is to talk about consciousness.

Definition 14.3.1. (Spatialization)
The term *spatialization* denotes a mode of consciousness of the Abstract Mind defined as the occurrence of following conditions for the Epistemology of the Abstract Mind:

- *understanding* is a terminal object

- *meaning* is a transformation that originates from an initial object of understanding

Definition 14.3.1 tells us that the Abstract Mind spatializes when there *exists* a basic, intuitive or preterintuitive understanding (terminal) of an object of thought, and when the semantogenesis *begins*. Spatialization is, in this sense, primarily associated to the *exploration* of meaning.

Definition 14.3.2. (Despatialization)
The term *despatialization* denotes a mode of consciousness of the Abstract Mind defined as the occurrence of following conditions for the Epistemology of the Abstract Mind:

- *understanding* is a initial object
- *meaning* is a transformation that originates from an terminal object of understanding

Definition 14.3.2 describes the condition where the Abstract Mind ceases to spatialize, and brackets the object of thought it has become aware of. This condition is described via the *existence* of an intuitive or preterintuitive understanding (initial), and via the *end* of semantogenesis. Despatialization is, in this sense, associated to a *definition* of meaning in which the Abstract Mind reframes an instance of understanding and reorients intentionality.

Definition 14.3.3. (Contravariance)
The term *contravariance* denotes a mode of consciousness of the Abstract Mind defined as the transition to and from spatialization and despatialization, as defined in 14.3.1 and 14.3.2 respectively, and characterized by the attainment or loss of a condition of *awareness*.

The term contravariance is used in this context with its category-theoretical meaning, and denotes the abstract operation of reversing the direction of the arrows of a diagram.

Definition 14.3.3 tells us that the achievement of a full spatialization or despatialization are conducive to contravariance whenever the particular configuration of the corresponding epistemology acquires meaning with respect to other (known) **concrete** concepts. In other words, contravariance is a meta-mode of consciousness, because it signals that a specific configuration of the Epistemology of the Abstract Mind has meaning in other $\mathcal{M}$-*signatures*.

14 Epistemology Of The Abstract Mind III

Spatialization And Objects Of Thought

So far in the text we have used the term "object of thought" as an intuitive reference to what an Abstract Mind grasps at a particular point in time. The characterization of the modes of consciousness just introduced, allows us to give a more precise interpretation of what an object of thought really is. Object of thought and spatialization are synonyms. They *mean* the same thing, but they *are not* the same thing. They are both references to what an Abstract Mind grasps, but spatialization defines how we become aware of an *abstract* "object of thought", whereas the object of thought—not wrapped in quotation marks—is the *experience* of that "object of thought" as a result of the dynamics of the modes of consciousness. Object of thought and "object of thought" coincide. It should be clear that we are *forced* to make this distinction between an abstract entity and our experience of it, the moment we introduce the mind-world dichotomy. The modes of consciousness define the structure of the mental experience, and with it, how the Abstract Mind handles these imaginary entities called objects of thought. The correspondence between an abstract object of thought, and the experience of that object of thought via the dynamics of the modes of consciousness that give rise to spatialization, is used in the construction of the epistemology of actual $\mathcal{M}$-*signatures*, as we will see in Sections 15.5, 16.5 and Section 17.5.

14.4 A Basic Description Of Consciousness

The descriptions of consciousness we give in Part IV are not meant to be comprehensive. They are intended to capture only those nonrepresentational aspects of consciousness that we regard as the most basic and, in a way that we will explain in Chapter 18, the most universal and revealing features of the nature of consciousness. Of course, we know that consciousness has subjective features, and that intentionality appears to have a role in the representational way in which we relate to the world through our thoughts. We regard subjectivity, intentionality and the qualitative and representational character of consciousness as epiphenomenal features of a fundamental reality of consciousness that, we are inclined to believe, can be sufficiently described and understood via two of its most distinctive features: *unity* and *continuity*. We will use this format to describe the experience of consciousness in the three

14.4 A Basic Description Of Consciousness

$\mathcal{M}$-*signatures* we present in this book.

The Unity Of Consciousness As The Coherence Of The Conscious Experience

The unity of consciousness denotes the intrinsic cohesive nature of all forms of conscious experience, and is described in Interaction Theory via the underlying *coherence* of all conscious states. In the model of Abstract Mind described by Interaction Theory, what is commonly referred to as the experience of the *unity* of consciousness is apprehended only intuitively via the metaphysical structure of thought defined by a given $\mathcal{M}$-*signature*, in the exact same way in which an Organizing Principle is comprehended intuitively. A given **concrete** concept defines "unity", and how the mind is able to grasp, experientially, intuitively and preterconceptually, anything. Consequently, the unity of consciousness, like any concept, experience and epistemology, has different *meanings* while being a reference to the same ontologically independent feature of consciousness: the unity of consciousness is, therefore, conveyed by an Actuation Mechanism.

The Flow Of Consciousness As The Cohesion Of All Conscious States

Consciousness is experienced as an uninterrupted flow. The *continuity* of consciousness is what is generally referred to as the flow or stream of the conscious experience. Continuity is a dynamic feature that characterizes, together with unity, the fundamental structure of the phenomenology of consciousness, and that gives rise to the experience of the flow of consciousness. The experience of consciousness never resembles stillness: even in the absence of thought, the unity and stillness of consciousness are perceived as an uninterrupted flow. This apparently counterintuitive notion becomes less impervious to intuition once we notice that the flow of consciousness is an atemporal, nonspatial kind of flow that can be vividly grasped without the need of time and space as representational aids. Consciousness is perceived as an uninterrupted flow via the same metaphysical structure of thought through which we construct our experience of anything else. To comprehend how exactly we come to experience consciousness as a flow, we need a framework to describe how a certain abstract notion of continuity gives rise to the experience of the flow of consciousness. What is continuous in the flow of consciousness? and What is continuity with respect to? To answer these questions, we need to resort to a higher level of abstraction and adopt a $\mathcal{M}$-*signature*-specific characterization of cohesion. Cohesion, as defined in Section 17.1, is in fact the basic structural feature necessary to define continuity, because it defines the dimensions of a certain spatialization of an object of thought along which continuity is defined. From these observations, it follow that while continuity and cohesion have different definitions and phenomenologies in each $\mathcal{M}$-*signature*, they give rise to different experiences of the same flow of consciousness.

The Invariance Of The Synthetic Structure of Intentionality As The Epistemic Principle Of The Conscious Experience

Recall Postulate 8.7.2: the Synthetic Structure of Intentionality is an invariant of consciousness. Since in Interaction Theory the Synthetic Structure of Intentionality model is used to define a conception of intelligibility, it follows that *each* state of consciousness encodes a conception of intelligibility: each state of consciousness represents a type of epistemic access to reality. This idea will be further explored in Part IV where we examine three examples of Symbolic Intuition of Reality.

14.5 The Archetypical Structure Of Communication

Communication is that which elicits a change of an Abstract Mind state.

It is impossible to talk about awareness, knowledge, comprehension and meaning without mentioning their culmination: *communication*. Here we want to sketch the *archetypical structure of communication* in Interaction Theory, which is a blueprint for the forms of communication that occur in an arbitrary Symbolic Intuition of Reality, but also between Abstract Minds in different cognitive states (Definition 11.4.1), and constitutes the essential framework on which any model of *communication* and *computation* is conceived in Interaction Theory.

In Interaction Theory the term *communication* denotes any pro-

14.5 The Archetypical Structure Of Communication

cess that results in a change of a specific Abstract Mind conscious state. Contrary to some the definitions of communication that require the presence a sender that intentionally conveys meaning to a receiver through the use of a shared symbol system, Interaction Theory does not require intentionality, does not require an agent that initiates the communication, does not require a shared symbol system, and doesn't even require a message to be shared. To begin to comprehend why in Interaction Theory communication is defined entirely via the dynamical properties of conscious states, let us ignore for a moment the Abstract Mind, and concentrate on three examples.

1. A waiter forgets a "Reserved" sign on a table. You walk into a restaurant and see the table with the "Reserved" sign. What do you think? Obviously, you think that that table is reserved, even though that is not the reason why the "Reserved" sign is on that table. Here a specific meaning about that table is conveyed unintentionally. It doesn't matter why the "Reserved" sign is on that table, the only element of the transaction between your attention and that table that conveys meaning is that the "Reserved" sign is on that table when you walk into the restaurant. In this example intentionality plays no role, and meaning is conveyed existentially: in space and time.

2. You are ambling on the street absentmindedly when you see or read something. It doesn't even catch your attention, but, suddenly, you remember something important, maybe, something you had been trying to call to mind for a long time. And not only do you remember that thought: that thought is now vividly clear in your mind and is perfectly anchored to the knowledge and awareness you need to reason about it. How can a random event that does not occur *for you* communicate something *to you*, and not to someone else? A random event becomes meaningful to you by virtue of helping you recall a memory, by helping you think of something with unusual clarify and focus. Unlike the "Reserved" sign, this random event doesn't even rely on a shared symbol system to convey meaning: it conveys meaning by virtue of *resonating* with a certain configuration of thoughts and memories in your mind, to which you are particularly receptive in the state of consciousness you are in when that event occurs.

3. You are talking with a friend about a particularly sensitive issue when there is a long silence. Even silence, and in general the complete absence of data or messages, conveys meaning, and can trigger feelings and ideas and memories in whoever is part of that transaction, voluntarily or involuntarily. The absence of uttered words becomes meaningful by virtue of being part of the shared semantic framework that is the conversation between you and your friend: to which you and your friend are *receptive* by virtue of having *tuned* your thoughts and feelings and emotions to the same sensitive topic.

The key concept that we can distill from these examples is that communication, like computation, is entirely *observer dependent*. What elicits meaning and thoughts and feelings and memories in an observer depends entirely on what *that* observer is receptive to at *that* particular moment, by virtue of being in a certain conscious state. Without states of consciousness and Symbolic Intuitions of Reality, the World remains a stage full of data that can only be transferred or acquired but never turned into meaningful messages, let alone, experiences. The turning of data into meaningful messages, or, more pompously, into information, is an observer dependent, subjective manifestation of consciousness and intentionality that consists in a form of *receptive resonance* between data and an Abstract Mind conscious state. We can crystallize this reflection in the following definition.

Definition 14.5.1. (Archetypical Structure of Communication) *Communication* is that which elicits or results in a change of a well-defined Abstract Mind state.

A corollary of these reflections, and of the definition of the archetypical structure of communication just given, is that *it is impossible to not communicate*. In other words, *communication is completely unavoidable*, being an integral part of the phenomenology of consciousness, and is, therefore, part of the unity of the reality that every living creatures incessantly cocreates.

14.6 How Data Becomes Message

Data and message denote respectively the initial and final stages of the supervenience of meaning in a mind-world transaction.

The definition of The Archetypical Structure of Communication we gave in the previous section (Definition 14.5.1), raises two questions: How does communication occur? and, in particular, How does data become message? Based on the definition just mentioned, answering these questions means, respectively, explaining how the transitions between conscious states occur in the Abstract Mind, and, within the context of those transitions, how "something" we call *data* becomes "something else" we call *message*, and how this transmutation constitutes the experience we commonly refer to as the occurrence of communication, and of other kinds of transactions where data becomes meaning without us necessarily becoming aware of it.

As usual in this research, the analysis reaches a point where the terms we borrow from ordinary language become inadequate to express the richness and complexity of the metaphysical realities we are uncovering, and we have to make up for this inadequacy whenever we have the right language and context to do so. The terms we have borrowed in this and in the previous section to introduce an archetypical theory of communication are data and message, and it is now time to explain what these terms denote in Interaction Theory. We have also smuggled an assumption in the wording of a question, for "How does data become message?" suggests that data is transmuted into something different. Why did we not say that data *gives rise to* a message?

To explain "data", we need the description of "Modes Of Consciousness As Spatialization" we gave in Section 14.3, because communication in Interaction Theory is defined via the transitions between Abstract Mind states, which in Section 14.3 are described in terms of the dynamical behavior of spatialization, despatialization and contravariance. To explain "message", we need the notions of protosemantic transformation and semantogenesis (Definitions 12.6.1 and 12.6.2 respectively) we introduced in "Meaning And Semantogenesis" (Section 12.6), because the conditions under which the Abstract Mind changes state are defined by certain configurations of the Epistemology of the Abstract Mind parametrized by *comprehension* and *meaning*.

In Interaction Theory, what we ordinarily call "data" is a certain representation of an object of thought or perception that, as such, becomes intelligible in the same fundamental way in which the Abstract Mind becomes *aware* of an object of thought. Data is, therefore, a reference to a type of spatialization in the context of a *specific* Interaction Model, and denotes the fact that a certain object of thought is part of a specific mind-world transaction described by *that* Interaction Model and its epistemology. Thus, as seen in "The Structure Of Awareness" (Section 14.2), when the mind-world transaction corresponds to a Minimal Interaction Model, "data" means *object of thought*—recall awareness is the self-evidence of an object of thought: the full correspondence of spatialization with epistemology (Definition 14.2.1)—when the mind-world transaction corresponds to a general Interaction Model, "data" means the spatialization where that Interaction Model is defined.

With these ideas in mind, we can think of data and message as two states of a single mind-world transaction, where data is a reference to the type of spatialization that defines the structure of that specific mind-world transaction, and message is a reference to a semantogenesis of a specific Interaction Model for that transaction. The distinction between data and message, therefore, while useful for explanatory purposes, is of a purely linguistic nature: it does not connote any ontological distinction between data and message. We crystallize this observation in the following definition.

Definition 14.6.1. (Data and Message)
The terms *data* and *message* denote respectively the initial and final stages of the supervenience of a semantic framework on the spatialization of a specific Interaction Model.

We proceed now to examine the types of changes of Abstract Mind states, and the corresponding types of communication they give rise to.

Figure 14.2 is a close-up view of the Abstract Mind. The region between dashed lines represents the surface of thought currently lit by consciousness. It is the stage where a Symbolic Intuition of Reality makes its appearance, and signals that in this area the

14.6 How Data Becomes Message

Abstract Mind is aware of concepts according to the epistemology defined by a certain conception of intelligibility—by a certain **concrete** concept. In the subliminal part, the Abstract Mind has various forms of intuitions that continuously bubble into existence and evaporate without being fully conceptualized, and this frantic activity, which we might call "internal dialogue", even though it may occur without words, takes place below the threshold of awareness. In the higher consciousness part, the mind is able to embrace other metaphysical structures of thought, as described by the $\mathcal{M}$-*signature* model and, for this reason, is able to achieve higher types of knowledge—higher with respect to the current Symbolic Intuition of Reality. The term higher consciousness is by no means some kind of overdrive mode. The nomenclature used in Figure 14.2 to denote the three levels of awareness is reflective of a view according to which the normal level of awareness is based on one $\mathcal{M}$-*signature* at a time. This is a simplification made for explanatory purposes only. In reality, the human mind *is indeed* a multi-$\mathcal{M}$-*signature* object that spontaneously and simultaneously functions at all levels of reality.

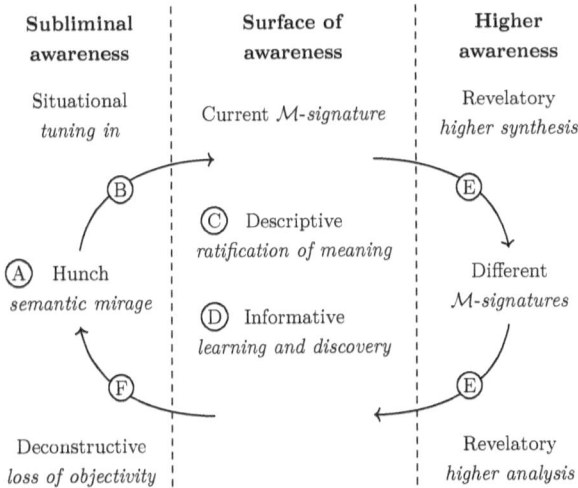

Figure 14.2: The 6 types of communication in Interaction Theory.

Hunches

A *hunch* (Figure 14.2-A) is a transition between Abstract Mind states of the same $\mathcal{M}$-*signature* that does not require or result in the **reconstruction of the success condition** (Section 11.3). Hunches are, therefore, changes between Abstract Mind states that occur outside of the pattern that defines cognitive states—recall the "Classification Of Interaction Models" in Section 11.4. They constitute forms of subliminal, unfinished and unformed communication, such as suggestions, allusions, presentiments, and in general all kinds of unformed intuitions about a particular type of matter or feeling or emotion.

In hunches and presentiments *meaning is not formed* because the Abstract Mind, being in a pre-conceptual or pre-cognitive state, does not have the cognitive structures underpinning the genesis of meaning, which are based, as we have seen in Section 12.6, on a specific equational system that models an instance of **reconstruction of the success condition**. The meaningfulness of a hunch is in its very emergence, which heralds the intuitive and emotional resonance of certain unexpressed mental states, and announces that an unidentified object of thought or perception is appearing on the horizon of consciousness.

Hunches and presentiments produce the experience of *semantic mirages*. In the absence of meaning, the sense of direction imparted to consciousness by the grasping of meaning is replaced by evocations of varying intensity, duration and clarity that may, collectively, give the illusion of an overarching meaning, not so much of the hunch per se, but of the patterns that seem to appear and disappear in the storm of memories and feelings and emotions that the hunch brings to mind. Hunches and presentiments, like mirages in the desert, may signal the presence of real objects that are misplaced by an optical illusion. Hunches and presentiments are, in every regard, and in every meaningful way, conceptual illusions that compel us to recognize them as such in order to reveal their true content and significance.

Situational

Situational communication (Figure 14.2-B) occurs when a transition between Abstract Mind states gives rise to a Minimal Interaction Model. Situational communication is a subliminal type of

14.6 How Data Becomes Message

communication where two important things occur: the Abstract Mind adjusts itself to a specific $\mathcal{M}$-*signature*, and, as a result, becomes *aware* of an object of thought according to the structure of the Minimal Interaction Model for that $\mathcal{M}$-*signature*. Situational communication, therefore, gives rise to the *cognitive states* of an Abstract Mind (Definition 11.4.1).

In situational communication *meaning is formed as objectivation*. Meaning is anchored to the elementary object of thought the Abstract Mind becomes aware of, by tuning in to a specific type of intelligibility, and, for this reason, its content is the very mode of existence embodied by the objectivation that gives rise to the awareness of an object of thought—recall "How Intelligibility Defines Objectivity" (Section 10.6).

Situational communication produces the experience of *tuning in to a specific reality*, and is perhaps most commonly detected through the physiological reactions it triggers, such as the distinctive sensation of getting in the zone, and other sensations associated with mental clarity and calm attention. As the Abstract Mind enters a state of situational perception, the formation of a Minimal Interaction Model is accompanied by an archetypical "internal dialogue"—wrapped in quotation marks because it is entirely intuitive and transcends language and imagination. This transition produces a very distinctive experience that we recognize, first intuitively and then rationally, as a sequence of epiphanies where the mind tunes in to the slice of reality it is becoming aware of.

Descriptive

Descriptive communication (Figure 14.2-C) occurs when the Abstract Mind transitions to and from the same Interaction Model. Descriptive communication is, in general, transient, because the epistemology of an Interaction Model has limited configurations—recall the discussion in "Modes Of Consciousness As Spatialization" we gave in Section 14.3 and the role played by comprehension and meaning—and preludes to two types of more "stable" transitions: back to pure awareness (Minimal Interaction Model), or to an informative state (new Interaction Models).

Descriptive communication *solidifies meaning as norm*. Meaning is or can be reproduced, rediscovered or reconstructed within the language defined by the Interaction Model where the knowledge of the current object of thought is developed. This discovery ascribes to meaning the status of norm, convention, and, to a certain extent, a degree of veridicality by virtue of its clear, immediate usefulness.

Descriptive communication produces the experience of *ratifying the knowledge and comprehension* of an object of thought. This is the experience captured by the notion of epistemic continuity (Definition 13.3.1), it is the feeling, and the intuitive understanding that something is "intrinsically right" about a certain thought or argument, but in a pretercognitive fashion. It is the experience through which, for example, we ascribe to physical reality a mathematical nature each time our mathematical knowledge—where the Interaction Model is the OSP Pattern (Chapter 9)—predictively confirms experience.

Informative

Informative communication (Figure 14.2-D) occurs when the Abstract Mind transitions between distinct Interaction Models within the same $\mathcal{M}$-*signature*. Informative communication is by far the most common type of Abstract Mind state transition, because its structure is, so to speak, a stable point of dynamic equilibrium where the most common cognitive functions are present and accessible in the mind's internal dialogue.

In informative communication *meaning is conveyed relationally or existentially* through the schema of the Epistemology of the Abstract Mind.

Informative communication produces the experience of *learning* according to the metaphysical structure of thought defined by the current $\mathcal{M}$-*signature*, as described in Chapters 12 and 13.

Revelatory

Revelatory communication (Figure 14.2-E) occurs when the Abstract Mind changes $\mathcal{M}$-*signature*, and signals a radical change in the way the intellect functions. Recall that the Abstract Mind does not actually change state like an electrical switch, it activates one or more states simultaneously. The name "revelatory" is meant to suggests the radical nature of this type of communication that does not simply elicit the creation of meaning, it shakes a system of thought at its foundations and changes the purpose and structure

14.6 How Data Becomes Message

of an entire epistemic enterprise. Revelatory communication, for this reasons, relates to the higher-order mind-world transactions where a foundation of human thought is no longer built on one single conception of intelligibility, but on higher-dimensional relations between types of intelligibility.

In revelatory communication *meaning is transcended and transmuted by intuitions and discoveries about the unity and cohesion of knowledge and Nature*. Revelatory communication is about what connects different Symbolic Intuitions of Reality, and, for this reason, gives rise to higher types of knowledge and explanations. Revelatory communication is always relative to a **concrete** concept, and is, therefore, of two types, *synthetic* and *analytical*, depending on the position of the current **concrete** concept in the transition.

Revelatory communication produces the intuitive experience of a *rearrangement of thoughts and beliefs and of the perceived purpose of one's own intellectuality*. It is, for this reason, the most powerful type of communication, because it shakes one's own structure of thought at its core, and can change irreversibly the way one sees, knowns and comprehends the World, and the way one gives meaning to her experiences and her existence.

Deconstructive

Deconstructive communication (Figure 14.2-F) occurs when the Abstract Mind transitions from a cognitive state to a non-cognitive state—a pre-cognitive or pre-conceptual state as described in "Classification Of Interaction Models" in Section 11.4. The name deconstructive, suggests that this transition is somehow related to creativity by virtue of bringing the mind, even if only temporarily, to a blank slate condition of pure potential.

In deconstructive communication *meaning is conveyed through the intuitive disassociation from the type of intelligibility in use*. The abandonment of a cognitive state is signaled by the disengagement of the Abstract Mind from a state of awareness of an object of thought, and, consequently, from the corresponding understanding of its objective status, value and significance in the narrative of thought. The loss of objectivity of an object of thought is a direct consequence of the disengagement of the Abstract Mind from the type of intelligibility at the core of the $\mathcal{M}$-*signature* where deconstructive communication unfolds.

Deconstructive communication produces a range of experiences that can vary from a mild reduction in the belief of a normative view about a certain conception of knowledge, to the erosion or disintegration of the very foundations of thought upon which a certain epistemic project was initiated. The power and intensity and depth and importance of deconstructive communication is comparable to that of revelatory communication. As a matter of fact, deconstructive communication has in many regards an anti-revelatory potential in that it forces to come to grips critically with a system of thought in its entirety.

14.7 Metaphysics Of Language And Computation

A language is a technology to communicate how the World is seen and understood by an Abstract Mind in a cognitive state.

After having defined "The Archetypical Structure Of Communication" (Section 14.5) as that which elicits an Abstract Mind state transition, and having explained "How Data Becomes Message" through a classification of the Abstract Mind state transitions (Section 14.6), we want to explore the archetypical structure of language, because *communication occurs through language.*

By the archetypical structure of language, we mean that which defines its purpose, and the metaphysical structure of its messages. The *purpose* of a language is what a language is designed to convey at the most fundamental level, and beyond the particularizations of grammar, rule systems, or of any systematization or representation or interpretation of its fundamental constituents. The *metaphysical structure of its messages* is that which defines the structure and limits of its expressivity, that is, what a language can intrinsically convey of a World, by virtue of being designed in a certain way.

To explore the archetypical structure of language in a given $\mathcal{M}$-*signature*, we start from the observation that communication occurs through language, and introduce a first, intuitive definition of language that we call *synthetic*, because it is centered on the symbol-meaning tension that manifests itself in the dichotomy of the World *seen* versus the World *understood* by an Abstract Mind.

14.7 Metaphysics Of Language And Computation

Definition 14.7.1. (Language – *synthetic*)
A *language* is a technology to communicate how the World is *seen* and *understood* by an Abstract Mind in a cognitive state.

We can appreciate a first, intuitive example of the archetypical structure of *ordinary language*, by observing the World seen by an Abstract Mind in an Identity-based cognitive state—a state where the **concrete** concept of *identity* is reconstructed (see also Definition 11.4.1). The World seen by an Abstract Mind in an Identity-based cognitive state, is the World where we ordinarily think of things such as making a sandwich, sending a human mission to Mars, paying taxes, gossip, elections and so on and so forth. That World is made of *processes*. Through the processes that we see, we *comprehend* how the manifestations of principles of permanence, difference, constancy, succession, variance, evolution and change give rise to the *identity* of the things we see in the World—we will examine these ideas in detail in Chapter 15. Ordinary languages—called Identity-based languages in Interaction Theory—convey a World made of processes through their fundamentally dichotomic structures, consisting of subject-object, verb-predicate and subordinate clauses, and their messages use spatiotemporal and causal references to convey how principles of permanence, difference, constancy, succession, variance, evolution and change form the identity of the objects in the World.

To explain what connects the seen and the understood to an Abstract Mind cognitive state, we revisit the metaphysical structure of explanations, because explanations, more than any other conceptual artifact, are deeply and firmly rooted in the way a **concrete** concept defines intelligibility and objectivity. We expressed this rationale in "The Metastructure Of Explanations" (Section 13.7), where we observed that

> "*[t]o comprehend the fundamental structure of explanations in Interaction Theory, and to establish the limits of the interpretations of what we see and comprehend when we look at the World through the conceptual lens of a coherent explanation, we need to explore what connects an explanation to the mechanism through which a given **concrete** concept defines intelligible thought, and understand how this nexus defines the conceptual structures that differentiate an*

explanation from any other utterance or thought. "

We expressed this nexus in "How **concrete** Concepts Define Intelligibility" (Section 10.5), and in the diagram of Figure 10.3 we discovered the fundamental way in which we see the World through a **concrete** concept. There the distinction between the phenomenology and epistemology of a **concrete** concept is captured by the notions of Actuation Mechanism and Definitional Model respectively.

In particular, in "The Metastructure Of Explanations" (Section 13.7) we observed that

> "*[w]hat differentiates an explanation from any other expression, is that an explanation is concerned with how a certain entity referenced in the explanans acquires or changes its definitional structure, whereas definitions stipulate that certain entities constitute the spatialization where that Actuation Mechanism exists as phenomenology of the Organizing Principle. It follows that, at the most fundamental level, explanations must necessarily be concerned with how an instance of Actuation Mechanism acquires or changes its definitional structure, or, in more intuitive terms, must be concerned with the genesis of the definitional structure of an Actuation Mechanism.* "

Thus, the fundamental mode of existence of the things in the World is the "seen", and defines the *purpose* of a language. Given a **concrete** concept, what is seen of the World is described by the Actuation Mechanism: its definitional structure determines the basic constituents of the World, and ultimately what *any* language in a given $\mathcal{M}$-*signature* conveys.

The features of existence, that is, how existence is manifested, are the "understood", and define the *structure of messages*. What is understood of the World seen through a **concrete** concept is defined by the Definitional Model, and determines *how* any language in a given $\mathcal{M}$-*signature* conveys the modes and features of existence of the things in World described by its Actuation Mechanism.

With these ideas in mind about the archetypical structure of language, we can connect the seen and the understood to an Abstract Mind cognitive state, and crystallize this relation in the

14.7 Metaphysics Of Language And Computation

following analytic definition of language.

Definition 14.7.2. (Language – *analytic*)
Given a **concrete** concept $\mathcal{C}$ and its corresponding Actuation Mechanism and Definitional Model, a *language for* $\mathcal{C}$ is a method to communicate how the features of existence defined by the Definitional Model constitute the manifestations of $\mathcal{C}$ in a World where the fundamental mode of existence is defined by the Actuation Mechanism.

This brief metaphysical investigation on the archetypical structure of language would be woefully incomplete without the mention of a topic strictly related to communication: *computation*. To redefine computation in the cosmology of Interaction Theory, we begin by examining the metaphysical structure of communication in the Identity-based languages, and then map its fundamental structure in the language of Interaction Theory.

In the ordinary Identity-based language, computation is synonym of *stepwise unfolding of a well-defined process*. Computation and process are, in fact, intimately related notions: they both relate input and output, and they are both based on the assumption that the solution to a certain problem, or the occurrence of a certain event, can be reduced to a well-defined, repeatable, ordered set of steps that, when occurring in a certain order and according to certain rules, gives rise to a certain output. Process and computation are, in essence, two views of what connects causally two entities that we call input and output. When a structured set of well-defined steps and rules that gives rise to a certain output is considered as a united whole (Figure 14.3-A), we talk about *process*, when we consider the execution of those steps according to those rules, we talk about *computation* (Figure 14.3-B).

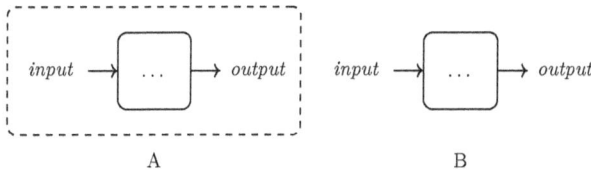

A   B

Figure 14.3: Process and computation.

These initial observations about process and computation raise a fundamental question: Why process? Where does the conceptualization of an objects of thought or a state of affair as process come from? and, most importantly, What does this specific conceptualization assume and imply about the object of thought or state of affair it is about? There is indeed a radical assumption underlying the notion of process. It is the assumption that the answer to a specific problem can be abstracted and synthesized into a solution to solve the same or similar problems, because it is in the nature of that problem to crop up always identical to itself, and, for this reason, the *same answer* will continue to apply to all the future instances of the that problem. It is the idea that certain causes are constant, and that therefore give rise to the same effects under the same initial configuration, and that, for this reason, the processes that connect those causes to their effects are *invariants* of the World. In this sense, process, and, a fortiori, computation, are the expression of the logic leap from *answer* to *solution*, and constitute an important example of an intrinsically non evolutionary worldview.

Answers have no requirement of reusability. Solutions, in contrast, not only must answer a specific question or solve a specific instance of a problem, they must continue to do so by virtue of being based on principles and knowledge that transcend the knowledge needed to answer a specific question or solve a specific problem. In this sense, answers are to solutions what tools are to machines. Solutions are, so to speak, conceptual machines to solve or address certain classes of problems that we are able to conceptualize as processes.

With these reflections in mind, we now turn our attention to the metaphysical structure of process and computation, and reinterpret these ideas in the broader context of the structure of an arbitrary $\mathcal{M}$-*signature*. We note that in the analysis above, process and computation are mere conceptual placeholders whose role is to denote how the World is seen by an Abstract Mind in the ordinary Identity-based cognitive state. Thus, given an arbitrary **concrete** concept $\mathcal{C}$, and in light of Definition 14.7.2, *process* is an example of how the Definitional Model for $\mathcal{C}$ describes a World where the fundamental mode of existence is defined by the Actuation Mechanism for $\mathcal{C}$, and *computation* is the corresponding example of how that description occurs. In other words, computation and language share the same archetypical structure. This is why, for

14.7 Metaphysics Of Language And Computation

example, conversations are often seen as forms of computation, because they consist of sequences of exchanges that, if handled constructively and intelligently—these are the rules of verbal computation—result in the achievement of a shared goal.

We distill these observations in the following definition which is based on the analytical definition of language (Definition 14.7.2):

Definition 14.7.3. (Computation – *metaphysical*)
Given **concrete** concept $\mathcal{C}$, a *computation in* $\mathcal{C}$ is a model of language for $\mathcal{C}$.

We conceptualize and comprehend computation in the ordinary language as a certain set of steps and rules to achieve a specific goal *because* the World we see through the **concrete** concept of *identity* is made of *processes*. What Definition 14.7.3 tells us is that the same relation between Definitional Model and Actuation Mechanism holds in any **concrete** concept, and that therefore, once the mode of existence of the things in the World seen through a given **concrete** concept is known, language and, a fortiori, computation, have *necessarily* a certain archetypical form, being the expression of the ontological relation between the Definitional Model and Actuation Mechanism for that **concrete** concept.

For example, in a World where the mode of existence—the seen—is atemporal and nonspatial we think in terms of canons and shapes, not processes, and the features of existence—the understood—are, therefore, of a relational kind that transcends and generalizes causality in many ways—we will see in detail this mechanism in Part IV in the description of non-standard Symbolic Intuitions of Reality. To solidify the core ideas we have discovered in this section about the deep structure of language and communication, we use the visual metaphor of the light spectrum (Figure 14.4), to shows how the World is seen and understood depending on what **concrete** concept defines what's intelligible.

14 Epistemology Of The Abstract Mind III

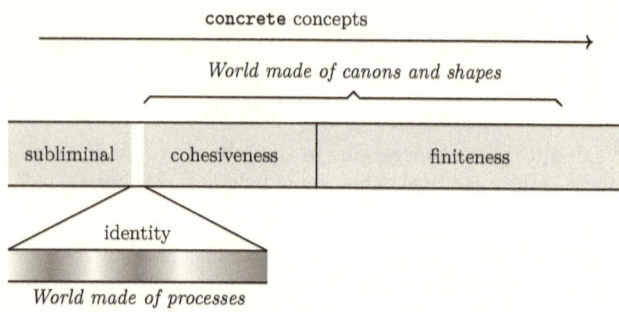

Figure 14.4: Illustration of the archetypical linguistic representations of the World as a spectrum of light.

14.8 Conceptions And Foundations Of Mathematics

A conception of mathematics is an intuition, intrinsic in the Symbolic Intuition of Reality, about how what we see explains what we comprehend.

In "The Common Mathematical Intuition Of Reality" (Chapter 9), we saw how the ordinary human mind's ability to mathematize is described by an Interaction Model of the Identity-based *M-signature* called the OSP Pattern. This raises some natural questions: How should we think of our ability to mathematize in the context of Interaction Theory? Is mathematics as we know it today the manifestation of the only possible type of mathematical intuition of reality? Can we define mathematical thought in other *M-signatures*?

These questions, and several other permutations of the theme of interpreting mathematization in the context of an arbitrary *M-signature*, are not questions about mathematization and mathematics: these are questions about a *conception of Mathematics*.

This book is called "Toward A New Conception Of Mathematics", and not "Toward A New Foundation Of Mathematics", precisely for this reason, because in Interaction Theory we first need to define what we mean by the term "mathematics" in the context of a specific *M-signature*, and only then can we think of

14.8 Conceptions And Foundations Of Mathematics

how to *construct* one: only then can we think of a *foundation* of mathematics.

One way to derive a conception of mathematics in Interaction Theory that generalizes the ordinary conception of mathematics along the dimensions of thought introduced by Interaction Theory, that is to say, a conception of intelligibility in the form of the $\mathcal{M}$-*signature* model, and a conception of knowledge in the form of the Cognitive Architecture model, consists in thinking of mathematics as a *language*. One of the key ideas we presented in "Metaphysics Of Language And Computation" (Section 14.7), is that a language (Definition 14.7.2):

> "*is a method to communicate how the features of existence defined by the Definitional Model constitute the manifestations of [a* `concrete` *concept] C in a World where the fundamental mode of existence is defined by the Actuation Mechanism.*"

Embedded in this definition is the mysterious tension between what we see (Actuation Mechanism) and what we comprehend (Definitional Model) that shapes our intuition of reality, and that forms our mental attitudes toward the pursuit of knowledge as a mission to reconcile symbol and meaning. We see the World through the lens of a `concrete` concept that defines our Symbolic Intuition of Reality, and we comprehend what we see according to the fundamental structure of knowledge for that specific Symbolic Intuition of Reality, which is based on a Minimal Interaction Model built around the Definitional Model.

With these ideas in mind, if we revisit what we discovered about the foundations of mathematics in the "Generative Lens" (Section 4.1), and examine the basic structure of the concepts used as foundations of mathematics, a pattern begins to surface. The basic concepts used as a foundation of modern mathematics: set (ZFC), category (ETCS and AST), etc. represent a medium to convey how what we *see* in the World through the concept of identity corresponds to what we *comprehend* of it.

In the Identity-based $\mathcal{M}$-*signature*, the World we see is made of entities that evolve in time and space, that is, the World is made of *processes*. We comprehend those processes through the notion of *change*, the identity of the things we see in the World. For example, the process $A \to B$, *explains* how B acquires its identity.

To convey this fundamental intuition about how what we see explains what we comprehend, we use, for example, a gadget called *intuitive set*, or *naive set*, which we can imagine as a "bag of points", together with some natural operations, the intuitive operations on sets such as union, intersection, etc. By explaining *change* via the natural operations on sets, we are able to use what we see (processes) to develop our knowledge of a World where things acquire their ontological status through change.

This reflection on the relation between an intuitive notion of set and the fundamental structure of Identity-based intelligibility, leads us to a turning point of this analysis: we have reduced the problem of reinterpreting the modern conception of mathematics in the context of Interaction Theory to the problem of explaining the relation between Identity-based intelligibility and the archetypical notion used as a foundation of mathematics, namely, a set—recall "The Deep Structure Of The Concept Of Set" (Section 9.8).

But in "Epistemic Continuity, Truth And Reality" (Section 13.3) we examined this very problem from the perspective of explanations. There, instead of talking about foundations of mathematics we were talking about the more abstract relation between explanandum and explanans. What connects intelligibility to a foundation of mathematics, explained in terms of the dichotomy between symbol and meaning, is a correspondence of epistemic continuity. Let us materialize this idea in the following:

Definition 14.8.1. (Conception of mathematics)
A *conception of mathematics* is an intuition about the existence of a fundamental relation of epistemic continuity, intrinsic in the mechanism through which a **concrete** concept defines intelligibility, by means of which what we see (Actuation Mechanism) explains what we comprehend (Definitional Model).

Thus, in a given $\mathcal{M}$-*signature*, "mathematics" denotes the fundamental intuition about what connects symbol and meaning according to a view of the World defined by a certain type of intelligibility. It follows that:

Definition 14.8.2. (Foundation of mathematics)
A *foundation of mathematics* is a language (Definition 14.7.2) that models a conception of mathematics (Definition 14.8.1).

14.8 Conceptions And Foundations Of Mathematics

A foundation of mathematics, therefore, is at its core a language to convey a fundamental relation of epistemic continuity, a relation that is necessarily built upon the models of the Definitional Model.

Outline Of A Foundation Of Mathematics

Here we use the definition of a foundation of mathematics in an arbitrary $\mathcal{M}$-*signature* we have just given (Definition 14.8.2) to sketch the steps to construct a foundation of mathematics. We will follow these steps in Part IV to sketch $\mathcal{M}$-*signature*-dependent conceptions and foundations of mathematics.

- We identify the qualities of the mode of existence defined by the Actuation Mechanism associated to a given **concrete** concept.

- We identify a *prototypical concept* that models those qualities. This is the concept upon which we begin to develop a conception of mathematics.

- A conception of mathematics is an intuition about the existence of a fundamental relation of epistemic continuity, intrinsic in the $\mathcal{M}$-*signature* defined by the given **concrete** concept, by means of which what we see in the form of (manifestations of the) Actuation Mechanism explains what we comprehend as (models of the) Definitional Model.

- A foundation of mathematics is a language to model of the intuition at the basis of a conception of mathematics based on the prototypical concept defined above and on its natural operations—operations that preserve its definitional integrity.

Bibliographical Notes

The range of subjects that converges in the definition of consciousness I give here is vast, and as I write this bibliographical note I realize how many sources and reflections and discussions I do not remember. Superficially, there are various themes of the Psychology of Invention, Psychology of Communication, Cognitive Psychology, and, to a certain extent, the themes motivating the tension between psychological nativism and empiricism. The references on these subjects are the classics, I mention only Watzlawick, Bavelas, and Jackson 1967.

14 Epistemology Of The Abstract Mind III

The bulk of the work presented in this chapter is in Lo Vetere 2015a, where I put forward the idea that the modes of consciousness are archetypical forms of spatialization, and in Lo Vetere 2013c, where there is an embryonal idea of the generalized models of computation based on the dynamical properties of the Abstract Mind introduced here.

But as I dig deeper into the relations between the structure of spatialization and the modes of consciousness I describe here, I am forced to come to grips with the themes of a domain that is neither philosophical nor mathematical nor psychological: it's mystical, and in a brutally practical way. Consciousness is the very fabric of human existence. And in this research I have certainly been indirectly influenced in more ways than I can remember, by my interpretations of Buddhist psychology and its sophisticated cosmology of the mind, by various philosophical texts of the Hindu tradition, and by a host of treatises on the fine structure of the mind.

PART IV

Three Symbolic Intuitions Of Reality

What does a mental reality look like?

Introduction

> Knowledge is indivisible. When people grow wise in
> one direction, they are sure to make it easier for
> themselves to grow wise in other directions as well.
> On the other hand, when they split up knowledge,
> concentrate on their own field, and scorn and ignore
> other fields, they grow less wise - even in their own
> field.
>
> Asimov 1983

> Single-mindedness is all very well in cows or
> baboons; in an animal claiming to belong to the
> same species as Shakespeare it is simply disgraceful.
>
> Huxley 1929

In this part we use the framework introduced in Part III to construct three $\mathcal{M}$-*signatures* and their corresponding Symbolic Intuitions of Reality.

Each Symbolic Intuition of Reality constitutes the basic structure of a self-contained foundation of human thought where it is possible to define a conception of Mathematics and an epistemological project. For this reason, as it will become apparent in the next three chapters, the vastness of the subjects involved in the description of just a single Symbolic Intuition of Reality is such that a choice had to be made as to what to emphasize. I chose to emphasize the *experiential* aspects of a Symbolic Intuition of Reality to make it easier to comprehend what the World looks like from some of the radically unintuitive perspectives that we will discover.

Less emphasis has been put on the mathematical representations and implications of the Symbolic Intuitions of Reality.

Key Concepts Of Part IV

1. The World we see through the lens of *identity* is made of processes, and we comprehend it by division. It is a World where objectivity is defined by principles of permanence, difference, constancy, succession, variance, evolution and change, where things exist as change, and where the qualities of existence are multiplicity, variety, range and diversity.

2. The World we see through the lens of *finiteness* is made of encompassments, and we comprehend it by inclusion. It is a World where objectivity is defined by principles of addition, subtraction, inclusion, exclusion, finitude, limitlessness and encompassment, where things exist as forms of proximity, and where the qualities of existence are multitude, gathering, conglomeration and aggregation.

3. The World we see through the lens of *cohesion* is made of morphogeneses, and we comprehend it by similarity and structural continuity. It is a World where objectivity is defined by principles of similarity, analogy, homomorphy, uniformity, wholeness, heterogeneity, affinity, chirality and symmetry, where things exist as forms, and where the qualities of existence are homogeneity, uniformity, regularity and homology.

Chapter 15

Identity-based $\mathcal{M}$-*signatures*

One can never know for sure
what a deserted area looks like.

George Carlin

The Identity-based $\mathcal{M}$-*signature*, also called the *default $\mathcal{M}$-signature*, because it gives rise to the ordinary mode of thought, defines the metaphysical structure of thought in which you drive your car, send a human mission to Mars, pay taxes, prepare a speech, fix your toaster and define Mathematics as we know it today. When the human mind functions in the default metaphysical structure of thought, we see the world through hierarchies of concepts that are the evolution of a single, irreducible Symbolic Intuition of Reality: we call this intuition *distinctiveness*.

To introduce the Identity-based $\mathcal{M}$-*signature*, we follow the outline described in Section 10.7. We begin with a description of the **concrete** concept of identity, and characterize the basic structure of Identity-based thought, which in Interaction Theory is called Identity-based intelligibility. The mind-world transactions that give rise to the Symbolic Intuition of Reality of distinctiveness are captured by the Minimal Interaction Model. The chapter culminates with the description of the Identity-based epistemology, which is a characterization of the experience of reality created by the Symbolic Intuition of Reality of distinctiveness, and captured by the meaning of the terms knowledge, comprehension, meaning,

15 Identity-based $\mathcal{M}$-signatures

explanation and consciousness. One of the key aspects of this chapter, is that it is meant to be interpreted literally. More precisely, it is a description in which there is a faithful correspondence between the elements of the Identity-based *M-signature* that we use to describe the Identity-based *M-signature* in the language of the Identity-based *M-signature*, and the entities of the Identity-based *M-signature* described in this way. This fact, which might seem a triviality or a vicious circle, is neither trivial nor vicious in a logical sense, and masks many practical and theoretical subtleties that will emerge vigorously in the description of the two non-standard *M-signatures* (Chapters 16 and 17), where the faithful correspondence between language and elements of an *M-signature* radically falls apart, and requires the interpretation framework of Interaction Theory to be understood correctly.

The origin of this faithful correspondence between the entities being described, and the elements of the language used to describe those entities, is to be sought in the relation between *identity* and *spatialization*. On the one hand, by spatializing an object of thought, we ascribe to that object the properties of a space made of distinct and definite entities. On the other hand, to identify each one of those distinct and definite entities, we rely on an Organizing Principle (Section 10.5) of distinctiveness. It is, in other words, thanks to the metaphysical intuition of distinctiveness that the human mind—and consequently the Abstract Mind in its default state—is able to discern each entity of a space as a distinct and definite entity. This is why in the non-standard *M-signature*, where identity is not available as a **concrete** concept, but only as an **abstract** concept, the ordinary notion of spatialization needs to be revisited and expanded, and this process has, as we will see, profound epistemological consequences.

15.1 What Is Identity?

One way to characterize identity is to say that it is a concept used to indicate that two things are the same. This definition is, as it will emerge from this discussion, a very rough approximation to what identity really is, because it hides identity behind a definition of sameness, which begs for questions such as In what way are two things the same? which lead to a vast array of philosophical problems, such as the dimensions along which we should study identity—e.g. quantitative, qualitative, ontological, and in general

15.1 What Is Identity?

any dimension consistent with the type of spatialization we adopt to represent the objects of thought to which we apply a notion of identity. Examples of the conceptual labyrinths that originate from these approaches to the study of identity—which I think it is fair to say, belong to the classic and modern philosophical traditions—range from Leibniz's indiscernibility of identicals—if A is identical to B then everything true about A is true about B—and its converse, the identity of indiscernibles—if everything true about A is true about B, then A is identical to B, to glamorous puzzles such as the Ship of Theseus—Is an object still identical to itself after we replace all of its parts? The list goes on with increasing levels of abstractions created by the different types of spatializations given to identity.

We have indicated two approaches to tackle the problem of defining identity. When in Chapter 5 we discussed the metaphysical nature of concepts, the notion of identity served as an example to introduce the framework of definitional interference. There, we gave, implicitly, a definition of identity as a certain conceptual placeholder for whatever definitional interference pattern preserves the definitional integrity of a given object of thought.

The framework of definitional interference, however, is too abstract to provide the practical characterizations of concepts we need in Interaction Theory, because it is designed to characterize *any* concept: its relevance is primarily theoretical. So in Section 10.5 we introduced a coordinate system—Organizing Principle, Actuation Mechanism and Definitional Model—to characterize the concepts we are interested in in Interaction Theory.

The leitmotif about the description of **concrete** concepts we examine here and in the next two chapters, is that **concrete** concepts are the extension to the mental world of physical acts. With these ideas in mind, the definition of identity we want to give here is: *identity is the extension to concepts of the physical act of grabbing an object.*

To begin to understand why we invented identity and how its pervasive use in practically evey mental process shapes our ability to think, we need to look, once again, at how the human mind works: we need to look at two examples of phenomenology of identity. The first example is about our own ability to perceive thoughts, the second example is about our ability to elicit thoughts through language. An intuitive observation about the phenomenology of identity, is that *any thought is invariably defined by the distinct presence of the object the thought is directed at.* The archetypical experience of identity is a fundamental state of consciousness that consists of all those mental states in which we discover our own thoughts as distinct and definite entities. It is the intuition that there is an "I" observing those thoughts, that those thoughts are distinct and separate from the "I" observing them, and that that "I" participates somehow in the evolution of those thoughts. This experience occurs in the preconceptual and pre-linguistic stages of psychological evolution, and is associated—and in part masked by—the physical sensations triggered by thoughts. What's remarkable about thoughts, as a manifestation of identity, is that our experience of them is never exclusive, and this is, perhaps, one of the most revealing clues about the nature of identity and about the Identity-based metaphysical structure of thought. This example shows that identity has an intrinsic character of *duality*.

The duality of identity is revealed when we observe how identity works, at a cognitive level, in communication. In written and spoken communication, the subject-object dichotomy that we experience in an archetypical form when we think, becomes a communication strategy whenever we delegate the completion of the subject-object construction to the reader or to the listener. This happens in certain forms of poetry, where the sentences are left incomplete by breaking the subject-predicate structure in order to trigger the reader's imagination, or in koans. The same techniques are used in mass communication, and especially in advertising, where vagueness, and very often the lack of an object or a subject, are used to elicit in the potential customers select associations of ideas with the product or service being advertised. This example shows the tendency of the mind to "fill the gap" left in a thought whenever we break duality.

To penetrate the nature of identity that begins to surface from these first observations about its phenomenology, we need to examine the type of intelligibility that originates from identity.

15.2 Identity-based Intelligibility

Identity-based intelligibility is a structure of thought in which the human mind has epistemic access to the world via a principle of distinctiveness, which extends to concepts the physical act of grabbing an object.

The definition of Identity-based intelligibility that follows, forms the conceptual link between the philosophical and psychological domains, where we have developed a framework to explain the workings of the Abstract Mind, and the mathematical domain, where we will begin to translate philosophical concepts into mathematical structures.

In the Identity-based mode of thought, the **concrete** concept of identity defines the fundamental structure of thought through which an intellect modeled by the Abstract Mind interacts with a reality. This is the general pattern described in Section 10.5, which applies to any $\mathcal{M}$-*signature*: it is the pattern through which a **concrete** concept defines intelligibility, and with it, an entire structure of thought, and an entire mind regarded as its own capacity to deliberately create thoughts. We may choose to believe—but not doing so has no influence on the line of reasoning we are presenting here—that reality is intelligible to an Identity-based mode of thought as a result of having a certain degree of mutual intelligibility with the Identity-based structure of human thought. This can be already considered a form of panpsychism—direct panpsychism, reality has some features of the mind, or reverse panpsychism, the mind has some features of physical reality—which we will not explore here. In general, every discourse about $\mathcal{M}$-*signatures* and intelligibility leads to various nuances of panpsychism, as does physicalism.

15 Identity-based $\mathcal{M}$-signatures

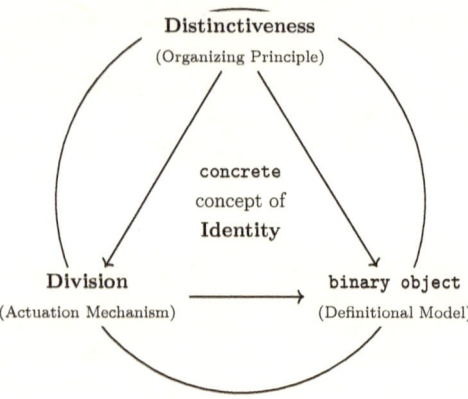

Figure 15.1: The structure of Identity-based intelligibility.

The next three sections describe with the tools of Organizing Principle, Actuation Mechanism and Definitional Model how the ordinary Identity-based mode of thought originates from the **concrete** concept of identity, and the corresponding mechanism through which a notion of **success** is constructed.

Organizing Principle Of Identity

*Distinctiveness is an unintelligible metaphysical intuition about the potential of the **concrete** concept of identity.*

The potential expressed by identity is *distinctiveness* (Figure 15.1). Two intertwined facts define identity: an object that is abstractly identified acquires a quality of distinctiveness, and there is no intelligible notion of identity that does not ascribe to the object being identified a quality of distinctiveness. Identity and distinctiveness always and only coexist: they have the same mode of existence but different qualities. The very essence of identity is to make that which acquires an identity distinct—but not necessarily separate—from other entities. Identity loosely defines *other* by defining *that*.

15.2 Identity-based Intelligibility

Actuation Mechanism Of Identity

*Division defines the phenomenology of
distinctiveness, and distinctiveness becomes
intelligible to the human mind as division.*

In the Identity-based mode of thought, the Actuation Mechanism describes how an Organizing Principle of *distinctiveness* manifests itself as *division* (Figure 15.1).

Distinctiveness becomes intelligible and epistemically accessible via the notion of division. The separation of the objects of thought from one another defines the mental experience of distinctiveness, which we might improperly refer to as the experience of the "identity of those objects". The project of describing the phenomenology of distinctiveness, is the project of giving an account of how division becomes an archetypical form of identity. This is, therefore, yet another project at the intersection of philosophy and mathematics. So far, the discussion has covered the philosophical aspects of this project by stating the ontological status and the epistemological relations between the Organizing Principle and the Actuation Mechanism. To describe the mathematical part of this project, we need to explain how the Actuation Mechanism of division is codified in the zoo of mathematical concepts that represent the various manifestations of identity. To this end, it is useful to review briefly the mathematical forms of the philosophical concept of identity.

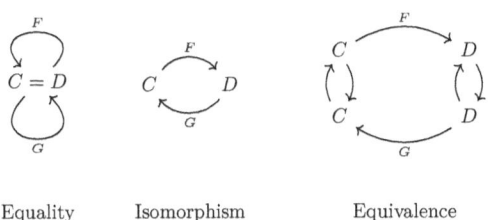

Equality   Isomorphism   Equivalence

Figure 15.2: Examples of mathematical representations of the philosophical notion of identity.

The mathematical representations of the abstract concept of identity, originate from various ways to interpret and generalize this concept. The mathematical forms of *equality* capture a strict ontological correspondence between mathematical entities. In Figure 15.2, the = sign in $C = D$ signifies that "everything about C corresponds to everything about D"—which is a hugely useless approach in theoretical mathematics and in most practical applications—where "everything" is defined within the same definitional framework in which C and D are defined. There is the mathematical notion of *isomorphism*, which expresses the idea that it is possible to travel from C to D and return to C, in such a way that the beginning and the end of the trip satisfy a certain relation *defined* in C that does not necessarily require two objects to be equal, because their being both in C and satisfying a certain relation generalizes equality—in the context defined by C satisfying that relation is practically indistinguishable from identity. There is the mathematical notion of *equivalence*, which expresses the idea that two objects C and D can be replaced by one another in a defined set of contexts. There are more sophisticated versions of equivalence called *adjunctions*, which extend and refine the idea of two objects behaving in the same way in a set of specific contexts, and that we will examine later.

Definitional Model Of Identity

A `binary object` defines the epistemology of distinctiveness by modeling division.

The Definitional Model of *division* is a philosophical device called **binary object** (Figure 15.1) that codifies the prototypical structure of **success** for the metaphysical structure of thought defined by the Identity-based *M-signature*.

The **binary object**:

- is a *form of awareness*—as seen in "The Structure Of Awareness" (Section 14.2),

- defines the structure of (de)spatialization in the Identity-based *M-signature*—see Definition 14.2.1 and "Consciousness And Spatialization" (Section 14.3),

- is used in Interaction Theory as a reference to formal models of *division*,

- forms the basis of *any* construction of the **success condition** in the Identity-based $\mathcal{M}$-*signature*—see also Section 8.3,

- is defined by the following type of *generative knowledge* (see Section 12.1) $\{\bullet, \rightarrow\}$,

- constitutes the core of the Minimal Interaction Model for the Identity-based $\mathcal{M}$-*signature*.

The name **binary object** is meant to evoke the intrinsic dichotomy at the basis of the structure of the Identity-based mode of thought. A structure of thought defined by identity is entirely determined by its intrinsic impulse, necessity and attitude of the mind to detach experience from thought, knowledge and understanding. Paradoxically, the Identity-based thought connects us to reality by separating us from the reality is it meant to connect us to. This fundamental characteristic of the Identity-based structure of thought has a very intuitive analogy with two common and antithetical human experiences: knowing and feeling. Knowing rationally is mostly a manifestation of Identity-based thought. Feeling is a type of direct, intuitive knowledge based on communion rather than identity: it really does not involve of rely on any form of distinction.

Identity-Based Generative Knowledge

Generative knowledge in the Identity-based $\mathcal{M}$-*signature* consists of the set of archetypes $\{\bullet, \rightarrow\}$. We find these elements in the representations of **binary object**.

The representation of **binary object** we introduce here is called the **identiton**.

15.3 The **identiton**

The **identiton** is a *form of awareness*—as seen in "The Structure Of Awareness" (Section 14.2)—that we *describe* as an atomic and irreducible pattern that models the notion of *division*—the Actuation Mechanism of *distinctiveness*—and that defines the *structure of spatialization* in the Identity-based $\mathcal{M}$-*signature*.

15 Identity-based $\mathcal{M}$-*signatures*

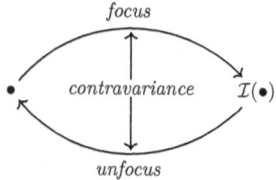

Figure 15.3: The **identiton**

Compare Figure 15.3 with Figure 14.1 in "Consciousness And Spatialization", where the **concrete** concept is $\mathcal{I}$: *identity*. The form of awareness codified by an **identiton** is a set of "sleights of mind" that symbolize the irreducible modes in which the conscious mind operates at a fundamental level in the Identity-based $\mathcal{M}$-*signature*. These modes of operation are: *focus, unfocus,* and *contravariance*.

To introduce these three sleights of mind, we interpret the diagram in Figure 15.3 as follows:

- the • represents an *object of thought*,

- $\mathcal{I}(\bullet)$ is a one-vertex diagram that represents the *spatialization* of the object of thought •,

- the arrows • $\to \mathcal{I}(\bullet)$ and $\mathcal{I}(\bullet) \to$ • connect • and $\mathcal{I}(\bullet)$ according to a set of patterns that are mathematical representations of the philosophical notion of identity,

- the arrow between arrows called "contravariance" denotes a certain symbolic operation to switch to and from • $\to \mathcal{I}(\bullet)$ and $\mathcal{I}(\bullet) \to$ •.

The structure of this correspondence codifies the philosophical notion of **binary object**, the Definitional Model of identity, through a variety of mathematical forms of sameness—like equality, isomorphism, equivalence, adjunction, fibration, homotopy, cobordism, etc. and their generalizations. These forms of mathematical sameness are expressed diagrammatically via the structure and properties of the arrows between • and $\mathcal{I}(\bullet)$. To examine this correspondence, it is therefore necessary to interpret $\mathcal{I}$. Thus, we ask: What is the difference between an object of thought •, and a one-vertex diagram $\mathcal{I}(\bullet)$?

15.3 The **identiton**

Interpretation Of $\mathcal{I}$

By turning an object of thought into a one-vertex diagram we reframe the abstract notion of "object of thought" in the frame of reference of perception, consciousness and intentionality.

Figure 15.4: Examples of one-vertex diagrams

One-vertex diagrams (Figure 15.4) are diagrams with one vertex and an arbitrary number of arrows to and from that vertex. On the one hand, an object of thought • is an abstract reference to an ontologically subjective entity that exists in one's consciousness. On the other hand, the *same* object of thought becomes part of the intentional cognitive processes that create the *experience* of that object of thought $\mathcal{I}(\bullet)$ via the changes in our perception of that object that—in Interaction Theory—we call the *awareness of that object of thought*.

Adjoints And Identity Through Change

The concept we introduce here informally, called *adjunction*, is center stage in modern mathematics, and is necessary to understand how the **identiton** defines its own version of spatialization. Adjunction is a construction of Category Theory that generalizes *equivalence*—recall the discussion about equivalence in the Actuation Mechanism (Section 15.2)—which means that it is a way to characterize the conditions for which two mathematical gadgets are *interchangeable*. The core idea of adjunction is to codify a conception of *identity through change*, which generalizes the philosophical notion of identity by describing not so much what identity *is* abstractly, but how it *becomes* a form of identity by *functioning* in a certain way in a certain context. To get a feel for how this idea is captured mathematically, let's consider the diagram in Figure 15.5, where **C** and **D** are certain entities made of objects connected by arrows, where objects and arrows have a certain meaning in **C** and **D** respectively that is irrelevant in this discussion. To define the conditions for which we can think of **C** and **D** as interchangeable entities, we want to be able to say

15 Identity-based M-signatures

things like "there is a certain correspondence between what we can do in **C** and what we can do in **D**". The things we can do in **C** (or **D**) are expressed by the *arrows* between objects in **C** (or **D**). The metaphor of arrows as "things we can do" or "tasks" or "processes" will be useful very soon.

Let's imagine now to have defined two sets of rules, F and G, which allow us to assign objects and arrows of **C** to objects and arrows of **D** respectively, and objects and arrows of **D** to objects and arrows of **C** respectively. To identify what rule is used, we prepend the name of the rule to the object or arrow to which we apply the rule. For example in Figure 15.5, if we pick an object C of **C** and apply F to it, we denote with FC its corresponding object in **D**. Similarly, if we pick an arrow q in **D** and apply G to it, we denote with Gq its corresponding arrow in **C**.

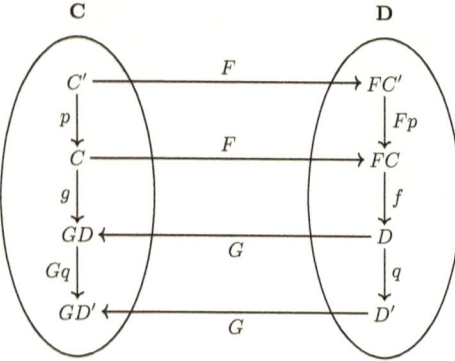

Figure 15.5: The structure of an adjunction (in Category Theory).

With these ideas in mind, we say that the rules F and G are an *adjunction*, that is, under these rules **C** and **D** are interchangeably equivalent, if what is *conducive* to a certain task g in **C**, corresponds under F to what is *conducive* to a task f in **D**, and what *derives* from a task f in **D** corresponds under G to what *derives* from a task g in **C**. The conditions of "being conducive to" and "derive from" can be expressed synthetically as follows: the type of correspondence between f and g (see Figure 15.5), is such that whenever in **C** a certain task p is conducive to g, the combination of tasks Fp followed by f in **D** corresponds via G to

15.3 The `identiton`

the combination of tasks p followed by g in **C**. Similarly, when a task q of **D** derives from f, the composite task f followed by q corresponds under G to the compositions of tasks g followed by Gq in **C**. When F or G exist, we say that F is *left adjoint* to G, and similarly that G is right adjoint to F.

Another useful way to think of F and G is as one as the best approximation to the other, and vice versa. That is to say, G as the best approximation to the set of rules that undo whatever F does, and similarly F as the best approximation to a the set of rules that undo what G does.

The Structure Of The `identiton`

In the next three sections we offer an overview of the three modes of operation of the `identiton`. This description is meant to be interpreted as an example of "The Abstract Mind-as-sound Analogy" we presented in Section 11.5, where the analogy is between the `identiton` and the sound produced by the buzzer circuit. Focus, unfocus and contravariance are a way to describe the basic, irreducible, atomic experience of the Abstract Mind, in an Identity-based state coming in contact with an object of thought.

Spatialization As The Experience Of Focusing On An Object Of Thought

Spatialization is the "focus" sleight of mind, and its mathematical analogy are certain generalizations of the category-theoretical notion of left adjoint.

Focus is the way in which an Abstract Mind in the Identity-based state grasps an object of thought. The term focus is used here literally to evoke the capacity to single out an object of thought from a background of mental perceptions where the mind-world transactions take place. Focus is at the core of the phenomenology of identity. It is the experience of the human mind coming in contact with an object of thought, and producing the experience of *that* object of thought as a result of this particular type of interaction with that object.

The act of focusing on an object of thought is the precursor event of the mental experience of coming in contact conceptually with an object of thought—what is commonly referred to as the experience of an object of thought, or the grasping of a thought. The mathematical construction of left adjoint becomes a representation of this particular mode of the mind by characterizing the conditions of (left) interchangeability between an object of thought •, and the mental experience of that object $\mathcal{I}(\bullet)$. These conditions of interchangeability are of course the conditions on the arrows described in the previous section. The structure of the Identity-based spatialization is, therefore, crystallized by the existence of a left adjoint, which defines how we can think of a certain entity in terms of another entity, but not necessarily the other way around. This condition of half-interchangeability is a refined form of identity which is not symmetrical—by analogy, it is *as if* a left adjunction defined certain conditions for which B=A does not follow necessarily from A=B, thus generalizing in this way the naïve symmetrical notion of identity, and many of its generalizations we encounter in the philosophical literature on the subject. The definition of the conditions of (left) interchangeability between and object of thought and the experience of it are, therefore, the structure of Identity-based spatialization.

Despatialization As The Experience Of Neutral Perception

Despatialization is the "unfocus" sleight of mind, and its mathematical analogy are certain generalizations of the category-theoretical notion of right adjoint.

To unfocus is meant here to convey the experience of returning to a state of neutral perception, in which the mental experience of thoughts is an unbiased, vividly unfocused awareness. Unfocus symbolizes the movement towards the initial state of awareness from which each perceptual experience originates, but it is *not*, in itself, the particular state of awareness it can be. The characterization of this mode of the mind via a right adjoint codifies the conditions for which a given mental experience $\mathcal{I}(\bullet)$ of an object of thought • becomes interchangeably intelligible with the object of thought from which it originates. As pointed out in the introduction to these three sleights of mind, the `identiton` is an atomic, irreducible process that captures the basic structure of the Identity-based mind-world transactions. Despatialization is

15.3 The `identiton`

meant to be understood as an illustration of a feature or quality of this atomic process rather than as an actual part of it. Hence, the distinction between the transition to and from an object of thought • and the experience of it $\mathcal{I}(\bullet)$ is purely descriptive, not ontological.

Contravariance Is The Mental Perception Of An Object Of Thought By Division

Contravariance—in fact any contravariance we encounter in the treatment of $\mathcal{M}$-*signatures*—is the closest thing to a process we can imagine, but unlike normal processes that are made of intermediate steps, contravariance is atomic and irreducible. In this sense, contravariance is an interpretation of the **binary object** as a process. In particular, Identity-based contravariance is the atomic process of grasping a concept by division. It is the sound, the acoustic signature of a given $\mathcal{M}$-*signature*—recall the analogy used in the introduction to these sleights of mind. In the Identity-based $\mathcal{M}$-*signature*, contravariance denotes the completion of an adjunction, where left and right adjoint define a full correspondence between an object of thought • and the mental experience of the same $\mathcal{I}(\bullet)$. In this sense, Identity-based contravariance is also the most intuitive and most basic form of awareness—recall Definition 14.2.1. The sound analogy gives us also another insight into the function of Identity-based contravariance. The continuous transition between spatialization and despatialization symbolized by contravariance can be thought as the ability to progressively develop a concept through an iterative process in which, starting from a rudimentary form of identity, the concept is explored and comprehended. To understand why, we note that the condition for an Identity-based (de)spatialization to come to an end is diagrammatically the same—same in category-theoretical terms—as the condition of understanding (Section 12.5 and Section 14.1).

Definition 15.3.1. (`identiton` – *informal*)
A `identiton` is the pattern defined in Figure 15.3 where $\mathcal{I}$ denotes a one-vertex diagram and $\bullet \to \mathcal{I}(\bullet)$ and $\mathcal{I}(\bullet) \to \bullet$ denote respectively category-theoretic models of left and right adjoints, and where contravariance denotes the category-theoretic operation of inverting arrows.

15 Identity-based $\mathcal{M}$-signatures

15.4 The Minimal Interaction Model

The Minimal Interaction Model of the Identity-based $\mathcal{M}$-signature:

- *is* an atomic, irreducible model of the fundamental makeup of the Identity-based mind-world interplay (Definition 11.5.2)

- *is* a Type-A Interaction Model that models the reconstruction of identity (Definition 11.5.1),

- *is* a prototypical model of *division*—the Actuation Mechanism of *distinctiveness*,

- *is built* via the application of the Knowledge-Reality Interface (Section 13.8) to a model of `binary object`—an `identiton`,

- *defines* the conceptions of Identity-based knowledge and epistemology.

Note

This description of the Minimal Interaction Model of the default $\mathcal{M}$-signature is based on an analytical view of the components of the Synthetic Structure of Intentionality, that consists in regarding $\mathcal{M}$-signatures and Cognitive Architectures as distinct and ontologically independent entities. This is an interpretation of the Synthetic Structure of Intentionality that sees in the linguistic distinction between $\mathcal{M}$-signature and Cognitive Architecture—motivated purely by explanatory purposes—a reflection of the ontological nature of those entities.

With these ideas in mind, the foundation of a Minimal Interaction Model is *always*, and, necessarily, a process of spatialization consistent with the ambient $\mathcal{M}$-signature.

Structure Of The Minimal Interaction Model

The `reconstruction of the success condition` takes place in the spatialization of an object of thought as described in the steps of Figure 15.6—compare with Figure 11.4.

A. The `reconstruction of the success condition` begins here and consists in applying to the elements of a spatialization of the object of thought a model of `binary object` based on

15.4 The Minimal Interaction Model

the **identiton**. The result of this step is a minimal framework to describe the object of thought via a set of equational properties.

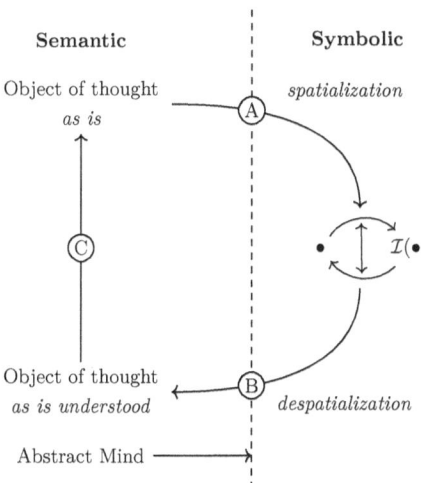

Figure 15.6: Structure of a Minimal Interaction Model in the Identity-based $\mathcal{M}$-*signature*.

B. This is the second half of the **reconstruction of the success condition**, and represents the construction of the of the **Identity-based knowledge**, and consists in applying to the elements of the framework defined in **A**, various models of *division* based on the **identiton**. This step produces statements of the form $A = B$, $A \cong B$, $A \Leftrightarrow B$, etc. which relate any two entities A and B via **identiton**-based models of *division*. Note that in the Minimal Interaction Model the equational relations between ontologies do not rely on any relation other those based on the **identiton**; they are, at least definitional and at most tautological.

C. This step symbolizes the construction of **higher forms of knowledge**. The inferences constructed in **B** constitute a higher spatialization of the initial object of thought, and represent *how* we see and *what* we see of that initial object of thought through the equational lens produced by this Interaction Model—what is called Local Identity (Section 8.7). By applying iteratively the schema of Figure 15.6 to the image produced by the Local Identity, we obtain, in this analogy, forms of identity along two dimensions (Figure 15.7): the admissible types of spatialization, and the models of identity.

$$
\begin{array}{c}
\xrightarrow{\textit{spatializations}} \\
\textit{models of identity} \downarrow
\begin{array}{cccc}
A = B & A' = B' & A'' = B'' & \ldots \\
A \cong B & A' \cong B' & A'' \cong B'' & \ldots \\
A \Leftrightarrow B & A' \Leftrightarrow B' & A'' \Leftrightarrow B'' & \ldots \\
A \dashv B & A' \dashv B' & A'' \dashv B'' & \ldots \\
\ldots & \ldots & \ldots & \ldots
\end{array}
\end{array}
$$

Figure 15.7: Local Identities produced by a Minimal Interaction Model in the Identity-based $\mathcal{M}$-signature.

where A' and B' represent higher order equational properties between lower order equations such as $A = B$, $A \cong B$, $A \Leftrightarrow B, \ldots$, and A'' and B'' represent higher order equational properties between $A' = B'$, $A' \cong B'$, $A' \Leftrightarrow B', \ldots$ and so on. This cognitive machinery produces equations, equations of equations, equations of equations of equations and so on, according to hierarchies of **identiton**-based constructs.

From The Minimal Interaction Model To An Interaction Model

The construction of an arbitrary Type-A Interaction Model consists in adding to the Minimal Interaction Model any number of transformations compatible with the equational properties defined by the models of **identiton**. These transformations are obtained by iterating the Minimal Interaction Model with itself (Figure 15.6-C) and by adding to it any transformation consistent with the spatialization to which the models of **identiton** are applied.

15.5 The Structure Of Knowledge By Division

The Identity-based epistemology

- *is* an interpretation of the terms *knowledge, comprehension, meaning, explanation* and *consciousness* as defined in Chap-

15.5 The Structure Of Knowledge By Division

ters 12, 13 and 14, based on the mechanism through which the **concrete** concept of *identity* defines what is *intelligible*—as described in Sections 15.2 and 10.5, and

- *consists* in the definition of a correspondence between the concepts introduced in Chapters 12, 13 and 14 and certain mathematical constructions of Category Theory based on the **identiton**.

To interpret the terms knowledge, comprehension, meaning, explanation and consciousness based on the type of intelligibility defined by the **concrete** concept of *identity*, means to interpret the terms that make up the definitions of Chapters 12, 13 and 14 according to the meaning that those terms pick up in the Identity-based $\mathcal{M}$-*signature*. For example, Definition 12.3.1 states that acquiring knowledge

"... *is the process of relating a specific set of objects of thought via models of the* **concrete** *concept that defines the* $\mathcal{M}$-*signature where the dialectic about those objects of thought takes place.* "

We have seen in the discussion about the Minimal Interaction Model (Section 15.4) that knowledge is produced by relating the objects of thought via the **identiton**. Thus, to define the process of acquiring knowledge, we need to define the *objects of thought* that we intend to relate through the **identiton**. These are the most elementary "unit of thought" that an Identity-based Abstract Mind can conceive. As it turns out, defining the object of thought is also a necessary step to interpret the other definitions given in Chapters 12, 13 and 14.

The informal construction of the Identity-based epistemology that follows is articulated in these 6 steps:

1. Define the Identity-based object of thought. This answers the question: What is the experience of an object of thought in the Identity-based $\mathcal{M}$-*signature*?

2. From 1, we derive the actual form of Identity-based knowledge. This answers the question: What is the experience of knowledge in the Identity-based $\mathcal{M}$-*signature*? It is a characterization of the practical and conceptual event of consciously knowing an object of thought *before* the development of any complex cognitive apparatus such as beliefs,

contemplation or justification, and any derived notion of truth. This form of knowledge is prototypical and forms the backdrop against which more sophisticated *descriptions* of knowledge are developed, such as for example the descriptions we find in the traditional philosophical accounts of knowledge as justified true belief.

3. From 2, we derive the interpretation of the diagram of comprehension (Figure 15.8). This answers the question: What does it mean to comprehend in the Identity-based $\mathcal{M}$-*signature*? What exactly happens when we understand something? and What are the conditions of, or conducive to, comprehension through division?

4. From 3, we derive the structure of the morphisms between Interaction Model, and define the conceptual machinery at the basis of the protosemantic and semantogenetic processes. This answers the question: How do we seek and create meaning in the Identity-based $\mathcal{M}$-*signature*?

5. From the points above, we outline the basic elements of the Identity-based explanatory system.

6. From the points above, we outline the archetypical structure of the conceptions of Identity-based communication and mathematics.

7. From 1, we deduce a description of the two key features of consciousness, unity and flow, and offer a characterization of these in the ordinary alert, problem-solving state of consciousness.

Since the Identity-based $\mathcal{M}$-*signature* happens to be the ordinary mode of thought, the mode of thought in which we run errands, define mathematics, go to Mars and play tennis, the brief description of the Identity-based epistemology is also a model of the ordinary way of seeking and creating knowledge with the tools of meaning, understanding and explanation.

Identity-based Object Of Thought

An object of thought in the Identity-based $\mathcal{M}$-signature can be thought of as an adjoint-preserving endomorphism.

15.5 The Structure Of Knowledge By Division

Endomorphism means morphism to and from the same object. Recall the discussion "Consciousness And Spatialization" in Section 14.3, where we saw how an object of thought coincides with the *experience* of that object of thought produced by the Abstract Mind via a certain form of *spatialization*. Those ideas are used here to describe what an object of thought is in the Identity-based $\mathcal{M}$-*signature*.

In the Identity-based $\mathcal{M}$-*signature* an object of thought coincides with the experience that we convey with "this is..." or "that is not...". This correspondence is not linguistic, is experiential. The use of language is purely representative of this correspondence. It is necessarily so, because the basic structure of the Identity-based thought revolves around various manifestations the form of spatialization described by the **identiton**. The only way the Abstract Mind grasps concepts in the Identity-based state is via identity, which implies that "object of thought" is solely and exclusively a model of **binary object**. As seen in the description of the identiton, these take various mathematical forms which we have described via left and right adjoints.

An intuitive way to crystallize these ideas is with this diagram

$$F$$

where the • is a mathematical gadget with enough structure to allow for the definition of adjoints—or generalizations of the category-theoretical notion of adjoint—and where F is a morphism from and to • characterized by the property of preserving adjoints.

Identity-based Knowledge As Knowledge By Division

Identity-based knowledge can be thought of as an adjunction-preserving morphism between objects of thought.

In the Identity-based $\mathcal{M}$-*signature*, knowledge is produced by relating the ontologies that form a spatialization of an object of thought via various mathematical constructs based on the **identiton** (Definitions 15.3.1 and 12.3.1). These constructs represent *instances* of the Minimal Interaction Model. The hierarchical

15 Identity-based $\mathcal{M}$-signatures

structure of Identity-based knowledge is described by the below correspondence.

1. The Identity-based *conception* of knowledge is represented by the `identiton`.

2. A *prototypical definition* of knowledge is the atomic cognitive process defined by the Minimal Interaction Model (Section 15.4).

3. Each *definition* of knowledge corresponds to the way in which an Interaction Model implements the **reconstruction of the success condition** based on 2.

To give an Identity-based interpretation of the process of acquiring knowledge described in Definition 12.3.1 we need to characterize two objects of thought are related to each other via the `identiton`.

The diagram above illustrates the type of relation that we want to characterize, where the • are two, possibly distinct objects of thought in the *same $\mathcal{M}$-signature*, and F is a morphism between them. The condition that both objects of thought in the diagram above are part of the same $\mathcal{M}$-signature, that is, the condition that they belong to the same mode of consciousness, implies that whatever F is, it must necessarily preserve the structure of those objects of thought. This condition translates into the requirement that F must be a morphism that *preserves adjoints*.

Knowledge By Division

The philosophical interpretation of Identity-based knowledge as an adjunction-preserving morphism between objects of thought is that Identity-based knowledge is knowledge by *division*. Knowledge by division captures the essence of a Cognitive Architecture that codifies a conception of knowledge as the intellectual act of splitting an object of thought into parts.

15.6 Identity-based Comprehension

Identity-based comprehension can be thought of as an adjunction-preserving morphism between two instances of Identity-based knowledge.

In the Epistemology of the Abstract Mind (Section 12.5) the process of understanding is defined as an object $\bullet^*$ together with an arrow g such that any other arrow f factors via g as in the diagram of comprehension (Figure 15.8).

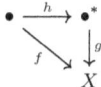

Figure 15.8: Diagram of comprehension.

This condition is captured by the arrow equation below, where $\circ$ is a certain operation that assign one arrow to a pair of adjacent arrows—adjacent meaning that the tip of one arrow coincides with the bottom of the other arrow.

$$f = g \circ h$$

We use the definitions of Identity-based object of thought, and Identity-based knowledge to interpret the diagram above: $\bullet$ are *objects of thought*, and $\rightarrow$ are *types of knowledge*.

Thus, the problem of defining the process of understanding in the Identity-based $\mathcal{M}$-*signature*, is the problem of defining the meaning of the $=$ sign in the equation above. But we know from the discussions about Actuation Mechanism and Definitional Model (Section 15.2), and from the Minimal Interaction Model (Section 15.4), that $=$ signifies that f and $g \circ h$ are related via a model of **binary object**. It follows that understanding in the Identity-based $\mathcal{M}$-*signature* is an **identiton**-based relation between the arrows f and $g \circ h$, that is, it is a (left)right adjoint, or both. Another way to put this, in light of the definition of Identity-based knowledge, is to say that understanding is a certain system in which we come to know f via $g \circ h$.

This conclusion suggests that to define forms of understanding in the Identity-based $\mathcal{M}$-*signature*, we need to switch from "object thinking" to "arrow thinking", being $f = g \circ h$ an adjunction between arrows, not objects. Formally, besides some technicalities, we really do not care whether the symbols we manipulate represent objects or arrows. The distinction between objects and arrows is a convenient way to explain certain basic ideas, but it becomes a hindrance in other contexts.

These observations, and the notion of Identity-based knowledge seen in the previous section, lead to the informal notion that Identity-based comprehension is an adjunction-preserving morphisms between types of Identity-based knowledge: one type of knowledge is represented by the arrow f, another type of knowledge is $g \circ h$. Note that the information in the diagram above that f and g point to the same object (of thought) X, and that f and h have the same origin (object of thought), is part of the *definition* of the adjunction-preserving morphism between the knowledge systems where f and $g \circ h$ are defined respectively.

15.7 Meaning And Semantogenesis

Here we use the definitions of protosemantic transformation (Definition 12.6.1), semantogenesis (Definition 12.6.3), and the intuitive description of a semantogenesis given in "Meaning And Semantogenesis In Interaction Theory" in Section 12.6 to sketch an Identity-based semantogenesis.

Recall from Definition 12.6.3 that a semantogenesis is a mathematical object that classifies types of understanding.

An Identity-based semantogenesis is, therefore, a morphism

$$\Gamma : \mathcal{U} \to \mathbf{P}$$

where $\mathcal{U}$ symbolizes a class of Identity-based *modes of understanding*—which are protosemantic morphisms—and $\mathbf{P}$ symbolizes a space of *values*. Here we describe an example of Identity-based semantogenesis without the notion of protosemantic transformation.

Recall that $\mathcal{U}$ is an object where the diagrams of comprehension

15.7 Meaning And Semantogenesis

Figure 15.9: The diagram of comprehension.

are defined, and where the condition of understanding $f = g \circ h$ for the arrows in Figure 15.9 is based on a certain definition of $=$ given by the Interaction Model in which $\mathcal{U}$ is defined, and according to the Identity-based conception of knowledge defined in this chapter. An example of $\mathcal{U}$ is **P**, where $=$ has the ordinary meaning, and where f, g and h are functions between sets. If in $\mathcal{U}$ we interpret the sets as *data* and the arrows as *strategies* activated by that data, we can think of the comprehension diagram as an illustration of the condition that a certain $\bullet^*$ represents the minimum data to activate a strategy (g) to achieve X.

How does Γ create meaning? The range of possible meanings is codified by the range of values that Γ can take in **P**, and this range of values is a representation of the *purpose* of the modes of understanding considered. Recall that (Section 12.6) the mode of existence of meaning is the mode of existence of the properties that define a classifier (Γ). The definition of a purpose via the structure of **P** and via $\mathcal{U}$, and consequently via the set of values that $\mathcal{U}$ can take, is therefore an implicit representation of the modes of existence of a particular instance of meaning. The most elementary purpose has a binary structure—a unary purpose, one that has only one possible answer, just one value $\mathcal{U}$ can take, does not convey any meaning per se because *any* strategy part of *any* mode of understanding gives the same outcome.

Semantogenesis As Measure Of Modes Of Understanding By Division

The observations made above can be summarized by saying that the creation of meaning in an Identity-based Symbolic Intuition of Reality coincides with classification of ways of understanding according to a certain purpose described by **P**.

The Emergence Of The Ordinary Notion Of Information

The correspondence between meaning and information pointed out in Section 12.6 reemerges in this interpretation of semantogenesis: the more a meaning becomes specific—as a manifestation of a finer form of knowledge by division—the more a certain type of understanding is relevant to a context, the higher its information content. Compare this observation with the measure of entropy given in Shannon's Theory of Information (see also Chapter 20) or in other modern theories of information. We will revisit these ideas in Chapter 20.

15.8 Explanatory Systems

Here we begin to characterize the structure of an Identity-based explanatory system by answering briefly the following questions. Section 15.9 outlines the structure of Identity-based explanations.

- What are explanandum and explanans?
- What are the accepted truths and self-explanatory ontologies?
- What is epistemic continuity?
- What is a reality?

Explanandum And Explanans

In the Identity-based structure of thought, the Abstract Mind relies on models of the `binary object` to articulate concepts: it differentiates, it knows by division. But the Identity-based structure of thought is also where we have defined ordinary language. Consequently, there is a full correspondence between the terms explanandum and explanans and what they represent in the Identity-based $\mathcal{M}$-signature, because the language we use to describe the Identity-based explanatory system is the same language we use to describe the knowledge and the explanations it produces. Explanandum and explanans are therefore direct references to the high-level structure of the Identity-based mind-world interplay: of which denote the semantic and symbolic parts respectively (Figure 15.6).

15.8 Explanatory Systems

Accepted Truths And Self-explanatory Ontologies

Recall Definition 13.2.1: *"An accepted truth is a spatialization of an explanandum, in which a reality is defined. "* By virtue of the definition above, the three types of spatializations described in Section 15.3 are the accepted truths in the Identity-based $\mathcal{M}$-signature. They are accepted truths because they are where the Identity-based thoughts takes place. Perhaps, in a less intuitive way, this observation applies to contravariance. The cognitive act of switching between the spatialization and despatialization described in Section 15.3 is itself an accepted truth in a fundamental way, in that it is a self-evident, basic, integral, organic feature of the Identity-based structure of thought. For the same reasons, self-explanatory ontologies are the minimal, irreducible unities of Identity-based thought, which, by virtue of Definition 15.3.1, and of the definition of the Identity-based object of thought, are mathematical models of the notion of adjoints. The notion of truth follows from the above and from Definition 13.3.2. In summary:

- The Identity-based accepted truths are: focus, unfocus and contravariance.

- The Identity-based self-explanatory ontologies are the various mathematical models and generalizations of adjoints.

Epistemic Continuity

Recall Definition 13.3.1: *"Epistemic continuity is the belief that an explanandum has an intrinsically protosemantic cohesiveness that corresponds to the symbolic cohesion inherent in the structure of an explanatory system $\boldsymbol{P}$ to a degree consistent with the explanans produced by $\boldsymbol{P}$. "* The definition of epistemic continuity conveys the fundamental belief that our explanations, the way we construct them, give us epistemic access to the things they explain. This belief is articulated in Interaction Theory via the notion of epistemic continuity, which captures the idea that the syntactic and semantic cohesion of our explanations is a reflection, a direct and reliable reference to the kind of truth that holds together the things that we explain. It is with this understanding of the basic mechanism of epistemic continuity that we tackle the problem of characterizing this central belief of an Identity-based explanatory system. As it turns out, the intrinsic coherence of the explanations

15 Identity-based $\mathcal{M}$-signatures

we use to interpret the structure and coherence that our explanations make us see in the physical world, is a reflection of the kaleidoscopic manifestations of *causality*. The syntactic version of causality takes the forms of various forms of logic systems—called *internal logic* in the parlance of Category Theory—upon which we construct our explanations. The semantic version of causality, the part that validates the quality and the depth of our explanations, is experimental, in the physical domain. At a conceptual level, the validation of Identity-based knowledge occurs via the *existence* of causal relations, which are expressions of forms of division between explanans and explanandum. The existence of a relation of division between explanans and explanandum, expressed via the **identiton**, is the type of coherence needed in order to validate our Identity-based knowledge of reality. In summary: *Identity-based epistemic continuity is predicated on the belief that the existence of a relation of identity through change between explanandum and explanans—described mathematically by various category-theoretic forms of adjunction—is a faithful and reliable expression of the intrinsic order of the Identity-based explanandum.*

Reality

The semantic part of a mind-world transaction is a *reality*. The same notion is expressed in the language of Interaction Theory in Definition 13.3.3: *"A reality is the experience created by an Interaction Model. "*

The semantic content of a reality is the collection of all explanations based on the symbolic content of that reality, interpreted according to the schema of epistemic continuity. The semantic content of an unexplained reality consists only of the accepted truths. To understand why, recall Definition 13.2.1: *"An accepted truth is a spatialization of an explanandum, in which a reality is defined. "*

From Definition 13.3.3, it follows that the most elementary kind of reality is the experience that originates from the most elementary Interaction Model, namely, the Minimal Interaction Model. As seen in this chapter in the description of a Identity-based object of thought, an **identiton** represents how an Abstract Mind in an Identity-based state grasps thoughts at a fundamental level, and coincides with the fundamental experience of thought described in Section 14.1 between a type of spatialization and the experience of an object of thought. The experience of an Identity-based object of thought is therefore the experience of a Minimal Interaction Model.

To get a feel for what this means practically, we construct an intuitive example that captures the basic idea of a certain morphism that preserves the cognitive structure at the basis of the Identity-based epistemology. Suppose that a certain Identity-based Object of Thought is the set $K = \{1, 2, 3\}$. To transform K into itself—we are looking for an endomorphism—we permute its elements, and obtain $\{2, 1, 3\}, \{1, 3, 2\}, \ldots$ Roughly, the *experience* of K is fully described by the permutations of K because each permutation preserves the *knowledge by division* which is at the basis of the Identity-based epistemology. Intuitively, knowledge by division is preserved because each element x of K can be thought of as the set of permutations that do not alter x. For example, we can think of 2 as the set of permutations $\{1, 2, 3\} \to \{1, 2, 3\}$ and $\{1, 2, 3\} \to \{3, 2, 1\}$. Our knowledge by division of 1, 2 and 3 has the same structure of our knowledge by division of the permutations of K, and this correspondence is what we call our *experience of K*. In modern mathematics, the intuition that we can think of objects as endomorphisms—automorphism more correctly—is crystallized by what is known as Galois Theory, and its generalizations.

15.9 Process And Causal Efficacy

The central themes of Identity-based explanations are the genesis of identity through change, and how the World appears through the notion of process.

Here we use the rationale presented in "The Metastructure Of Explanations" (Section 13.7), and the definitions of *lens* and *image* (Definitions 13.7.1 and 13.7.2 respectively), to outline the structure of Identity-based explanations, and to describe the limits of the interpretations of what we see and comprehend when we look at the World through an Identity-based explanation.

Recall from Section 15.2 that *division*, the Actuation Mechanism of identity, is the phenomenology of distinctiveness, the Organizing Principle of identity. To comprehend the structure of Identity-based explanations we follow the line of reasoning we introduced in Section 13.7: we need to explore what connects an Identity-based explanation to the mechanism through which the **concrete** concept of identity defines intelligible thought, and understand how this nexus defines the conceptual structures that differentiate an Identity-based explanation from any other Identity-based utterance or thought. In "The Deep Structure Of Explanations" (Section 13.7), we discovered that for an utterance to be an explanation, it must necessarily, directly or indirectly, describe the genesis of the definitional structure of the Actuation Mechanism for the *M-signature* where the explanation is formulated, which in this case is: *division*. With these ideas in mind, and with the concepts of *lens* and *image*, we can describe the structure of an Identity-based explanations as follows.

Identity, Process And Change

The genesis of the definitional structure of division is codified by the notion of *change*. Change is the *lens* of identity, and expresses at an archetypical level the mechanism through which the notion of division originates. The term change is used here to capture the fundamental quality of becoming, and to signify how through change an instance of identity acquires its definitional structure.

Through change we see *process*. Process is the *image* of identity, and characterizes what we see when we look at the World through the lens of change. The World described by Identity-based explanations is *made of* processes because that is how we *comprehend* change.

Through process we comprehend *identity*. We *grasp* the various manifestations of *identity* in the World through the way the notion of change is portrayed as process in our Identity-based explanations.

Objectivity

The term *objectivity* denotes how an Abstract Mind in an Identity-based state *apprehends* the fundamental mode of existence of the things in the World.

Objectivity is defined by *division*, the Actuation Mechanism of identity, and characterizes the condition that resolves the epistemic tension between the semantic and symbolic parts of a mind-world

15.9 Process And Causal Efficacy

transaction. Thus, objectivity is formulated as a *reference* to a model of **binary object**—such as, for example, the **identiton**.

A claim is *objective* by virtue of transcending the features that define the epistemic tension where we develop our metaphysical, theoretical and practical comprehension of an object of thought as identity through change. This is why the profundity of objective Identity-based knowledge is gauged by how much of it can be *identified* with its object.

From the perspective of Identity-based thought, *objectivity* is the natural and ultimate mode of existence of facts and ideas that marks the end of the epistemic project: that end is called *ultimate truth*.

The World Explained Through Identity

The *requirement* of an Identity-based explanation is to objectively discern, reveal and describe the *processes* that hold together the World, and in this rests its predictive character.

The *features* of reality revealed by Identity-based explanations are principles of permanence, difference, constancy, succession, variance, evolution and change.

Through an Identity-based explanation, the World appears intrinsically designed as a *machine*, whose workings we can understand and describe through the development of our knowledge of its underlying processes.

Identity-based explanations describe the World in terms of archetypical *dichotomies*, which are instances of the Organizing Principle of distinctiveness.

An Identity-based *explanation* is a statement about the distinctive processes that define the objects in the World intended as a reference to instances of those objects.

Ontic And Epistemic Contents Of An Explanation

Identity-based explanations are *referential*: the objects of thought or perception, and the states of affair they describe are direct references to the objects and states of affair captured by the language of the **identiton**.

15 Identity-based $\mathcal{M}$-*signatures*

Something "is" by virtue of being the *reference* according to which a **identiton** rearranges the objects of thought the way it does.

To *construct* Identity-based explanations we look for relations of division—as described by the **identiton**—between entities that we represent in the ordinary Identity-based language.

The most basic Identity-based explanation is the explanation of a self-evident reality, namely, an *object of thought*, and is described by the condition of epistemic continuity. Recall that Identity-based epistemic continuity is predicated on the belief that the existence of a relation of identity through change between explanandum and explanans—described mathematically by various category-theoretic forms of adjunction—is a faithful and reliable expression of the intrinsic order of the Identity-based explanandum.

Identity-based explanations are *gauged* by how much of the World they describe in terms of *causal efficacy*.

The diagram in Figure 15.10 maps some of the points we have just presented, the description of the **identiton** (Section 15.3) and of the Identity-based epistemology (Section 15.5) in the Knowledge-Reality Interface, where we recognize spatialization, despatialization, and contravariance (Figure 15.10-A, B and C).

15.9 Process And Causal Efficacy

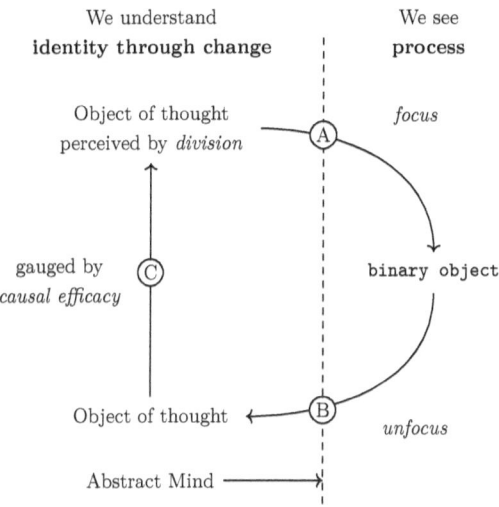

Figure 15.10: How an Abstract Mind in an Identity-based mode of consciousness sees and understands reality.

In Interaction Theory, the term *causal efficacy* denotes a measure of how much an Identity-based explanation describes a reality—a set of known truths—in terms of the minimum number of causal relations between the elements of that reality, as described by the **identiton**. The term efficacy is used in this context to denote a measure of certain extensive properties through which division is represented, and does not have any causal connotation. Epistemic continuity applied to an Identity-based explanatory framework tells us that the greater the causal efficacy, the more fundamental or universal the process through which we comprehend a reality.

The ultimate form of causal efficacy leads to what is sometimes called a Theory of Everything, that is, a theory that with just a few equations explains the laws of Nature, every known natural process and, ultimately, every aspect of reality as we know it. The representation of a causal relation is of course codified by the various mathematical forms of **identiton**, in the form of the equational systems that define the mathematical gadget where the explanation is formulated.

15 Identity-based $\mathcal{M}$-*signatures*

From Change To Identity To Process

The discussion about the phenomenology and representation of identity via the **identiton** (Section 15.3), gave us the opportunity to explore the notion of identity with a level of abstraction that transcends the philosophical interpretations of this elusive concept. But it is, perhaps, only in the analysis of how we connect to the World through an Identity-based explanatory framework, that we begin to reveal an even deeper structure of this notion. This time, the depth is not metaphysical, as in the anatomy of "Identity-based Intelligibility" (Section 15.2), is conceptual. By asking what connects change, identity and process, we are forced to explain an origin of identity that is not phenomenological, but psychological. Change is an integral element of the mechanism through which we experience and comprehend identity because our primary mode of thought is Identity-based. There is, perhaps, a very intuitive and very practical way to convey the quintessential relation between change and process. Process is the manifestation of organized change. It is, so to speak, the application of a principle of change for a specific purpose. The very notion of process, in fact, is based on individual tasks that realize specific changes. In all this, causality is merely a signpost used to mark that a certain process has reached a certain goal. But identity has no causal quality, and has no requirement of causal efficacy. We can describe the nature of identity by comparing change to an intent, and the achievement of that intent to a definition of process based on that change. By identity we mean the evolutionary process of an elementary mode of consciousness that relates us humans with the World through various processes. The reason why we denote identity as the evolution from an intent of change to a process based on that intent is, I think, rooted in the way this transformation allows us to manifest our intentionality in the World. But it is essential, at this point, to note that in this analysis, change, identity and process play no particular role, only the conceptual structure through which they manifest intentionality does, as we will see in "From Proximity To Finiteness To Encompassment" and "From Form To Cohesiveness To Morphogenesis" in Sections 16.9 and 17.9 respectively.

Fully Explainable Realities

The theme of fully explainable realities is a generalization of the argument presented in Section 8.4 "The Limits Of Formal Systems". In particular, it is a generalization of Lawvere's Fixed Point Theorem to the context of an Identity-based $\mathcal{M}$-*signature*: a reality is fully explainable if it is fixed-point-free, in the sense of this generalization—this is a work in progress. A fully explainable reality is free from paradoxes because every explanation of that reality is provable within the system in which that explanation is built.

15.10 Identity And The Language Of Change

In "The Archetypical Structure Of Communication" (Section 14.5) we introduced the idea that communication is that which results in a change of Abstract Mind state, and in "How Data Becomes Message" (Section 14.6) we used this rationale to categorize communication based on the types of Abstract Mind state transitions.

To describe the structure of Identity-based communication—the communication that occurs when the Abstract Mind is in an Identity-based state—we interpret the classification of the Abstract Mind state transitions using identity as a reference Abstract Mind state, and derive from this analysis the structure of Identity-based languages.

15 Identity-based $\mathcal{M}$-signatures

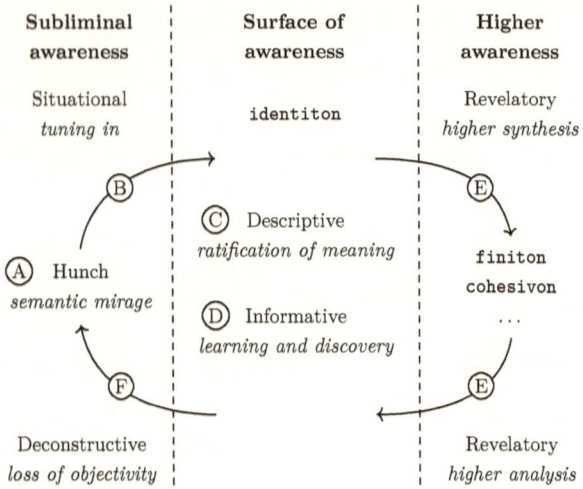

Figure 15.11: The 6 types of Identity-based communication.

Figure 15.11 is a close-up view of the Abstract Mind in the Identity-based state. The region between dashed lines represents the surface of awareness currently lit by consciousness, and signals where the mind-world transactions are defined by a conception of intelligibility based on the **concrete** concept of *identity*, and where, therefore, the Symbolic Intuition of Reality emerges as the experience of grasping concepts as described in "The Structure Of Knowledge By Division" (Section 15.5).

In the brief description of Figure 15.11 that follows, we ground the basic ideas of Identity-based communication to the experience that they produce, because this greatly helps to comprehend how certain subtle fluctuations of consciousness are related to these types of communication. To this end, for each type of communication in Figure 15.11, we explain when it occurs, how meaning is created, and the experience produced by the genesis of meaning.

A *hunch* in the Identity-based $\mathcal{M}$-signature (Figure 15.11-A) is a transition between Abstract Mind states that does not involve the reconstruction of identity. Hunches do not produce meaning because, when they occur, the cognitive structures underpinning

398

15.10 Identity And The Language Of Change

semantogenesis are absent. Hunches produce, instead, a form of false awareness called *semantic mirage*, that in the Identity-based $\mathcal{M}$-*signature* corresponds to the experience of seeming intuitions about the distinctiveness of an object of thought, and consists in the illusion that the features of permanence, difference, constancy, succession, variance, evolution and change that characterize the experience of that object of thought, signify that the identity of *that* object of thought is actually being revealed through the changes in our awareness of it. For these reasons, epistemic continuity (Section 13.3) does not apply to hunches.

Situational communication (Figure 15.11-B) occurs when a transition between Abstract Mind states gives rise to a Minimal Interaction Model for the Identity-based $\mathcal{M}$-*signature*. The emergence of a Minimal Interaction Model in the Identity-based $\mathcal{M}$-*signature* signals that the mind is spatializing through the **identiton**. This produces a sense of *tuning in* to what one is thinking of—see also Contravariance in the **identiton** (Section 15.3)—and is the experience of becoming aware of an object of thought (Definition 14.2.1), that in the Identity-based $\mathcal{M}$-*signature* corresponds to the grasping of an object of thought as a type of process through the features that define *knowledge by division* (Section 15.5). Situational communication marks also the beginning of an embryonal form of Identity-based objectivity, where intuition, knowledge, comprehension and meaning are guided by the Identity-based form of epistemic continuity—see also "Epistemic Continuity" in Section 15.8.

Descriptive communication (Figure 15.11-C) occurs when the Abstract Mind transitions to and from the same Interaction Model within the Identity-based $\mathcal{M}$-*signature*. Descriptive communication is the most basic form of abstraction, and produces the experience of becoming aware of the current Interaction Model—as described in Section (14.2)—by objectifying it according to the structure of *knowledge by division* (Section 15.5). Descriptive communication ratifies meaning via the forms of Identity-based epistemic continuity where principles of permanence, difference, constancy, succession, variance, evolution and change explain the qualities of multiplicity, variety, range and diversity of the objects and states of affair in the World. These produce examples of how semantogenesis measures different modes of understanding by division of the "same" object of thought accessed through that Interaction Model.

15 Identity-based M-*signatures*

Informative communication (Figure 15.11-D) occurs when the Abstract Mind transitions between distinct Interaction Models in the Identity-based M-*signature*. Informative communication gives rise to the experience of *learning by division*, as explained in Section 15.5, and creates meaning by revealing the content of an object of thought through various forms of causal relations between different mental representations of that object. Linguistically, whenever applicable, informative communication conveys how principles of permanence, difference, constancy, succession, variance, evolution and change make up a World where things exist as processes.

Revelatory communication (Figure 15.11-E) occurs when the Abstract Mind changes M-*signature*, and signals a deep change in the way the intellect functions. A synthetic type of revelatory communication denotes a transition *to* a non-Identity-based M-*signature*, and gets its name from the fact that the activation of a different M-*signature* produces the experience of reframing an entire system of thought, and not just specific types of knowledge. In synthetic revelatory communication, meaning is created by translating the structure and purpose of a non-Identity-based epistemic project into the language of Identity-based epistemology. This is, for example, the interpretation of a non process-based view of the World in terms of the concepts of the Identity-based epistemology, and conveys the intuitive experience of thinking of an object from the perspective of how the knowledge of that object is acquired and understood in terms of permanence, difference, constancy, succession, variance, evolution and change, rather than from the narrative of the frame of reference within which that object was originally known and conceived. Similarly, an analytic type of revelatory communication denotes a transition *from* a non-Identity-based M-*signature* to the Identity-based M-*signature*. Its name suggests that the experience associated to this transition is that of how thinking from a perspective not based on multiplicity, variety, range and diversity primes the mind to appreciate new details of what is already known. Analytic revelatory communication produces meaning by reframing or interpreting the narrative of an Identity-based epistemological project into the language of non-Identity-based M-*signature* without changing the structure and purpose of that epistemic project. This is, for example, the interpretation of a process-based view of the World in terms of non-Identity-based M-*signatures*.

Deconstructive communication occurs when the Abstract Mind

15.10 Identity And The Language Of Change

disengages the Identity-based Symbolic Intuition of Reality, and enters a pre-cognitive or pre-conceptual state. This type of communication corresponds to a *deconstruction* in the sense that, as a result of this transition, the reconstruction of identity no longer takes place. In deconstructive communication, the Abstract Mind suspends its awareness of an object of thought by division and of the corresponding epistemology—as modeled by the **identiton** (Section 15.3), and as explained in "Process And Causal Efficacy" (Section 15.9)—and this interruption results in the disintegration of the current Symbolic Intuition of Reality. Deconstructive communication does not produce meaning as such, because it suspends the cognitive structures underpinning semantogenesis. As a result, the identity of an object of thought is no longer grasped via the qualities of multiplicity, variety, range and diversity, and this produces a class of experiences that can be described as a loss of objectivity.

Worlds, Languages And Utterances

An Identity-based language is a method to communicate how permanence, difference, constancy, succession, variance, evolution and change make up a World where things exist as processes.

In Section 14.7 we defined an archetypical language as a method to communicate how the World is seen and understood by an Abstract Mind in a cognitive state (Definition 14.7.1).

We know how the World is seen and understood by an Abstract Mind in an Identity-based state. In "Process And Causal Efficacy" (Section 15.9), we described the basic features of reality as it appears through the lens of identity. There, we used the ideas of "The Metastructure Of Explanations" (Section 13.7) to outline the fundamental mode of existence of the things in an Identity-based World, and to describe how that World can be described through the features of its objects, and of the relations between those objects. To introduce Identity-based languages, we revisit here some of those ideas—also summarized in Figure 15.10—to explain two essential dimension of this problem:

- How the World *seen* through identity defines the *purpose* of Identity-based languages.

15 Identity-based $\mathcal{M}$-signatures

- How the modes of existence of the things in a World *understood* through identity, and the features of that World, define the *structure* of Identity-based utterances.

Let us recall that the World *seen* by an Abstract Mind in a Identity-based state defines how things *exist*. In "Identity, Process And Change" we observed that the World we *see* through identity is made of processes.

> "*[t]hrough change we see process. Process is the image of identity, and characterizes what we see when we look at the World through the lens of change. The World described by Identity-based explanations is made of processes because that is how we comprehend change.*"

Process is, therefore, how things *exist* in the World seen through identity, and this defines the fundamental purpose of Identity-based languages as technologies to convey how certain fundamental processes make up the World.

The features of World *understood* by an Abstract Mind in an Identity-based state define the fundamental *features of existence*. In "The World Explained Through Identity", we observed that

> "*[t]he features of reality revealed by Identity-based explanations are principles of permanence, difference, constancy, succession, variance, evolution and change.*"

These features dictate that the types of *sentences* of an Identity-based language are *necessarily* about how diverse referential relations of causality hold together Identity-based realities, by virtue of forming types of processes. This is how, at the most fundamental level, Identity-based languages relate us to the World via epistemic continuity.

> "*[t]hrough process we comprehend identity. We grasp the various manifestations of identity in the World through the way the notion of change is portrayed as process in our Identity-based explanations.*"

From these observations, it follows that Identity-based languages function by reference.

15.11 Conception And Foundations Of Mathematics

A conception of Identity-based mathematics is a fundamental intuition about how what we see as process explains what we comprehend as manifestations of identity through change.

Here we apply the ideas we presented in "Process And Causal Efficacy" (Section 15.9), in "Identity And The Language Of Change" (Section 15.10), and in "Conceptions And Foundations Of Mathematics" (Section 14.8) to interpret the fundamental intuition of epistemic continuity at the basis of Identity-based Mathematics. The steps below outline the adaptation of the line of reasoning we presented in Section 14.8 to the structure of Identity-based languages and intelligibility.

The process-explains-change intuition at the basis of the modern, Identity-based conception of mathematics, constitutes a mental attitude and a metaphysical sensitivity that allows us to grasp and use the symbol-meaning epistemic tension in a way that we can illustrate as follows.

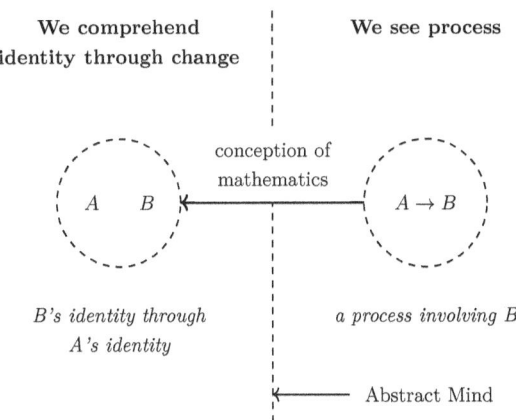

Figure 15.12: Illustration of the Identity-based conception of mathematics as the fundamental intuition about the relation of epistemic continuity between process and identity through change.

15 Identity-based $\mathcal{M}$-signatures

- The qualities of *division* are multiplicity, variety, range and diversity, and form the substratum where the principles of permanence, difference, constancy, succession, variance, evolution and change that make up an Identity-based reality, exist as *process*.

- A *prototypical concept* and a model of multiplicity, variety, range and diversity is the *intuitive notion of aggregate or set*. Intuitive means that set is grasped as a whole, and its elements are defined by the very grasping and not by a membership relation.

- A *conception of mathematics* (Figure 15.12) is an intuition about the existence of a fundamental relation of epistemic continuity, intrinsic in the Identity-based intuition of reality, by means of which what we see as process *explains* what we comprehend as manifestations of identity through change.

- *Process explains change* through the intuitive, prototypical notion of set, and its natural operations—union, intersection, etc.—that, together, constitute an archetypical language to convey how permanence, difference, constancy, succession, variance, evolution and change *explain* multiplicity, variety, range and diversity.

- Identity-based mathematics is a language of *process*. It is the language of formal systems.

- A *foundation of mathematics* is a language to model of the intuition at the basis of the conception of mathematics based on the prototypical notion of set and its natural operations.

- The distinctive features of Identity-based mathematical reasoning are abstract notions of number, function, maximum, minimum, derivation, inversion, and their many generalizations, and characterize the natural ways to construct processes.

Identity-based Computers

An archetypical Identity-based computer, the ordinary computer, what in Interaction Theory we should call a *change computer*, is a device or a system that translates in its dynamical properties the features of Identity-based languages. Identity-based computers, therefore, are parametrized processes.

The instructions to set their parameters are called programs, and Identity-based programs and programming languages are, therefore, designed to model the features of permanence, difference, constancy, succession, variance, evolution and change.

In this sense, and for this archetypical reason, the elementary units of process are Boolean operations, or pre-configured logic functions such as elementary and advanced arithmetic and logic functions, and in general any reusable function, because "function" is a model of process.

15.12 Features Of Consciousness

The three modes of consciousness modeled by the **identiton** form the basic structure of the Identity-based conscious experience of any mental or perceptual reality: they *are* the subjective reality. As described in Section 14.1, and in particular in the characterization of awareness of Definition 14.2.1, in Interaction Theory, what is referred to as the conscious experience of a certain object of thought $\mathcal{O}$ is a reference to a configuration where the epistemology of $\mathcal{O}$—as described in Chapters 12, 13 and 14—coincides with the phenomenology of the Actuation Mechanism associated to the **concrete** concept in which $\mathcal{O}$ emerges as part of a dialectic of thought. In the previous sections we have outlined the basic structure of the Identity-based epistemology. Here we want to describe briefly the two main properties of the general notion of consciousness as they appear in the Identity-based $\mathcal{M}$-*signature*. Thus, we ask: What are the unity of consciousness and the flow of consciousness in the default $\mathcal{M}$-*signature*? Let's remember that "is" means experience, since the experience of the unity of consciousness and of the flow of consciousness are themselves unity and flow of consciousness.

Unity Of Consciousness

In the Identity-based $\mathcal{M}$-signature, the unity of consciousness is experienced via forms of division.

This might seem a paradox. How is it possible to experience the unity of all conscious phenomena via forms of division? The term division here denotes the Actuation Mechanism of identity (Figure 15.1). How can we reconcile the idea of the unity of consciousness

15 Identity-based $\mathcal{M}$-signatures

with the notion of division that seems to negate unity altogether? The *unity* of consciousness is an experience that is comprehended intuitively, via the binary, dualistic features of the Identity-based subjective reality. The characteristics of the Identity-based reality are used to comprehend, to perceive, to grasp the Organizing Principle at the origin of the things that come into existence in one's subjective conscious experience in the Identity-based $\mathcal{M}$-*signature*—that is, via *distinctiveness*. The term division, refers to the way the Abstract Mind grasps and constructs the subjective experience, *not to the content of the experience*. Consequently, the statement that in the Identity-based $\mathcal{M}$-*signature* the unity of consciousness is experienced via forms of division means that in the every aspect of the conscious experience it is possible to consistently recognize through a dichotomy the "I" that is both the creator and the recipient of the conscious experience. This is unity through division.

Flow Of Consciousness

In the Identity-based $\mathcal{M}$-signature, the flow of consciousness is experienced as distinctiveness through division.

The experience of consciousness as an uninterrupted, all-pervading event, is the manifestation of how an underlying conception of continuity creates a coherent and cohesive experience of each conscious experience. In the Identity-based mode of thought, this feature of continuity of the conscious experience is interpreted as continuity with respect to the structure of thought defined by the **identiton**. What is perceived as continuous, therefore, is continuous by virtue of having certain characteristics that make it appear so to an Abstract Mind in the Identity-based state. Thus, to comprehend how we perceive the mental, perceptual and spatiotemporal bubbling of each experience as a coherent flow, we must examine what makes a sequence of experiences continuous. We are looking for what makes two conscious experiences stick together coherently, and appear as if one grows out of the other that precedes it. These spatiotemporal metaphors are of course purely instrumental to anchoring the terms of this analysis to our physical experience for descriptive purposes. We know that the human experience of the flow of consciousness applies to mental

15.12 Features Of Consciousness

experiences as well, where the frame of reference of space and time does not always apply.

Intuitively, *continuity* denotes a very general form of identity through the changes that occurs according to the type of spatialization defined by the structure of the **identiton**.

Bibliographical Notes

The notion of identity has been and continues to be at the center of numerous philosophical investigations, and this presence is reflected in the vast literature on the subject. The take on identity presented in this chapter is, first and foremost, archetypical. It is not concerned with its logic (Black 1952, Hawley 2009, Russell 1940), or with the determination of standards by which to judge identity (Frege 1980, Wittgenstein 1958), or with the classical problems of identity over time (Lewis 1986, Hawley 2001, Theodore Sider 2001, Quine 1960) or over possible worlds (Kripke 1981, Lewis 1986) or with other subtle characterizations of the numerous problems posed by this intriguing concept. Instead, identity, like any other **concrete** concept in this research, is studied exclusively as the seed that gives rise to a specific metaphysical structure of thought. Identity is studied by virtue of being the origin of the ordinary Symbolic Intuition of Reality, the Symbolic Intuition of Reality of the alert, problem-solving state of consciousness.

The other important theme of this chapter is the manifestation of the abstract notion of identity in mathematics, and in particular its many category-theoretical representations. An introduction to adjunctions can be found in any introductory text to Category Theory, such as Lawvere and Schanuel 2009 and Steve Awodey 2008b, or the classic but more advanced MacLane 1998. A reference to advanced mathematical representations of identity can be found in the general theory of fibered categories, such as Bénabou 1985, Streicher 1999, Jacobs 2001 for a type theoretic application of fibered category theory, and Peter T Johnstone 2002a B1.3. The basic intuition underpinning fibrations is in Grothendieck 1959.

The descriptions of the **identiton** and of the Identity-based epistemology, and the interpretation of consciousness as a mode of spatialization are sketched in my unpublished work Lo Vetere 2015a. For a short bibliographical reference on consciousness see the note at the end of Chapter 14.

Chapter 16

Finiteness-based $\mathcal{M}$-signatures

Long ago, when shepherds wanted to see if two
herds of sheep were isomorphic, they would look for
an explicit isomorphism. In other words, they would
line up both herds and try to match each sheep in
one herd with a sheep in the other. But one day,
along came a shepherd who invented
decategorification. She realized one could take each
herd and 'count' it, setting up an isomorphism
between it and some set of 'numbers', which were
nonsense words like 'one, two, three, ...' specially
designed for this purpose. By comparing the
resulting numbers, she could show that two herds
were isomorphic without explicitly establishing an
isomorphism! In short, by decategorifying the
category of finite sets, the set of natural numbers
was invented.

Baez and Dolan 1998

In this chapter, we introduce an $\mathcal{M}$-*signature* that gives rise
to a metaphysical structure of thought based on the concept of
finiteness.

Finiteness is a central intuition of the ordinary perceptual experience of the world, that captures the observation that most
thoughts, objects, and states of affairs in the world, appear to

16 Finiteness-based $\mathcal{M}$-signatures

the human mind as having certain features of boundedness, determinacy and definiteness. Here we explore the basic structure of thought of a mind that functions in a Finiteness-based fashion, and describe how that mind sees the World, how it thinks, how it communicates, how it constructs concepts and explanations, and how it develops its knowledge of the World. A Finiteness-based mind thinks through hierarchies of concepts that are the evolution of a single, irreducible metaphysical intuition about the mind-world interplay: we call this intuition *boundedness*.

So far, we have used spatialization quite freely in our discussions, because we have never had to challenge the intelligibility of the elements of a spatialization. In other words, thinking about the elements of a space as distinct and definite entities has always been an accepted reality, a true fact deeply embedded in the metaphysical structure of thought in which we were asking questions about those very same distinct and definite entities. The typical example of what this basic presupposition about the structure of thought has allowed us to do, is thinking of an object of thought as a set of symbols, or as a space. This situation changes drastically when we abandon the comfort of the Identity-based $\mathcal{M}$-*signature*, and study the non-standard $\mathcal{M}$-*signatures*, because the very notion of spatialization changes radically. In the non-standard $\mathcal{M}$-*signatures*, the lack of a notion of identity in the form of a **concrete** concept, disintegrates the ordinary notion of spatialization because the distinct and definite entities become unintelligible, and with them, the mind's ability to read an atomic object of thought via the properties of a space in the usual way.

This change does not mean that the outline to describe $\mathcal{M}$-*signatures* introduced in Section 10.7 is not applicable to the non-standard $\mathcal{M}$-*signatures*; it means that we need to make a sharp distinction between the language we use to describe the structure of a Finiteness-based $\mathcal{M}$-*signature*—which continues to be the ordinary, Identity-based language of the default $\mathcal{M}$-*signature*—and the entities of the Finiteness-based $\mathcal{M}$-*signature* described by that language. We need, in other words, to *interpret* the Identity-based description of a Finiteness-based metaphysical structure of thought in a way that does not rely on the notions of identity. The effects of this interpretation are subtle, and are clearly visible in the description of the Finiteness-based epistemology, where the need to revisit the notions of knowledge, comprehension, meaning, explanation and consciousness without spatialization, reveals the magnitude of the conceptual landscape portrayed by Interaction Theory.

For example, one of the puzzles posed by the interpretation of an Identity-based description of a Finiteness-based $\mathcal{M}$-*signature* is the following. Recall the definition of knowledge given in the Epistemology of the Abstract Mind (Section 11.1):

> *"Knowledge is the process of relating a specific set of objects of thought via models of the* **concrete** *concept that defines the $\mathcal{M}$-signature where the dialectic about those objects of thought takes place. "*

Thus, we ask: If in a Finiteness-based $\mathcal{M}$-*signature* there are no distinct and definite ontologies as such, how do we relate distinct and definite ontologies via a **concrete** concept?

16.1 What Is Finiteness?

When we cannot truly say that a given object of thought or perception displays certain intelligible, distinctive characteristics of boundedness, determinacy and definiteness, we say that that object of thought or perception is *infinite*. Here we use the three properties of boundedness, determinacy and definiteness to interpret and describe how finiteness shapes our ability to interact with the physical and mental realities.

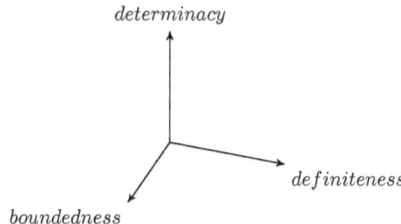

Figure 16.1: The 3+1 dimensions of infiniteness.

Finiteness draws a demarcation line between a peculiar feature of perceptual experience, and the mental realities that we create when we abstract and generalize that experience. One such generalization of finiteness is the concept of infinity. We construct a concept of infinity by altering the degree to which the three features that define finiteness exists or manifests themselves (Figure 16.1). Each type of infinity corresponds to a plane parallel to one of the three planes defined by boundedness-determinacy, boundedness-definiteness and determinacy-definiteness planes. In addition, when we collapse the three dimensions, we obtain a type of infinity which has none of the properties of finiteness, namely, the origin of the coordinate system of Figure 16.1. Linguistically, infiniteness is constructed by negating or weakening the meaning of a feature of finiteness.

Boundedness

The opposite of *boundedness* gives rise to infinities in space, time, number, and in general in any measurable—quantitative or qualitative—dimension of discourse or category of thought.

Determinacy

The opposite of *determinacy* gives rise to ontological infinities, and to forms of infiniteness in location, meaning and place—intended also as the context or milieu of a system of ideas, or a language or a culture.

Definiteness

The opposite of *definiteness* gives a type of conceptual and rational infiniteness, such as for example love, as that feeling that cannot be defined because it defies any definition, or a divine being, as that primordial or archetypical concept that cannot be defined other than by everything or nothing, etc.

Higher Infiniteness

Other constructions of infiniteness as abstractions of finiteness go beyond the modification of boundedness, determinacy and definiteness, and give rise to exotic concepts that are both finite and infinite, by attacking directly the definitional structure of finiteness. We will not discuss those generalizations of finiteness here. However, what needs to be noted about those generalizations of finiteness, is that they all share the strategy of weakening the definitional integrity of the features they abstract or generalize, which is an example of the general scheme of thought we described in Chapter 5.

16.2 Finiteness-based Intelligibility

In the discussion that follows, we cross once again the thin line that separates mathematics and metaphysics, and attempt to give an account of a radically different metaphysical structure of thought, which is unintelligible with the tools of the ordinary Identity-based thought. We will do this in a similar fashion to the Lego brick analogy we presented in the introduction to Chapter 10. The aim is to define intelligibility in the Finiteness-based metaphysical structure of thought, and how knowledge is acquired. These two elements fully characterize, as discussed in Chapters 10 and 11 the basic structure of a model of Abstract Mind.

16.2 Finiteness-based Intelligibility

Finiteness-based intelligibility is a structure of thought in which the human mind has epistemic access to the world via a principle of boundedness, which extends to concepts the physical property of being bounded with respect to one or more extensional properties.

The definition of Finiteness-based intelligibility that follows constitutes the connective tissue between the philosophical and psychological domains that have been the stage where we have developed a framework to define the structure of the Abstract Mind, and the mathematical domain, that will become the stage where we will begin to translate these ideas into mathematical structures.

In the Finiteness-based mode of thought, boundedness is *not* a linguistic construction, and does *not* extend to concepts the physical property of being bounded with respect to one or more extensive properties. This does not contradict what we said in the paragraph above. In the Finiteness-based mode of thought boundedness is the intrinsic, quintessential character of what it means to think in that mode of thought, in the exact same way in which distinctiveness is the prototypical way to form thoughts in the default $\mathcal{M}$-*signature*. When we attempt to express in an Identity-based language the meaning of boundedness, as it is understood in a Finiteness-based metaphysical structure of thought, we find an insurmountable barrier in the structure of natural language and in the structure of our own thoughts. Identity-based languages, that is, any natural and formal language, revolve entirely around various forms of dichotomies to generate utterances, such as "this and that", "subject and object", "is and isn't" etc. These dichotomies are a reflection of the effects of the notion of identity on the mind's ability to grasp concepts: identity *separates* the objects of thought, hence the dichotomies. This is what makes the Identity-based structure of thought utterly inadequate to describe anything that is not a manifestation of a binary principle. The cognitive and linguistic barriers we find in the description of the Finiteness-based mode of thought are an example of the limitations intrinsic to the study of the Abstract Mind we described with the Lego brick analogy in the introduction to Chapter 10. To overcome those limitations, in full accord with the Lego brick analogy, we can describe the structure of a Finiteness-based mind without having to *reason* in a Finiteness-based fashion.

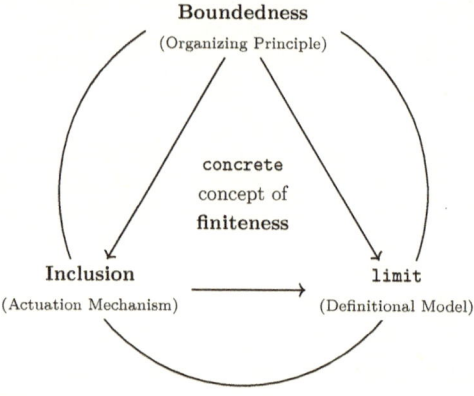

Figure 16.2: The structure of Finiteness-based intelligibility.

In the next three sections we use the notions of Organizing Principle, Actuation Mechanism and Definitional Model to describe the exact mechanism through which the `concrete` concept of finiteness gives rise to a Finiteness-based mode of thought.

16.2 Finiteness-based Intelligibility

Organizing Principle Of Finiteness

Boundedness is an unintelligible metaphysical intuition about the potential of the concrete *concept of finiteness.*

The capacity expressed by finiteness is *boundedness*, and is based on specific extensive properties of an object of thought (Figure 16.2). The term "extensive" suggests that this description of finiteness is given via a process of spatialization. The human mind spatializes concepts to be able to grasp them, and to think "inside" those concepts and not only *about* them (Section 10.1). In particular, the spatialization this section refers to, occurs in the default $\mathcal{M}$-*signature*, because this description of the Finiteness-based $\mathcal{M}$-*signature* is written in the default mode of thought. In this case, the function of the conceptual arena of thought is to carry or manifest the Organizing Principle of boundedness through which we are able to access and comprehend finiteness. Boundedness extends to concepts the physical intuition of boundedness in space, or of other instances or modalities of perceptual experience. The way this happens is described by the Actuation Mechanism.

Actuation Mechanism Of Finiteness

Inclusion defines the phenomenology of boundedness, and boundedness becomes intelligible to the human mind as inclusion.

In the Finiteness-based mode of thought, the Actuation Mechanism describes how an Organizing Principle of *boundedness* manifests itself as *inclusion* (Figure 16.2).

The description of an Actuation Mechanism for a given Organizing Principle of boundedness, is the description of how the Finiteness-based notion of intelligibility *becomes* conscious and intentional thought. This step is, like any other Actuation Mechanism, intrinsically model-theoretic, in that it demands us to produce a *model* of boundedness upon which we can formalize finiteness. In the discussion that follows, we present a characterization of mathematical finiteness via the set-theoretic notion of inclusion. These ideas will form the conceptual link between philosophical finiteness and mathematical finiteness when, in Section 16.3, we will immerse the example of mathematical finiteness we

16 Finiteness-based 𝓜-*signatures*

are about to see, in the broader context of a category-theoretical notion that generalizes inclusion along many dimensions.

There are many characterizations of finiteness in mathematics, the one we use here is, I think, the most intuitive. The basic idea is to use certain properties of the correspondences between sets, to characterize the intuitive idea that a certain collection has a finite number of objects.

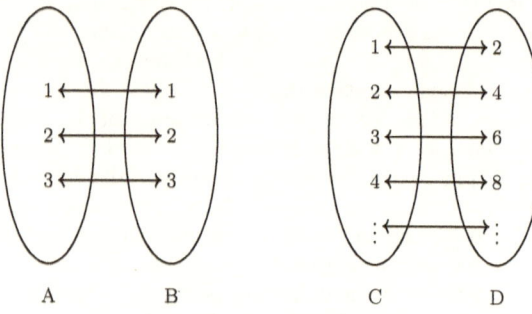

Figure 16.3: Bijection between finite sets (A,B), and between infinite sets (C,D).

In the finite sets of Figure 16.3-A and 16.3-B, it is possible to assign to each element of B an element of A, in such a way that distinct elements of A correspond to distinct elements of B, and only if B is *not* properly contained in A. This fails to be true when we consider the infinite sets of Figure 16.3-C (the set of all natural numbers) and 16.3-D (the set of all even natural numbers), where the construction of Figure 16.3-A and 16.3-B is still possible, *but D is* a proper subset of C. This fact can be used to characterize finiteness, and this characterization, due to R. Dedekind, defines finiteness via the properties of morphisms between sets—the arrows in Figure 16.3. The core of this characterization of finiteness is in the shift from "object thinking" to "arrow thinking", and the implications of this idea are profound. Let X and Y be sets. A morphism, or map, or function, is a rule to assign the elements of X to elements of Y in such a way that every element of X is assigned to at most one element of Y. Morphisms are indicated as $f : X \to Y$, or as $X \to Y$ when there is no need to identify the arrow. Given $f : X \to Y$, the element y of Y to which x is

16.2 Finiteness-based Intelligibility

assigned via f is called the *image* of x under f, and the totality of images of x under f is denoted by $f(X)$.

The properties of morphisms between sets involved in this characterization of finiteness are the following. A morphism $X \to Y$ is called:

- *injective*, if distinct elements of X correspond to distinct elements of Y via $\to$,
- *surjective* if every element of Y is the image of an element of X via $\to$,
- *bijective* if $\to$ is injective and surjective.

Dedekind's characterization of finiteness states that it is possible to define a bijection between a set X and a proper subset Y of X, only if X and Y are infinite. This line of thought shows that the characterization of finiteness can be reframed as the problem of classifying the morphisms between sets according to the definitions of injective and surjective morphism given above.

Another useful way to interpret injective morphisms is to observe that injective means surjective on equations. To see this, we consider an injective morphism $f : X \to Y$, and observe that equations in Y are certain morphisms $g : f(X) \to f(X)$ for which $f(x) = g(f(x)) = f(y)$ for y in $f(X)$. Thus, for every such equation defined by g, there are x and y in X for which $x = y$ because f is injective, which is itself an equation in X. If we compare this with the definition of surjective morphism we find that for every equation g in Y (via f), there is an equation in X.

Definitional Model Of Finiteness

A **limit** *defines the epistemology of boundedness by modeling inclusion.*

The Definitional Model of *inclusion* is a philosophical device called **limit** (Figure 16.2) that codifies the prototypical structure of **success** for the metaphysical structure of thought defined by the Finiteness-based $\mathcal{M}$-*signature*.

The **limit**:

- is a *form of awareness*—as seen in "The Structure Of Awareness" (Section 14.2),
- defines the structure of (de)spatialization in the Finiteness-based $\mathcal{M}$-*signature*—see Definition 14.2.1 and "Consciousness And Spatialization" (Section 14.3),
- is used in Interaction Theory as a reference to formal models of *inclusion*,
- is the basis of *any* construction of the `success condition` in the Finiteness-based $\mathcal{M}$-*signature*—see also Section 8.3,
- is defined by the following type of *generative knowledge* (see Section 12.1) $\{Diagram, \rightarrow\}$,
- constitutes the core of the Minimal Interaction Model for the Finiteness-based $\mathcal{M}$-*signature*.

The term `limit` captures the basic features of the experience of finitude, and is meant to convey the view that the definition of any concept conducive to some sort of abstract demarcation between entities is a manifestation of the Finiteness-based mode of thought.

Finiteness-Based Generative Knowledge

Generative knowledge in the Finiteness-based $\mathcal{M}$-*signature* consists of the set of archetypes $\{Diagram, \rightarrow\}$, where *Diagram* is a graph (bullets connected by arrows) and $\rightarrow$ is an element that connects • to *Diagram* in one direction, or vice versa. We find these elements in the representations of `limit`.

The representation of `limit` we introduce here is called the `finiton`.

16.3 The finiton

The `finiton` is a *form of awareness*—as seen in "The Structure Of Awareness" (Section 14.2)—that we *describe* as an atomic and irreducible pattern that models the notion of *inclusion*—the Actuation Mechanism of *boundedness*—and that defines the *structure of spatialization* in the Finiteness-based $\mathcal{M}$-*signature*.

16.3 The finiton

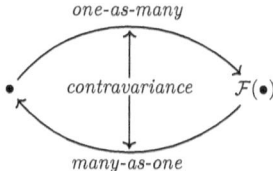

Figure 16.4: The finiton

Compare Figure 16.4 with Figure 14.1 in "Consciousness And Spatialization", where the concrete concept is $\mathcal{F}$: *finiteness*. The form of awareness codified by a finiton is a set of "sleights of mind" that symbolize the irreducible modes in which the conscious mind operates at a fundamental level in the Finiteness-based $\mathcal{M}$-signature. These modes of operation are: *one-as-many*, *many-as-one*, and *contravariance*.

To introduce these three sleights of mind, we interpret the diagram in Figure 16.4 as follows:

- the • represents an *object of thought*,
- $\mathcal{F}(\bullet)$ is a diagram—bullets connected by arrows—that represents the *spatialization* of the object of thought •,
- the arrows $\bullet \to \mathcal{F}(\bullet)$ and $\mathcal{F}(\bullet) \to \bullet$ are *bundles of arrows* that connect • to each vertex of $\mathcal{F}(\bullet)$, and each vertex of $\mathcal{F}(\bullet)$ to • respectively,
- the arrow between arrows called "contravariance" denotes a certain symbolic operation to switch to and from $\bullet \to \mathcal{F}(\bullet)$ and $\mathcal{F}(\bullet) \to \bullet$.

The purpose of the above correspondence is to encapsulate the philosophical notion of limit—the Definitional Model of finiteness—via a mathematical concept of "closeness to a diagram" that we express diagrammatically via certain conditions on the arrows $\bullet \to \mathcal{F}(\bullet)$ and $\mathcal{F}(\bullet) \to \bullet$. The closest object to $\mathcal{F}(\bullet)$ is •, and the closest diagram to • $\mathcal{F}(\bullet)$. Closeness to a diagram is a hugely important mathematical notion that, disguised in a vast array of mathematical forms, pervades the entire edifice of our mathematical knowledge of the world in surprising and unifying ways. As a first approximation, closeness to a diagram is to arbitrary mathematical gadgets what being a factor is to integers.

16 Finiteness-based $\mathcal{M}$-*signatures*

To situate the discussion about $\mathcal{F}(\bullet)$ and the characterization of closeness to a diagram, we need first to interpret $\mathcal{F}$.

Interpretation Of $\mathcal{F}$

$\mathcal{F}(\bullet)$ symbolizes the object of thought $\bullet$ in the context of perception, intentionality and consciousness. In the `finiton`, $\mathcal{F}(\bullet)$ is a diagram, that is, a mathematical object that we *represent* as a whole made of parts that we call vertices and edges. But $\mathcal{F}(\bullet)$ is still an *atomic and irreducible* object. We must not confuse the diagrammatic representation of $\mathcal{F}(\bullet)$ with its definitional identity. In other words, the use of vertices and edges to represent $\mathcal{F}(\bullet)$ is not to be interpreted as an indication or a suggestion of the existence of a finer structure of thought beyond $\mathcal{F}(\bullet)$. We can, however, ascribe a meaning to the vertices and edges of a diagrammatic representation of $\mathcal{F}(\bullet)$ by saying that with those symbols we crystallize the notion of a unity that, in the Finiteness-based structure of thought, is spatialized via multiple modes of existence or modes of appearance symbolized by vertices and edges.

Closeness To A Diagram

In Figure 16.5, let $\mathcal{F}(\bullet)$ be the one-vertex-and-no-arrows diagram $\{\bullet\}$. To express the idea that a given object $\bullet^*$ is the *closest object to* $\mathcal{F}(\bullet)$, we say that $\bullet^*$ has the property of being connected to $\mathcal{F}(\bullet)$ via an arrow g such that any other vertex connected to $\mathcal{F}(\bullet)$ via, say, f, factors via g, that is, $f = g \circ h$, where $\circ$ is a certain operation of composition of arrows, and h is an arrow that connects $\bullet$ to $\bullet^*$.

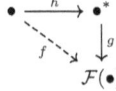

Figure 16.5: The limit of $\mathcal{F}(\bullet) = \{\bullet\}$

To generalize this characterization of "closeness to a diagram" to a diagram $\mathcal{F}(\bullet)$ with an arbitrary number of vertices and edges, we apply iteratively to each vertex of $\mathcal{F}(\bullet)$ the characterization of closeness of an object to the one-vertex-and-no-arrows diagram $\mathcal{F}(\bullet) = \{\bullet\}$ seen in the example above.

16.3 The finiton

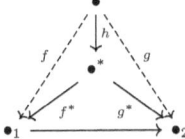

Figure 16.6: The limit of $\mathcal{F}(\bullet) = \{\bullet_1 \to \bullet_2\}$

For example, if $\mathcal{F}(\bullet) = \{\bullet_1 \to \bullet_2\}$, then $\bullet \to \mathcal{F}(\bullet)$ is the bundle of arrows $\{f, g\}$ of Figure 16.6. An object $\bullet^*$ connected to $\mathcal{F}(\bullet)$ via $\{f^*, g^*\}$ is the *closest* object to $\mathcal{F}(\bullet)$ if for any object $\bullet$ connected to $\mathcal{F}(\bullet)$ via, say, $\{f, g\}$, the following factorizations hold $f = f^* \circ h$ and $g = g^* \circ h$, where h is an arrow that connects $\bullet$ to $\bullet^*$. In the parlance of Category Theory, when $\bullet^*$ and the pair of arrows $\{f^*, g^*\}$ exist, they are called the *limit* of $\mathcal{F}$.

The Structure Of The finiton

In the next three sections we offer an overview of the three modes of operation of the **finiton**. This description is meant to be interpreted as an example of "The Abstract Mind-as-sound Analogy" we presented in Section 11.5, where the analogy is between the **finiton** and the sound produced by the buzzer circuit. Focus, unfocus and contravariance are a way to describe the basic, irreducible, atomic experience of the Abstract Mind, in a Finiteness-based state coming in contact with an object of thought.

Spatialization As Detailed Perception

Finiteness-based spatialization is the "one-as-many" sleight of mind, and its mathematical analogy are certain generalizations of the category-theoretical notion of limit.

The intuitive notion of "one that becomes many" is captured by the various mathematical representations of the concept of closeness of a vertex to a diagram. Since the definition of **finiton** depends on the shape of a diagram $\mathcal{F}$, the study of Finiteness-based spatialization is the study of various category-theoretical versions of closeness to diagrams, the most elementary of which is the *limit*. One-as-many is how an Abstract Mind in the Finiteness-based state grasps an object of thought. It is the very experience of

16 Finiteness-based $\mathcal{M}$-signatures

the object of thought (one) becoming a space of thought (many), which is nothing but a *detailed description* of "itself".

Recall the characterization of finiteness à la Dedekind introduced in Section 16.2 in the description of the Actuation Mechanism of finiteness. There, we gave a set-theoretic characterization of the notion of inclusion. In the parlance of Category Theory, the set-theoretic notion of inclusion is modeled by a subobject, which generalizes and improves in many ways the intuitive notion of subset. In the diagram below, let the arrows $B \to A$ and $C \to A$ represent injective morphisms—recall that a morphism $X \to Y$ is *injective*, if distinct elements of X correspond to distinct elements of Y via $\to$. A *subobject* is the class of the injective morphisms in A, characterized by the property that any two morphisms f, g of this class factor as $f = g \circ h$, where $g : B \to C$ is an isomorphism— an isomorphism is a generalization of equality such the notion of bijection we saw in the Actuation Mechanism in Section 16.2.

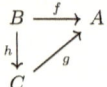

The way this definition of subobject generalizes the set-theoretic notion of inclusion is precisely via the ideas we presented in Section 16.2, where we switched from "object thinking" to "arrow thinking" and, as a result of this change, we replaced the intuitive notion of equality via certain properties or arrows. A close-up inspection of the definition of subobject should reveal how this definition is essentially a polished up version of the ideas we presented in Section 16.2, and how the basic idea of finiteness (à la Dedekind) is expressed as an isomorphism class of injective morphisms.

In particular, when $\mathcal{F}(\bullet)$ is the following diagram

its category-theoretic limit is a subobject.

We can think of the notion of closeness to a diagram as a generalization of the *type of finiteness* described by Dedekind in two ways:

16.3 The `finiton`

- it generalizes finiteness à la Dedekind via the notion of subobject, which expands the use of a set-theoretic-like notion of subset to category-theoretic generalizations of Set Theory

- it generalizes the set-theoretic notion of finiteness through the shape of $\mathcal{F}(\bullet)$

When we construct limits of arbitrary diagrams—and, a fortiori, generalizations of those limits—we discover a zoo of known and new mathematical objects in which the philosophical notion of finiteness as the boundedness-determinacy-definiteness trichotomy appears in dignified forms that greatly unify our vision and understanding of the hidden structure of mathematics. I think that there is a sense in which it is more appropriate to think of finiteness à la Dedekind as a very special case of other forms of finiteness disguised as ordinary mathematical objects that have been unified through the language of Category Theory in the past 70 years.

Other Forms Of Closeness To A Diagram

Among the many mathematical forms of closeness to a diagram, there is one that is, perhaps, the most important for its impressive descriptive and unifying power: it's called *Kan extension*. A Kan extension is a certain type of adaptation of the intuitive notion of closeness to a diagram to the context of categories.

Despatialization As Synthesis Of An Object Of Thought

Finiteness-based despatialization is the "many-as-one" sleight of mind, and its mathematical analogy are certain generalizations of the category-theoretical notion of colimit.

Despatialization in the Finiteness-based $\mathcal{M}$-*signature* captures the intuitive notion of "many that become one", and is expressed by the various mathematical representations of the concept of closeness of a diagram to a vertex. By swapping "vertex" with "diagram" in the discussion about how Finiteness-based spatialization is constructed mathematically, we obtain a mathematical description of despatialization. Alternatively, we can invert each arrow in the description of spatialization given above—recall that arrow inversion is a less fancy name for contravariance—in the category-theoretic parlance. Despatialization is here depicted as a space of thought (many), the diagram, becoming one, a vertex. Despatialization is many becoming one, as a result of a certain primary perceptual ability that synthesizes many into a new object of thought that carries in its ontological status some of the features of that space of thought. This intuitive transaction between the mind and the world is symbolic, allegoric, but also practical, as captured by various category-theoretical versions of the notion of *colimit*.

Contravariance Is The Mental Perception Of An Object Of Thought By Inclusion

In the Finiteness-based $\mathcal{M}$-*signature* spatialization and despatialization are related to each other in what is, perhaps the most intuitive way, both visually and conceptually, because those two sleights of mind can be visually anchored to an actual depiction of a space—a diagram where certain parts (vertices) are related to each other by arrows—and because diagrams are a direct and exact representation of the structure of a concept of space, namely a set of relations between entities. This correspondence is visual and mathematical, as we have seen. But in the Finiteness-based $\mathcal{M}$-*signature*, spatialization and despatialization are more than a convenient correspondence between diagrams and fundamental category-theoretic constructions that are found practically everywhere in modern mathematics: they are also modes of consciousness. They symbolize the experience of the grasping of an object of thought, and in this, I think, rests their most profound meaning. Limits and colimits are complementary ways to grasp objects of thought and articulate thoughts, and contravariance is the process of manipulating a concept via spatialization and despatialization in order to reveal its inner structure. This makes Finiteness-based contravariance a process-based representation of `limit`.

Definition 16.3.1. (`finiton` – *informal*)
A `finiton` is the pattern defined in Figure 16.4 where $\mathcal{F}$ denotes a diagram and $\bullet \to \mathcal{F}(\bullet)$ and $\mathcal{F}(\bullet) \to \bullet$ denote respectively various generalization of the category-theoretic notions of limit and colimit respectively, and where contravariance denotes the category-theoretic operation of inverting arrows.

16.4 The Minimal Interaction Model

The Minimal Interaction Model of the Finiteness-based $\mathcal{M}$-*signature*:

- *is* a Type-A Interaction Model that models the reconstruction of finiteness (Definition 11.5.1),
- *is* an atomic, irreducible model of the fundamental makeup of the Finiteness-based mind-world interaction (Definition 11.5.2),
- *is* a prototypical model of *inclusion*—the Actuation Mechanism of *boundedness*,
- *is built* via the application of the Knowledge-Reality Interface (Section 13.8) to a model of `limit`—a `finiton`,
- *defines* the conception of Finiteness-based knowledge and epistemology.

Note

In the description of the Minimal Interaction Model for the Finiteness-based $\mathcal{M}$-*signature* that follows, we assume that $\mathcal{M}$-*signature* and Cognitive Architecture are distinct entities of the model. This interpretation results in a conceptual frame of reference in which an Interaction Model is necessarily constructed on the spatialization of a certain object of thought. This choice is convenient for explanatory purposes, but it is by no means *necessary*.

Structure Of The Minimal Interaction Model

The **reconstruction of the success condition** takes place in the spatialization of an object of thought as described in the steps of Figure 16.7—compare with Figure 11.4.

A. The **reconstruction of the success condition**, which begins here, consists in applying to the elements of a spatialization of the object of thought a model of `limit` based on the `finiton`. The result of this step is a framework to describe the object of thought via a set of equational properties. In this case, though, the equational properties between the elements of a spatialization are *references* to cognitive processes that occur in a metaphysical structure of thought where the mind is not aware of the objects of thought the way it is ordinarily, and where the *only* terms of reference are objectless relations between arrows such as those described earlier in the Actuation Mechanism.

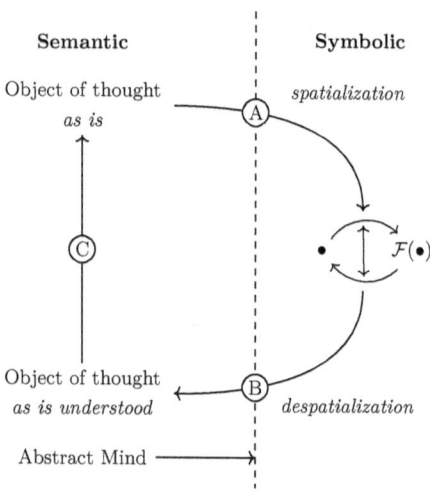

Figure 16.7: Structure of the Minimal Interaction Model in the Finiteness-based $\mathcal{M}$-*signature*.

B. This is the second half of the **reconstruction of the success condition**, and represents the construction of the **Finiteness-based knowledge**. It consists in applying to the elements of the framework defined in **A**, various models of *inclusion* based on the **finiton**. This step produces statements of the form $(Lim + Colim)A = B$ which relate any two entities A and B via **finiton**-based models of *inclusion*. Note that in the Minimal Interaction Model the equational relations between ontologies do not rely on any relation other those based on the **identiton**; they are, at least definitional and at most tautological. The notation "Lim + Colim" symbolizes a certain combination of a limit and a Colimit.

C. This step symbolizes the construction of **higher forms of knowledge**. The inferences constructed in **B** constitute a higher spatialization of the initial object of thought, and represent *how* we see and *what* we see of that initial object of thought through the equational lens produced by this Interaction Model—what is called Local Identity (Section 8.7). By applying iteratively the schema of Figure 15.6 to the image produced by the Local Identity, we obtain, in this analogy, forms of finiteness along two dimensions (Figure 16.8): the admissible types of spatialization, and the models of finiteness.

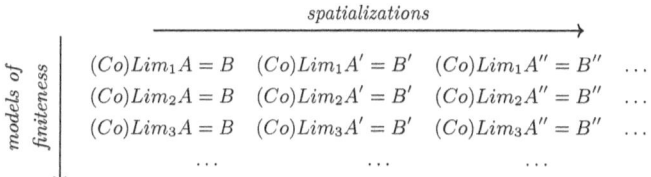

Figure 16.8: Local Identities produced by a Minimal Interaction Model in the Finiteness-based $\mathcal{M}$-signature.

where A' and B' represent higher order equational properties between lower order equations such as $(Lim + Colim)A = B$. This cognitive machinery produces equations, equations of equations, equations of equations of equations and so on, according to hierarchies of `finiton`-based constructs.

From The Minimal Interaction Model To An Interaction Model

The construction of an arbitrary Type-A Interaction Model consists in adding to the Minimal Interaction Model any number of transformations compatible with the equational properties defined by the models of `finiton`. These transformations are obtained by iterating the Minimal Interaction Model with itself (Figure 16.7-C) and by adding to it any transformation consistent with the spatialization to which the models of `finiton` are applied.

16.5 The Structure Of Knowledge By Inclusion

The Finiteness-based epistemology

- *is* an interpretation of the terms *knowledge, comprehension, meaning, explanation* and *consciousness* as defined in Chapters 12, 13 and 14, based on the mechanism through which the **concrete** concept of *finiteness* defines what is *intelligible*—as described in Sections 16.2 and 10.5, and

- *consists* in the definition of a correspondence between the concepts introduced in Chapters 12, 13 and 14 and certain mathematical constructions of Category Theory based on the **finiton**.

Here, to interpret the terms knowledge, comprehension, meaning, explanation and consciousness based on the version of intelligibility that characterized the Finiteness-based $\mathcal{M}$-*signature*, it is necessary to interpret the definitions of Chapters 12, 13 and 14 according to the meaning that those terms acquire in the this $\mathcal{M}$-*signature*. For example, Definition 12.5.1 states that the process of understanding is

> "... the determination of a minimal set of data that activates the strategy of an Observational Structure of Intentionality where the purpose of that apprehension is defined."

As we saw in Section 12.5, this definition can be modeled mathematically via various types of the category-theoretic condition of universality, and this demands us interpret the corresponding diagrammatic representations of this mathematical condition. As seen in the previous chapter, the definition of the Finiteness-based epistemology begins with the definition of the process of acquiring knowledge. To define the process of acquiring knowledge, we define the *objects of thought* that we relate through the **finiton**.

The informal process we follow to define the Finiteness-based epistemology consists of these 6 steps:

1. Define the Finiteness-based object of thought. Here the aim is to answers the question: What is the experience of an object of thought in the Finiteness-based $\mathcal{M}$-*signature*?

2. From 1, we derive the actual form of Finiteness-based knowledge, which explains the experience of knowledge in the Finiteness-based $\mathcal{M}$-*signature*. This form of basic knowledge is prototypical in that it defines at a cognitive level the practical and conceptual event of consciously knowing an object of thought *before* the development of any complex cognitive apparatus such as beliefs, contemplation or justification, and any derived notion of truth. Here we will begin

16.5 The Structure Of Knowledge By Inclusion

to notice the radical difference between the Identity-based metaphysical structure of thought and the Finiteness-based metaphysical structure of thought, and how this discrepancy is translated into a fundamentally different mental experience.

3. From 2, we derive the interpretation of the diagram of comprehension (Figure 16.9, and explain the structure of Finiteness-based understanding, the fundamental process through which the changes in the Abstract Mind's configuration give rise to what we consciously experience as apprehending (one's own) thoughts, not by grasping the individuality of those thoughts, but by clutching their mutual relations of abstract inclusion—as described by the **finiton** and by some of its mathematical representations.

4. From 3, we derive the structure of the morphisms between Interaction Model, and define the conceptual machinery at the basis of the Finiteness-based protosemantic and semantogenetic processes.

5. From the points above, we outline the basic elements of the Finiteness-based explanatory system.

6. From the points above, we outline the archetypical structure of the conceptions of Finiteness-based communication and mathematics.

7. From 1, we deduce a description of the two key features of consciousness, unity and flow, and offer a characterization of these in the Finiteness-based state of consciousness.

Finiteness-based Object Of Thought

An object of thought in the Finiteness-based
$\mathcal{M}$-signature can be thought of as a
(co)limit-preserving endomorphism.

Here we use the ideas presented in "Consciousness And Spatialization" in Section 14.3 about the relation between spatialization and objects of thought to describe an object of thought in the Finiteness-based $\mathcal{M}$-signature.

In the Finiteness-based $\mathcal{M}$-signature, the minimum "unit of

16 Finiteness-based *M-signatures*

thought" is an object of thought represented by the **finiton**. It represents how the Abstract Mind grasps any abstract object of thought. An Finiteness-based object of thought is, therefore, a Limit of a certain diagram $\mathcal{F}$. Much like in the Identity-based *M-signature* we think by division by forming thoughts like "this is...", "that is not..." etc., in the Finiteness-based *M-signature* the basic structure of thought is such that the Abstract Mind perceives *any* thought as a manifestation of finiteness in a form that we describe in this research via the mathematical models of **finiton**. We can, perhaps, delegate to an analogy the daunting task of giving a dramatization of the idea we are trying to convey. A Finiteness-based thought emerges not as a distinction between ontologies, but as what in the ordinary language we describe as form of inclusion. However, inclusion has to be understood in this context not a feature of a certain object of thought, but an actual mode of existence. This is truly an "existence by inclusion", much like in the Identity-based *M-signature* any experience of thought exists via a form of identity that we experience as forms of division.

An way to depict these ideas is with the following diagram

F

where the • is a mathematical gadget with enough structure to allow for the definition of (co)limits—or other generalizations of the category-theoretical notion of limit—and where F is a morphism from and to • characterized by the property of preserving (co)limits.

Finiteness-based Knowledge As Knowledge By Inclusion

Finiteness-based knowledge can be thought of as a (co)limit-preserving morphism between objects of thought.

In the Finiteness-based *M-signature*, knowledge is produced by relating the ontologies that form a spatialization of an object of thought via various category-theoretic constructions based on the **finiton** (Definitions 16.3.1 and 12.3.1). These constructions

16.5 The Structure Of Knowledge By Inclusion

represent *instances* of the Minimal Interaction Model. The hierarchical *structure* of Finiteness-based knowledge is described by the below correspondence.

1. The Finiteness-based *conception* of knowledge is represented by the **finiton**.

2. A *prototypical definition* of knowledge is the atomic cognitive process defined by the Minimal Interaction Model (Section 16.4).

3. Each *definition* of knowledge corresponds to the way in which an Interaction Model implements the **reconstruction of the success condition** based on 2.

To interpret the process of acquiring knowledge in an Finiteness-based fashion, as described in Definition 12.3.1, we need to describe how the objects of thought are related to each other via the **finiton**.

As seen in the Identity-based epistemology, the diagram above illustrates the construction we want to characterize, where • are objects of thought, represented in the ordinary Identity-based language, and F is a morphism between them. The remarks made in the previous chapter apply here for the same reasons. For a morphism to connect objects of thought defined in the same $\mathcal{M}$-*signature*, it must preserve their structure, which in this case translates into being a (co)limit-preserving morphism.

Knowledge By Inclusion

The philosophical interpretation of Finiteness-based knowledge as a (co)limit-preserving morphism between objects of thought is what we call knowledge by *inclusion*. Knowledge by inclusion denotes a Cognitive Architecture that codifies a conception of knowledge as the intellectual act of *representing* a spatialization by preserving its relation of closeness to another spatialization.

16 Finiteness-based *M-signatures*

16.6 Finiteness-based Comprehension

Finiteness-based comprehension can be thought of as a (co)limit-preserving morphism between two instances of Finiteness-based knowledge.

In the Epistemology of the Abstract Mind (Section 12.5) the process of understanding is defined as an object $\bullet^*$ together with an arrow g such that any other arrow f factors via g as in the diagram of comprehension (Figure 16.9).

Figure 16.9: Diagram of comprehension.

This condition is captured by the arrow equation below, where ∘ is a certain operation that assign one arrow to a pair of adjacent arrows—adjacent meaning that the tip of one arrow coincides with the bottom of the other arrow.

$$f = g \circ h$$

To interpret the diagram above, we use the definitions of Finiteness-based object of thought, and Finiteness-based knowledge: $\bullet$ are *objects of thought*, and $\to$ are *types of knowledge*.

As seen in the previous *M-signature*, the problem of defining the process of understanding in the Finiteness-based *M-signature*, is the problem of defining the meaning of the = sign in the equation above. The observations made in the discussions about Actuation Mechanism and Definitional Model (Section 16.2), and in the Minimal Interaction Model (Section 16.4), indicate that = codifies a relation between f and $g \circ h$ modeled by a **limit**. It follows that understanding in the Finiteness-based *M-signature* is an **finiton**-based relation between the arrows f and $g \circ h$, that is, it is a limit, a colimit, or both. Another way to picture this concept, in light of the definition of Finiteness-based knowledge, is to say that understanding is a certain system in which we come to know f via its inclusion (limit) in, or projection (colimit) on $g \circ h$.

These observations, and the notion of Finiteness-based knowledge seen in the previous section, lead to the intuitive notion that the forms of Finiteness-based understanding are (co)limit-preserving morphisms between types of Finiteness-based knowledge: one type of knowledge is represented by the arrow f, another type of knowledge is $g \circ h$. Note that the information in the diagram above that f and g point to the same object (of thought) X, and that f and h have the same origin (object of thought), is part of the *definition* of the (co)limit-preserving morphism between the knowledge systems where f and $g \circ h$ are defined respectively.

16.7 Meaning And Semantogenesis

Here we use the definitions of protosemantic transformation (Definition 12.6.1), semantogenesis (Definition 12.6.3), and the intuitive description of a semantogenesis given in "Meaning And Semantogenesis In Interaction Theory" in Section 12.6 to sketch an Identity-based semantogenesis.

Recall from Definition 12.6.3 that a semantogenesis is a mathematical object that classifies types of understanding.

A Finiteness-based semantogenesis is, therefore, a morphism

$$\Gamma : \mathcal{U} \to \mathbf{P}$$

where $\mathcal{U}$ symbolizes a class of Finiteness-based *modes of understanding*—which are protosemantic morphisms—and $\mathbf{P}$ symbolizes a space of *values*. Here we describe an example of Finiteness-based semantogenesis without the notion of protosemantic transformation.

Recall that $\mathcal{U}$ is an object where the diagrams of comprehension

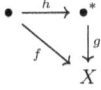

Figure 16.10: The diagram of comprehension.

are defined, and where the condition of understanding $f = g \circ h$ for the arrows in Figure 16.10 is based on a certain definition of $=$ given by the Interaction Model in which $\mathcal{U}$ is defined, and according to the Finiteness-based conception of knowledge defined in this chapter. An example of $\mathcal{U}$ is **P**, where = has the ordinary meaning, and where f, g and h are functions between sets. If in $\mathcal{U}$ we interpret the sets as *data* and the arrows as *strategies* activated by that data, we can think of the comprehension diagram as an illustration of the condition that a certain $\bullet^*$ represents the minimum data to activate a strategy (g) to achieve X.

To describe a Finiteness-based semantogenesis $\Gamma : \mathcal{U} \to \mathbf{P}$ for the example above as a certain morphism that assigns each comprehension diagram of $\mathcal{U}$ to a set in **P**, we might apply the definition of Finiteness-based understanding pedantically, and derive the description we are looking for. But this process is very abstract, and the important message it encodes would be easily obfuscated by the layers of abstraction needed to apply the definition of Finiteness-based understanding to the comprehension diagram. An intuitive way to describe the genesis of meaning in a Symbolic Intuition of Reality based on the `finiton` is via the notion of knowledge by inclusion we discovered earlier in this chapter. The function and the interpretation of **P** remain unchanged by the analogies we are about to introduce.

Since Finiteness-based knowledge is knowledge by inclusion, and since Finiteness-based understanding is the Finiteness-based knowledge that f is $g \circ h$—compare this with the definition of Finiteness-based knowledge and understanding—we can think of Finiteness-based understanding as a prototypical form of causality, where what we describe intuitively in the ordinary language as "A follows from B" is codified by the knowledge by inclusion of A *and* B. We want to use this intuition about Finiteness-based understanding to interpret how Γ creates meaning. Γ creates meaning by classifying the prototypical forms of causality symbolized by instances of Finiteness-based understanding according to a given schema **P**. What does **P** mean in this intuitive description of a Finiteness-based semantogenesis? **P** is still a representation of the purpose of this semantogenesis, and is instrumental in the definition of what type of causality fulfills a certain purpose.

Semantogenesis As Measure Of Modes Of Understanding By Inclusion

This brief analysis of meaning in a Finiteness-based Symbolic Intuition of Reality with the tools of Interaction Theory shows that it is indeed possible to not only conceive, but also relate to ordinary experience the deeply unintuitive metaphysical structures of thought of a Finiteness-based $\mathcal{M}$-*signatures*. Nothing like the creation of meaning provides a more vivid link to reality and experience. Finiteness-based meaning is a set of cognitive processes that guide our understanding of reality via various relations of inclusion, which in our ordinary conceptualization of experience are, perhaps, best understood via the analogy with causality and the analogy of set-theoretic inclusion—and how this gives rise to mathematical forms of finiteness.

Toward Finiteness-based Information

This discussion about Finiteness-based semantogenesis reveals another aspect of this Symbolic Intuition of Reality: the existence of a new type of information. The correspondence between meaning and information pointed out in Section 12.6 reemerges in this interpretation of semantogenesis: the more (or the less) inclusive a meaning, the more (or the less) a certain type of understanding is relevant to a context, the higher its information content. Here the informational content of a certain type of Finiteness-based understanding is measured via the same mechanism used to create knowledge, and is therefore a measure of inclusiveness, not exclusiveness like in the ordinary notion of information. We will revisit these ideas in Chapter 20.

16.8 Explanatory Systems

As seen in "The Principle Of Epistemic Relativity" (Section 13.1), the exploration of explanatory systems beyond the ordinary Identity-based $\mathcal{M}$-*signature* is based on the observation that there seems to be no indication that in Nature the **concrete** concept of identity is a privileged frame of reference to define mathematical thought and a scientific method. In this brief introduction to Finiteness-based explanatory systems we outline the scaffolding of a system of thought to construct explanations based on the paradigm of epistemic continuity described in Section 13.3 applied to the Finiteness-based metaphysical structure of thought. To begin to outline the structure of a Finiteness-based explanatory system, we answer the following questions. Section 16.9 outlines the structure of Finiteness-based explanations.

16 Finiteness-based $\mathcal{M}$-*signatures*

- What are explanandum and explanans?
- What are the accepted truths and self-explanatory ontologies?
- What is epistemic continuity?
- What is a reality?

Explanandum And Explanans

In the Finiteness-based structure of thought, the Abstract Mind relies on models of the `limit` to articulate concepts: it knows by inclusion. While explanandum and explanans remain references to the semantic and symbolic parts of the Finiteness-based mind-world interplay, the language we use to talk about them is Identity-based, therefore we need to interpret what they mean. To this end, we must remember what thinking in an Finiteness-based fashion means: it means that the Abstract Mind grasps concepts via various forms of inclusion. And this is precisely the relation between explanandum and explanans, it is the relation between the symbolic and the semantic parts of the Finiteness-based mind-world interplay. Hence the relation between explanandum and explanans—what an explanans says about an explanandum—is to be understood as a `finiton`-based relation between two or more Finiteness-based objects of thought, which captures properties that are common to a number of objects of thought greater than those directly involved in the construction of the Finiteness-based explanation—recall the "Structural Canon" of good explanations in Section 13.2.

Accepted Truths And Self-explanatory Ontologies

Recall Definition 13.2.1: *"An accepted truth is a spatialization of an explanandum, in which a reality is defined."* The three spatializations described in Section 16.3 are the accepted truths in the Finiteness-based mode of thought, because they are the stage where Finiteness-based thought takes place. Within the horizon defined by the accepted truths, the self-explanatory ontologies, are the minimal, irreducible unities of Finiteness-based thought, which, by virtue of Definition 16.3.1, and of the definition of the Finiteness-based object of thought, are mathematical models of the notion of (co)limit. The notion of truth follows from the above and from Definition 13.3.2. In summary:

16.8 Explanatory Systems

- The Finiteness-based accepted truths are: one-as-many, many-as-one and contravariance.
- The Finiteness-based self-explanatory ontologies are the various mathematical models and generalizations of (co)limit.

Epistemic Continuity

Recall Definition 13.3.1: *"Epistemic continuity is the belief that an explanandum has an intrinsically protosemantic cohesiveness that corresponds to the symbolic cohesion inherent in the structure of an explanatory system P to a degree consistent with the explanans produced by P."* The notion of epistemic continuity is predicated on the belief that the syntactic and semantic cohesion and coherence of our explanations corresponds to a similar type of cohesion and coherence that holds together the realities we explain in this way. To comprehend how to interpret this general schema in an Finiteness-based structure of thought, we must therefore ask: How do we characterize the cohesion and coherence of a Finiteness-based explanans? and How do we relate the coherence of a Finiteness-based explanans to what we believe is the intrinsic order that holds together a Finiteness-based explanandum? The coherence of a Finiteness-based explanans is syntactic and synthetic. In contrast, the coherence that (we have to believe) holds together a Finiteness-based explanandum is semantic: it is the coherence we recognize and validate when we can explain a Finiteness-based reality in the language of a Finiteness-based syntax. What's important to observe, at this point, is in what way the Finiteness-based syntax-semantics dichotomy differs from the Finiteness-based dichotomy. The Identity-based conception of knowledge, being based on various flavors of logic and causality, is predictive: you don't know something enough, rationally, if you can't make quantitative predictions about it. This perspective changes radically in the Finiteness-based mode of thought, where the conception of knowledge is based on forms of inclusion. As seen in the paragraph above about Explanandum and Explanans, Finiteness-based knowledge revolves around a form relation of (**finiton**-based) inclusion between explanans and explanandum. The validation of Finiteness-based knowledge occurs therefore not via causality or logic, but via the *existence* of a type of inclusion between explanans and explanandum. The existence of a relation of inclusion between explanans and explanandum, expressed via the

16 Finiteness-based $\mathcal{M}$-*signatures*

finiton, is the type of coherence needed in order to validate our Finiteness-based knowledge of reality. Thus, we might say that the structure of the correspondence underpinning the Finiteness-based form of epistemic continuity has the form of canons and schemes. In summary: *Finiteness-based epistemic continuity is predicated on the belief that the existence of a relation of closeness to a diagram between explanandum and explanans—described mathematically by various category-theoretic notions of limit—is a faithful and reliable expression of the intrinsic order of the Finiteness-based explanandum.*

Reality

The semantic part of a mind-world transaction is a *reality*. The same notion is expressed in the language of Interaction Theory in Definition 13.3.3: *"A reality is the experience created by an Interaction Model. "*

The semantic content of a reality is the collection of all explanations based on the symbolic content of that reality, interpreted according to the schema of epistemic continuity. The semantic content of an unexplained reality consists only of the accepted truths. To understand why, recall Definition 13.2.1: *"An accepted truth is a spatialization of an explanandum, in which a reality is defined. "*

From Definition 13.3.3, it follows that the most elementary kind of reality is the experience that originates from the most elementary Interaction Model, namely, the Minimal Interaction Model. As seen in this chapter in the description of a Finiteness-based object of thought, a **finiton** represents how an Abstract Mind in a Finiteness-based state grasps thoughts at a fundamental level, and coincides with the fundamental experience of thought described in Section 14.1 between a type of spatialization and the experience of an object of thought. The experience of a Finiteness-based object of thought is therefore the experience of a Minimal Interaction Model.

To convey an intuitive idea of what reality looks like to an Abstract Mind in a Finiteness-based state, suppose that an object of thought is the set of all natural numbers $N = \{1, 2, 3, \ldots \infty\}$. An endomorphism of N is, for example, the function that multiplies each element of N by an integer $m > 0$. Let's indicate with $m(N)$

the set of all integers multiple of m. For example, if $m = 2$, $2(N) = \{2, 4, 6, \ldots\}$, if $m = 3$, $3(N) = \{3, 6, 9, \ldots\}$ and so on. We observe that there is a correspondence between the elements of N and the functions $1(N), 2(N), 3(N), \ldots$, described by the function—it's a bijection—that assigns to each integer m the set $m(N)$. Recall the discussion about the Actuation Mechanism of finiteness (Section 16.2), and how the notion of closeness to a diagram generalizes the definition of finiteness à la Dedekind. In a Finiteness-based $\mathcal{M}$-*signature* the structure of our experience of N is *not* that of a set of distinct integers $1, 2, 3, \ldots$, as one might think in an Identity-based fashion: it is solely and entirely an experience defined by the relations of closeness to a diagram satisfied by $1(N), 2(N), 3(N), \ldots$, an example of which is the condition of (in)finiteness à la Dedekind. To comprehend why a relation of closeness to a diagram on $1(N), 2(N), 3(N), \ldots$ is an Finiteness-based experience of an object of thought, we observe that if a sequence of integers $a, b, c, \ldots$ satisfies the condition of finiteness à la Dedekind, so does the sequence $a(N), b(N), c(N), \ldots$. In other words, the only existential feature that counts is the existence of a certain condition of closeness to a diagram, and *not* which set of elements expresses that condition: this independence of a condition of closeness to a diagram from the object of thought—represented in an Identity-based fashion—is what we call the Finiteness-based experience of N.

16.9 Encompassment And Nearby Efficacy

The central themes of Finiteness-based explanations are the genesis of finiteness through proximity, and how the World appears through the notion of encompassment.

Here we use the rationale presented in "The Metastructure Of Explanations" (Section 13.7), and the definitions of *lens* and *image* (Definitions 13.7.1 and 13.7.2 respectively), to outline the structure of Finiteness-based explanations, and to describe the limits of the interpretations of what we see and comprehend when we look at the World through a Finiteness-based explanation.

Recall from Section 16.2 that *inclusion*, the Actuation Mechanism of finiteness, is the phenomenology of boundedness, the Organizing Principle of finiteness. To comprehend the structure

16 Finiteness-based $\mathcal{M}$-signatures

of Finiteness-based explanations we follow the line of reasoning we introduced in Section 13.7: we need to explore what connects a Finiteness-based explanation to the mechanism through which the **concrete** concept of finiteness defines intelligible thought, and understand how this nexus defines the conceptual structures that differentiate a Finiteness-based explanation from any other Finiteness-based utterance or thought. In "The Deep Structure Of Explanations" (Section 13.7), we discovered that for an utterance to be an explanation, it must necessarily, directly or indirectly, describe the genesis of the definitional structure of the Actuation Mechanism for the $\mathcal{M}$-signature where the explanation is formulated, which in this case is: *inclusion*. With these ideas in mind, and with the concepts of *lens* and *image*, we can describe the structure of a Finiteness-based explanations as follows.

Finiteness, Encompassment And Proximity

The genesis of the definitional structure of inclusion is codified by the notion of *proximity*. Proximity is the *lens* of finiteness, and expresses at an archetypical level the mechanism through which the notion of inclusion originates. The term proximity is used here to denote an abstract relation of closeness with respect to a given spatialization of an object of thought, and to signify how through proximity an instance of finiteness acquires its definitional structure.

Through proximity we see *encompassment*. Encompassment is the *image* of finiteness, and characterizes what we see when we look at the World through the lens of proximity. The World described by Finiteness-based explanations is *made of* encompassments because that is how we *comprehend* proximity.

Through encompassment we comprehend *finiteness*. We *grasp* the various manifestations of *finiteness* in the World through the way the notion of proximity is portrayed as encompassment in our Finiteness-based explanations.

Objectivity

The term *objectivity* denotes how an Abstract Mind in an Finiteness-based state *apprehends* the fundamental mode of existence of the things in the World.

Objectivity is defined by *inclusion*—the Actuation Mechanism of finiteness—and characterizes the condition that resolves the epistemic tension between the semantic and symbolic parts of a mind-world transaction. Thus, objectivity is formulated as a *reference* to a model of `limit`—such as, for example, the `finiton`.

A claim is *objective* by virtue of transcending the features that define the epistemic tension where we develop our metaphysical, theoretical and practical comprehension of an object of thought as finiteness through proximity. The profundity of objective Finiteness-based knowledge is gauged by how much of it can be *reapproached*, or reconnected to its object.

From the perspective of Finiteness-based thought, *objectivity* is the natural mode of existence of facts and ideas, from which we are "separated" only relationally, not epistemically or causally, and defines the scope of an epistemic project, namely: to describe how things are the way they are by virtue of how their intrinsic constitution is a manifestation of the inclusiveness of Nature, a concept corresponding in the Finiteness-based thought to the ordinary word *truth*.

The World Explained Through Finiteness

The *requirement* of a Finiteness-based explanation is to objectively discern, reveal and describe the types of *encompassment* that hold together the World, and in this rests its descriptive character.

The *features* of reality revealed by Finiteness-based explanations are principles of addition, subtraction, inclusion, exclusion, finitude, limitlessness and encompassment.

Through a Finiteness-based explanation, the World appears as intrinsically defined by *intensive and extensive properties*, properties that we can understand and describe through the development of our knowledge of various notions of quantity that represent fundamental types of encompassment.

Finiteness-based explanations describe the World in terms of archetypical *enfoldments*, which are instances of the Organizing Principle of boundedness.

A Finiteness-based *explanation* is a statement about the types of proximity that hold together the fabric of the World, intended as indirect references to instances of those types of proximity.

Ontic And Epistemic Contents Of An Explanation

Finiteness-based explanations are *existential*: the objects of thought or perception, and the states of affair they describe are indirect references to objects and states of affair captured in the language of the `finiton`.

Something "is" by virtue of being the *condition* according to which a `finiton` rearranges the objects of thought the way it does.

To *construct* Finiteness-based explanations we look for the conditions for which relations of proximity exist—as described by the `finiton`—between entities that we represent in the ordinary Identity-based language.

The most basic Finiteness-based explanation is the explanation of a self-evident reality, namely, an *object of thought*, and is described by the condition of epistemic continuity. Recall that Finiteness-based epistemic continuity is predicated on the belief that the existence of a relation of closeness to a diagram between explanandum and explanans—described mathematically by various category-theoretic notions of limit—is a faithful and reliable expression of the intrinsic order of the Finiteness-based explanandum.

Finiteness-based explanations are *gauged* by how much of the World they describe in terms of *nearby efficacy*.

The diagram in Figure 16.11 maps some of the points we have just presented, the description of the `finiton` (Section 16.3) and of the Identity-based epistemology (Section 16.5) in the Knowledge-Reality Interface, where we recognize spatialization, despatialization, and contravariance (Figure 16.11-A, B and C).

16.9 Encompassment And Nearby Efficacy

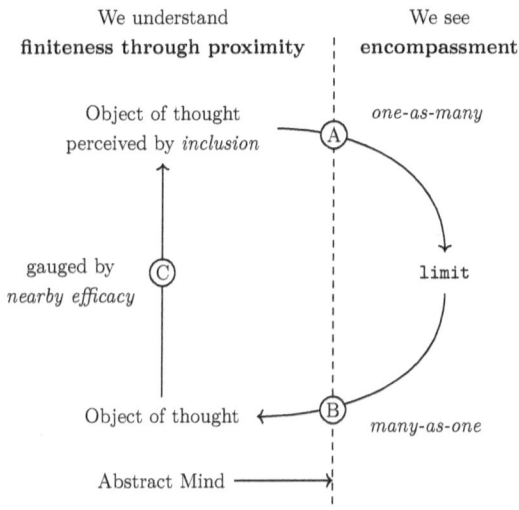

Figure 16.11: How an Abstract Mind in an Finiteness-based mode of consciousness sees and understands reality.

In Interaction Theory, the term *nearby efficacy* denotes a measure of how much a Finiteness-based explanation describes a reality—a set of known truths—in terms of the minimum number of proximity conditions according to which a `finiton` organizes the elements of that reality. The term efficacy is used in this context to denote a measure of certain extensive properties through which inclusion is represented, and does not have any causal connotation. Epistemic continuity applied to a Finiteness-based explanatory framework tells us that the greater the nearby efficacy, the more fundamental or universal the kind of encompassment through which we comprehend a reality.

From Proximity To Finiteness To Encompassment

An important element of the deep nature of Finiteness-based explanations is the relation between proximity and the overarching concept of finiteness. How does the abstract notion of "being close to" signify or enable or herald or set the scene for an entire mode of thought based on finiteness? Is perhaps proximity a prototypical version of finiteness, or is it the other way around? In the discussion about the **finiton** we have given a simple mathematical interpretation of the notion of proximity via the category-theoretic notion of (co)limit, and have shown how a relation of proximity to a certain diagram gives rise to an intuitive version of set-theoretic finiteness—finiteness à la Dedekind. From there, we have used the same category-theoretic apparatus to generalize the set-theoretic notion of finiteness as (co)limits of general diagrams. Hidden in the rarefied symbolism of the category-theoretic treatment of proximity is the nexus between proximity, finiteness and encompassment. We can, perhaps, convey what connects proximity, finiteness and encompassment at an archetypical, pre-conceptual, pre-cognitive level, by comparing them to the evolution of an intentional act. We can describe the nature of finiteness by comparing proximity to an intent, and the achievement of that intent to a condition of encompassment consistent with that form of proximity. Finiteness is, in this sense, the evolutionary process of an elementary mode of consciousness that relates us humans with the World through various forms of encompassment. This is, essentially, how and what Finiteness-based explanations explain the World.

Fully Explainable Realities

The notion of fully explainable reality generalizes Lawvere's Fixed Point Theorem (Section 8.4 "The Limits Of Formal Systems") to the context of a Finiteness-based $\mathcal{M}$-*signature*: a reality is fully explainable if it is fixed-point-free, in the sense of this generalization. Roughly, a fixed-point in a Finiteness-based theory is a fixed-point as in Lawvere's original formulation but in the equational system defined by the **finiton**—this is a work in progress.

16.10 Finiteness And The Language Of Proximity

In "The Archetypical Structure Of Communication" (Section 14.5) we introduced the idea that communication is that which results in a change of Abstract Mind state, and in "How Data Becomes Message" (Section 14.6) we used this rationale to categorize communication based on the types of Abstract Mind state transitions.

To describe the structure of Finiteness-based communication—the communication that occurs when the Abstract Mind is in an Finiteness-based state—we interpret the classification of the

16.10 Finiteness And The Language Of Proximity

Abstract Mind state transitions using finiteness as a reference Abstract Mind state, and derive from this analysis the structure of Finiteness-based languages.

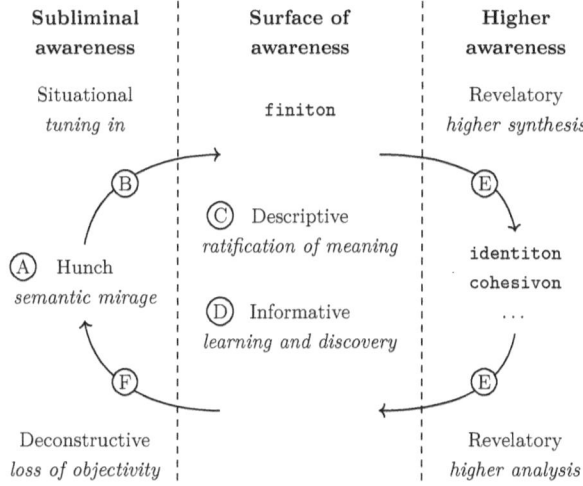

Figure 16.12: The 6 types of Finiteness-based communication.

Figure 16.12 is a close-up view of the Abstract Mind in the Finiteness-based state. The region between dashed lines represents the surface of awareness currently lit by consciousness, and signals where the mind-world transactions are defined by a conception of intelligibility based on the **concrete** concept of *finiteness*, and where, therefore, the Symbolic Intuition of Reality emerges as the experience of grasping concepts as described in "The Structure Of Knowledge By Inclusion" (Section 16.5).

In the brief description of Figure 16.12 that follows, we ground the basic ideas of Finiteness-based communication to the experience that they produce, because this greatly helps to comprehend how certain subtle fluctuations of consciousness are related to these types of communication. To this end, for each type of communication in Figure 16.12, we explain when it occurs, how meaning is created, and the experience produced by the genesis of meaning.

A *hunch* in the Finiteness-based $\mathcal{M}$-*signature* (Figure 16.12-A) is a transition between Abstract Mind states that does not involve the reconstruction of finiteness. Hunches do not produce meaning because, when they occur, the cognitive structures underpinning semantogenesis are absent. Hunches produce, instead, a form of false awareness called *semantic mirage*, that in the Finiteness-based $\mathcal{M}$-*signature* corresponds to the experience of seeming intuitions about the boundedness of an object of thought, and consists in the illusion that the features of addition, subtraction, inclusion, exclusion, finitude, limitlessness and encompassment that characterize the experience of that object of thought, signify that the identity of *that* object of thought is actually being revealed through the changes in our awareness of it. For these reasons, epistemic continuity (Section 13.3) does not apply to hunches.

Situational communication (Figure 16.12-B) occurs when a transition between Abstract Mind states gives rise to a Minimal Interaction Model for the Finiteness-based $\mathcal{M}$-*signature*. The emergence of a Minimal Interaction Model in the Finiteness-based $\mathcal{M}$-*signature* signals that the mind is spatializing through the `finiton`. This produces a sense of *tuning in* to what one is thinking of—see also Contravariance in the `finiton` (Section 15.3)—and is the experience of becoming aware of an object of thought (Definition 14.2.1), that in the Finiteness-based $\mathcal{M}$-*signature* corresponds to the grasping of an object of thought as a type of encompassment through the features that define *knowledge by inclusion* (Section 16.5). Situational communication marks also the beginning of an embryonal form of Finiteness-based objectivity, where intuition, knowledge, comprehension and meaning are guided by the Finiteness-based form of epistemic continuity—see also "Epistemic Continuity" in Section 16.8.

Descriptive communication (Figure 16.12-C) occurs when the Abstract Mind transitions to and from the same Interaction Model within the Finiteness-based $\mathcal{M}$-*signature*. Descriptive communication is the most basic form of abstraction, and produces the experience of becoming aware of the current Interaction Model—as described in Section (14.2)—by objectifying it according to the structure of *knowledge by inclusion* (Section 16.5). Descriptive communication ratifies meaning via the forms of Finiteness-based epistemic continuity where principles of addition, subtraction, inclusion, exclusion, finitude, limitlessness and encompassment explain the qualities of multitude, gathering, conglomeration and aggregation of the objects and states of affair in the World. These

16.10 Finiteness And The Language Of Proximity

produce examples of how semantogenesis measures different modes of understanding by inclusion of the "same" object of thought accessed through that Interaction Model.

Informative communication (Figure 16.12-D) occurs when the Abstract Mind transitions between distinct Interaction Models in the Finiteness-based $\mathcal{M}$-*signature*. Informative communication gives rise to the experience of *learning by inclusion*, as explained in Section 16.5, and creates meaning by revealing the content of an object of thought through various forms of relations of proximity—the abstract concept of closeness to a diagram used in the `finiton`—between different mental representations of that object. Linguistically, whenever applicable, informative communication conveys how principles of addition, subtraction, inclusion, exclusion, finitude, limitlessness and encompassment make up a World where things exist as forms of encompassment.

Revelatory communication (Figure 16.12-E) occurs when the Abstract Mind changes $\mathcal{M}$-*signature*, and signals a deep change in the way the intellect functions. A synthetic type of revelatory communication denotes a transition *to* a non-Finiteness-based $\mathcal{M}$-*signature*, and gets its name from the fact that the activation of a different $\mathcal{M}$-*signature* produces the experience of reframing an entire system of thought, and not just specific types of knowledge. In synthetic revelatory communication, meaning is created by translating the structure and purpose of a non-Finiteness-based epistemic project into the language of Finiteness-based epistemology. This is, for example, the interpretation of a view of the World not based on types of encompassment, in terms of the concepts of the Finiteness-based epistemology, and conveys the intuitive experience of thinking of an object from the perspective of how the knowledge of that object is acquired and understood in terms of addition, subtraction, inclusion, exclusion, finitude, limitlessness and encompassment, rather than from the narrative of the frame of reference within which that object was originally known and conceived. Similarly, an analytic type of revelatory communication denotes a transition *from* a non-Finiteness-based $\mathcal{M}$-*signature* to the Finiteness-based $\mathcal{M}$-*signature*. Its name suggests that the experience associated to this transition is that of how thinking from a perspective not based on multitude, gathering, conglomeration and aggregation primes the mind to appreciate new details of what is already known. Analytic revelatory communication produces meaning by reframing or interpreting the narrative of a Finiteness-based epistemological project into the language of non-Finiteness-based $\mathcal{M}$-*signature* without changing the structure and purpose of that epistemic project. This is, for example, the interpretation of a encompassment-based view of the World in terms of non-Finiteness-based $\mathcal{M}$-*signatures*.

Deconstructive communication occurs when the Abstract Mind disengages the Finiteness-based Symbolic Intuition of Reality, and enters a pre-cognitive or pre-conceptual state. This type of communication corresponds to a *deconstruction* in the sense that, as a result of this transition, the reconstruction of finiteness no longer takes place. In deconstructive communication, the Abstract Mind suspends its awareness of an object of thought by inclusion and of the corresponding epistemology—as modeled by the `finiton` (Section 15.3), and as explained in "Encompassment And Nearby Efficacy" (Section 16.9)—and this interruption results in the disintegration of the current Symbolic Intuition of Reality. Deconstructive communication does not produce meaning as such, because it suspends the cognitive structures underpinning semantogenesis. As a result, the identity of an object of thought is no longer grasped via the qualities of multitude, gathering, conglomeration and aggregation, and this produces a class of experiences that can be described as a loss of objectivity.

Worlds, Languages And Utterances

A Finiteness-based language is a method to communicate how addition, subtraction, inclusion, exclusion, finitude, limitlessness and encompassment make up a World where things exist as encompassments.

In Section 14.7 we defined an archetypical language as a method to communicate how the World is seen and understood by an Abstract Mind in a cognitive state (Definition 14.7.1).

We know how the World is seen and understood by an Abstract Mind in a Finiteness-based state. In "Encompassment And Nearby Efficacy" (Section 16.9), we described the basic features of reality as it appears through the lens of finiteness. There, we used the ideas of "The Metastructure Of Explanations" (Section 13.7) to outline the fundamental mode of existence of the things in a

16.10 Finiteness And The Language Of Proximity

Finiteness-based World, and to describe how that World can be described through the features of its objects, and of the relations between those objects. To introduce Finiteness-based languages, we revisit here some of those ideas—also summarized in Figure 16.11—to explain two essential dimension of this problem:

- How the World *seen* through finiteness defines the *purpose* of Finiteness-based languages.
- How the modes of existence of the things in a World *understood* through finiteness, and the features of that World, define the *structure* of Finiteness-based utterances.

Let us recall that the World *seen* by an Abstract Mind in a Finiteness-based state defines how things *exist*. In "Finiteness, Encompassment And Proximity" we observed that the World we *see* through finiteness is made of encompassments.

> *"[t]hrough proximity we see encompassment. Encompassment is the image of finiteness, and characterizes what we see when we look at the World through the lens of proximity. The World described by Finiteness-based explanations is made of encompassments because that is how we comprehend proximity."*

Encompassment is, therefore, how things *exist* in the World seen through finiteness, and defines the fundamental purpose of Finiteness-based languages as technologies to convey how certain manifestations of encompassment make up the World.

The features of World *understood* by an Abstract Mind in a Finiteness-based state define the fundamental *features of existence*. In "The World Explained Through Finiteness", we observed that

> *"[t]he features of reality revealed by Finiteness-based explanations are principles of addition, subtraction, inclusion, exclusion, finitude, limitlessness and encompassment."*

These features dictate that the types of *sentences* of a Finiteness-based language are *necessarily* about how diverse existential relations of proximity hold together Finiteness-based realities by virtue of forming types of encompassments. This is how, at the most fundamental level, how Finiteness-based languages relate us to the World via epistemic continuity.

> *"[t]hrough encompassment we comprehend finiteness. We grasp the various manifestations of finiteness in the World through the way the notion of proximity is portrayed as encompassment in our Finiteness-based explanations. "*

From these observations, it follows that Finiteness-based languages function by *evocation*, like symbols. It is essential to emphasize that the term "evocation" is used in this context to convey a concept that pertains to a different metaphysical structure of thought, and that, as such, we can grasp exclusively through metaphors, not through references. The metaphor of evocation helps us comprehend the structure of Finiteness-based languages from the point of view of the Identity-based structure of thought that we are using to describe these concepts in ordinary language: but that is where the meaning and content of this metaphor end: in an Abstract Mind in a Finiteness-based state there is no such thing as evocation.

16.11 Conception And Foundations Of Mathematics

A conception of Finiteness-based mathematics is a fundamental intuition about how what we see as encompassment explains what we comprehend as manifestations of finiteness through proximity.

Here we apply the ideas we presented in "Encompassment And Nearby Efficacy" (Section 16.9), in "Finiteness And The Language Of Proximity" (Section 16.10), and in "Conceptions And Foundations Of Mathematics" (Section 14.8) to interpret the fundamental intuition of epistemic continuity at the basis of Finiteness-based Mathematics. The steps below outline the adaptation of the line of reasoning we presented in Section 14.8 to the structure of Finiteness-based languages and intelligibility.

The encompassment-explains-proximity intuition at the basis of the Finiteness-based conception of mathematics, constitutes a mental attitude and a metaphysical sensitivity that allows us to grasp and use the symbol-meaning epistemic tension in a way that we can illustrate as follows.

16.11 Conception And Foundations Of Mathematics

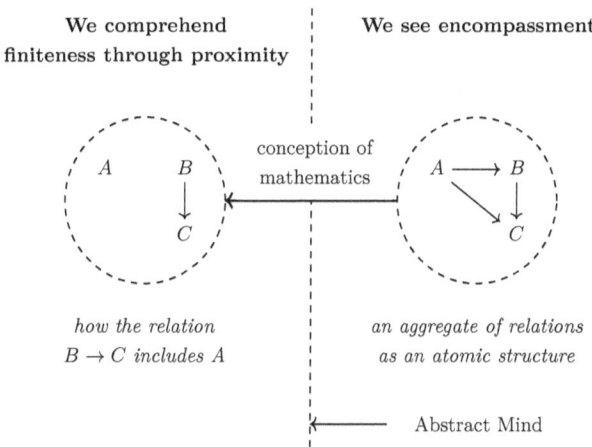

Figure 16.13: Illustration of the Finiteness-based conception of mathematics as the fundamental intuition about the relation of epistemic continuity between encompassment and finiteness through proximity.

- The qualities of *inclusion* are multitude, gathering, conglomeration and aggregation, and form the substratum where the principles of addition, subtraction, inclusion, exclusion, finitude, limitlessness and encompassment that make up a Finiteness-based reality, exist as *encompassment*.

- A *prototypical concept* and a model of multitude, gathering, conglomeration and aggregation is the *intuitive notion of a graph*. Intuitive means that a graph is grasped as a whole, and the relations it is made of, depicted as the arcs connecting two vertices, are defined by the very grasping and not through the vertices they connect.

- A *conception of mathematics* (Figure 16.13) is an intuition about the existence of a fundamental relation of epistemic continuity, intrinsic in the Finiteness-based intuition of reality, by means of which what we see as forms of encompassment *explains* what we comprehend as manifestations of finiteness through proximity.

- *Encompassment explains proximity* through the intuitive, prototypical notion of graph, and its natural attributes and organization—union, intersection, transposition, permutation, etc.—that, together, constitute an archetypical language to convey how addition, subtraction, inclusion, exclusion, finitude, limitlessness and encompassment *explain* multitude, gathering, conglomeration and aggregation.

- Finiteness-based mathematics is a language of *relation and structure*. It is the language of proportions and aesthetic and structural canons.

- A *foundation of mathematics* is a language to model of the intuition at the basis of the conception of mathematics based on the prototypical notion of graph and its natural operations.

- The distinctive features of Finiteness-based mathematical reasoning are abstract notions of density, separation, connectedness, inversion, and their many generalizations, and characterize the natural ways to construct relations.

Finiteness-based Computers

An archetypical Finiteness-based computer, what in Interaction Theory we should call a *proximity computer*, is a device or a system that realizes the features of Finiteness-based languages via the canons according to which its is built, and that realizes with its structure types of encompassments as described in Section 16.3. These can be thought of as objects that implement in their form or structure various mathematical forms of closeness to a diagram (Section 16.3) that give rise to numerous fundamental mathematical objects of Category Theory, such as (co)products, (co)equalizers, pullbacks and pushouts, etc. Finiteness-based computers are structures that allows for the parametrization of forms of encompassment.

The instructions to set their parameters are still called programs, and correspond to what in ordinary language we generally refer to as canons, such as architectural canons, numeric canons such as ratios, and aesthetic canons. Finiteness-based programs and programming languages are designed to model the features of addition, subtraction, inclusion, exclusion, finitude, limitlessness and encompassment.

Finiteness-based computers don't need to have moving parts.

They do not realize processes of any sort—recall the fundamental mode of Finiteness-based existence.

16.12 Features Of Consciousness

In the Finiteness-based $\mathcal{M}$-*signature*, the structure of any mental or perceptual subjective reality is defined by the three modes of consciousness modeled by the **finiton**. The characterization of the condition of awareness given in Definition 14.2.1 tells us that the basic features of the Finiteness-based conscious experience, that is, the basic features of Finiteness-based reality, are to be sought in the epistemology of an object of thought.

Unity Of Consciousness

In the Finiteness-based $\mathcal{M}$-signature, the unity of consciousness is experienced via forms of inclusion.

The unity of consciousness in the Finiteness-based $\mathcal{M}$-*signature* denotes the basic feature of any conscious experience of being consistent with the same overarching form of inclusion defined by the **finiton**. Unity is, therefore, synonym of coherence with respect to an archetypical template of inclusion that defines the metaphysical structure of any Finiteness-based experience. In this sense, unity does not imply or rely on a notion of part to be defined. Unity means that any experience of an object of thought is such that it is consistent with an underlying thought architecture through which the conscious experience evolves. The underlying structure of consciousness in this $\mathcal{M}$-*signature* is of course the **finiton**. The statement that in the Finiteness-based $\mathcal{M}$-*signature* the unity of consciousness is experienced via forms of inclusion means that in the every aspect of the conscious experience it is possible to consistently recognize through a form of inclusion, as modeled by the **finiton**, the creator and the recipient of the conscious experience. The term inclusion, refers to the way the Abstract Mind grasps and constructs the subjective experience, and does not denote the content of the experience. A way to convey a concept so impervious to intuition is to imagine that the constant presence of and reliance on relations of inclusion to connect to a reality conveys a sense of consistency with the archetypical principle with which any thought is grasped by a Finiteness-based Abstract Mind. The *existence* of a relation of inclusion as defined by the **finiton** is to a Finiteness-based Abstract Mind what identity is to the ordinary mode of thought. The main obstacle to imagining this situation is of course that we cannot think without identity. However, an intuitive way to characterize what the unity of consciousness should feel like in the Finiteness-based $\mathcal{M}$-*signature* could be to imagine that no matter what the experience, there is a constant, intuitive knowledge and feeling that *that* experience is in some relation of closeness to *all* the other experiences, in the way we have described the notion of closeness to a diagram in the description of the **finiton** (Section 16.3).

Flow Of Consciousness

In the Finiteness-based $\mathcal{M}$-signature, the flow of consciousness is experienced as boundedness through inclusion.

By flow of consciousness in the Finiteness-based $\mathcal{M}$-*signature*, we mean the quality shared by every conscious experience of being connected to the conscious experience that precedes it according to the canons that dictate how connectedness is comprehended by an Abstract Mind in the Finiteness-based state. Thus, to comprehend what features of a Finiteness-based experience make appear that experience as making part of a continuous flow, it is necessary to understand how continuity is characterized in the spatializations defined by the Finiteness-based. Continuity in the Finiteness-based $\mathcal{M}$-*signature* is continuity with respect to the ways in which inclusion operates in the objects of thought constructed via the **finiton**.

Roughly, *continuity* is used here as a mass noun to denote various form of finiteness allowed by the type of spatialization defined by the structure of the **finiton**.

Bibliographical Notes

Infinites are problematic, this is why in the history of Philosophy, Mathematics and Science we find various attempts, more or less explicit, to propose forms of *finitism*, a worldview according to which only finite entities exist. Very broadly speaking, infinities are problematic because the basic structure of human thought

16.12 Features Of Consciousness

and the Symbolic Intuition of Reality that originates from it, are intrinsically spatiotemporal. This means that geometrical and temporal infinities often spell ontological, epistemological and mathematical trouble because are intrinsically non-referential: we do not have any direct experience or indirect proof or evidence of infinity in space and time. The only way in which we grasp infinity is through the use of syntactic tools like the mathematical language: only through formal languages are we able to represent infinity, and to grasp its phenomenology and its practical and conceptual implications. For these reasons, many approaches to finitism share variations of the mantra of *constructivism*: constructability is the only proof of existence.

Geometrical infinities are conceptual puzzles that emerge from certain abstractions of reality that transcend and generalize our direct experience of physical relations between quantities. Examples of these puzzles are two-dimensional geometrical figures with finite area and infinite perimeter, such as Koch's snowflake (Figure 16.14), of which we show the first five iterations, from an equilateral triangle to a snowflake-like shape.

Figure 16.14: The first 5 iterations of Koch's snowflake.

Koch's snowflake is a fractal that has an area equal to $\frac{8}{5}$ the area of the equilateral triangle on the left, but infinite perimeter because its border is infinitely corrugated in such a way that its length diverges to infinity. The conceptual imbalance posed by the fractal above, originates from a counterintuitive relation between two kinds of quantity, area and perimeter, and can be grasped purely as a syntactical construction, not via direct experience: we see snowflakes in Nature, not Koch's snowflakes.

Temporal infinities lead to so-called supertasks. An example of supertask is the Ross-Littlewood paradox. There is an empty vase and an infinite supply of marbles. We want to add the marbles to the vase according to the following rules: (1) at each step add 10 marbles to the vase and remove 1 marble from the vase, (2) each step should take half the time of the previous step. How many marbles are in the vase when the task is finished? The Ross-Littlewood paradox challenges our direct experience of time and quantity with a counterintuitive relation between them. On the one hand, rule 2 tells us that the Ross-Littlewood task can be completed by a certain time. If for example the first step takes 60 seconds, the second step takes 30 seconds and so on, the total time to complete the Ross-Littlewood task is 2 minutes. On the other hand, rule 1 plays with infinity and quantity, and leaves us scratching our head asking whether the vase will be full of marbles or empty after two minutes. First it show us that we can count each marble we add to the vase, 1, 2, 3, 4, 5,..., which leads us to think that after two minutes, at the end of the task, the number of marbles in the vase must be infinite since we add 10 marbles for each marble we remove. Then challenges this very intuition with the notion that, eventually, every single marble we have counted 1, 2, 3, 4, 5,... when we added it to the vase will leave the vase, one marble at a time.

Infinities, intended as negations of the various forms of finiteness, appear problematic only if examined from the perspective of the Identity-based $\mathcal{M}$-signature, where the structure of thought does not allow the grasping of concepts without resorting to dichotomies (see "The Structure Of Knowledge By Division" in Section 15.5). This limitation of Identity-based thought can be fully appreciated by noting how in the Koch's snowflake example infinity breaks the principle of Identity-based epistemic continuity (see "Epistemic continuity" Section 15.5) by inviting us to reconcile two irreconcilable types of quantity, a finite area inside an infinite perimeter, that no longer reflect the intrinsic order through which we comprehend physical quantities. As explained in the description of the Finiteness-based epistemology, an Abstract Mind in a Finiteness-based state grasps concepts directly via forms of inclusion—as described by the `finiton`—and *not* via antithetic entities as in "this is this by virtue of not being that". This has radical and profound implications in the way $\mathcal{M}$-*signature*-agnostic concepts such as quantity are spatialized by a Finiteness-based intellect. For an Finiteness-based mind, a geometric figure is no longer a carrier of area and perimeter in the ordinary sense, it is a manifestation of the types of inclusion defined by the Finiteness-based. From a Finiteness-based perspective, the edges in Koch's snowflake that make up an infinite perimeter are a form of infinity that is grasped

16.12 Features Of Consciousness

holistically, in much the same way in which a Dedekind infinite set is grasped via the property of morphisms, without any reference to quantity as such.

The category-theoretic treatment of limits can be found in any introduction to Category Theory, such as Steve Awodey 2008b and MacLane 1998, and more general forms of finiteness are in Peter T Johnstone 2002b D5.2, D5.4. The interpretation I give here of finiteness via the category-theoretical notion of limit de facto extends the philosophical notion of finiteness to the construction of limits of arbitrary diagrams, and, in this way, generalizes the intuitive notion of finiteness à la Dedekind.

The **finiton**, the Finiteness-based epistemology, and the interpretation of consciousness as a mode of spatialization are sketched in my unpublished work Lo Vetere 2015a. For a short bibliographical reference on consciousness see the note at the end of Chapter 14.

Chapter 17

Cohesiveness-based $\mathcal{M}$-signatures

> In general we seem to make up for inadequacies of
> the human mind [...] by a search for 'order' or
> 'meaning' often pushed to absurd limits. The
> unceasing, obsessional search for regularities is
> certainly fundamental to human intelligence.
>
> David Ruelle

The presentation of the non-standard $\mathcal{M}$-signatures continues here with an introduction to the metaphysical structure of thought defined by the notion of *cohesiveness*.

The extension to concepts of the observation that like molecules stick together, is an abstract notion of cohesion. In this chapter, we describe how a theoretical mind that is wired to use cohesiveness to articulate concepts sees the World. A Cohesiveness-based mind thinks through hierarchies of concepts that originate from a single, irreducible metaphysical intuition about the mind-world interplay: we call this intuition *unity*.

The observations we made in Chapter 16 about the sharp distinction we need to make between the language we use to describe a non-standard $\mathcal{M}$-signature, and the entities of the non-standard $\mathcal{M}$-signature described by that language, apply clearly to the Cohesiveness-based $\mathcal{M}$-signature. What's peculiar about the Cohesiveness-based metaphysical structure of thought though, is that the notion of cohesiveness, as we will see, is essentially a generalization of the notion of spatialization, and this makes the philosophical and mathematical study of the Cohesiveness-based $\mathcal{M}$-*signature* particularly important from the practical and theoretical perspectives. In particular, in the Cohesiveness-based $\mathcal{M}$-*signature*, the structure of thought *is* the spatialization of an object of thought, being fully defined by the dimensions along which the elements of a spatialization of an object of thought can change, without any reference to those objects.

17.1 What Is Cohesion?

Cohesion comes from the Latin word *cohærere*, which means to stick together. The following three definitions offer three important interpretations of sticking together.

> *"Cohesiveness as the quality of forming a united whole."*
>
> Oxford American Dictionary

> *"Cohesion [...] or cohesive attraction or cohesive force is the action or property of like molecules sticking together, being mutually attractive. It is an intrinsic property of a substance that is caused by the shape and structure of its molecules [...]"*

https://en.wikipedia.org/wiki/Cohesion_(chemistry)

> *"Cohesion, in physics, the intermolecular attractive force acting between two adjacent portions of a substance, particularly of a solid or liquid. It is this force that holds a piece of matter together."*
>
> Encyclopaedia Britannica

The first definition uses unity and wholeness to describe cohesiveness. This approach adds to the observation of the sticking together of certain entities, a layer of abstraction completely independent of the concept of cohesion, implied by the notion that those entities form a whole *as a result of* their sticking together. Thus, in the first definition, sticking together appears to play a *causal* role, the effect of which validates the ascription of a quality or feature of unity and wholeness to the entities sticking together—recall "Observational Identities" Section 5.2. The sec-

17.1 What Is Cohesion?

ond definition introduces the idea that certain like entities stick together because sticking together is a property of the stuff they are made of. In this case, sticking together is described as the manifestation of the *nature or identity* of the entities at hand, and is therefore definitional—it is definitional with respect to the Observational Structure of Intentionality in which the definition of sticking together is constructed. The third definition situates cohesiveness within the larger context of the phenomenology of physical laws that explain the attractive forces we observe between two adjacent portions of a substance. In this case the sticking together of like molecules is an *effect* of intermolecular forces. The difference between the second and the third definition is subtle. In the second definition what makes the objects stick together may not be intrinsic to the nature of the objects, it can be for example an abstract of a physical property that the objects acquire, whereas in the third definition the sticking together is a direct and ontologically objective manifestation of the physical nature of the objects.

These three definitions pose a clear question about the nature of cohesiveness via the common reference model used to construct scientific explanations, namely: causality. For, in these definitions we seem to be dealing with two forms of cohesiveness, one, objective, being a physical property of the stuff certain entities are made of, or the manifestation of certain physical laws (2nd and 3rd definitions), and one subjective, being a feature of the observational process we apply to those entities (1st definition). This first, cursory analysis of cohesiveness indicates that there must be something deeper than the conceptual phenomenology of cohesiveness at work in our conception of cohesiveness, something that we cannot see clearly if we limit our analysis to the definition of cohesion.

To illustrate the structure of this conundrum, let's consider some water molecules and a wooden table (Figure 17.1). We ask: What is the difference between the cohesiveness of water molecules, and the cohesiveness of table parts? The cohesiveness we observe in water molecules denotes a physical property of water molecules for which they exhibit a certain force of mutual attraction. Cohesiveness in a table indicates that certain pieces of wood which we intentionally cut and assembled in such and such way, form a certain wooden artifact that we identify as a whole and agree to refer to as a table. The physical force of mutual attraction that keeps the water molecules together, corresponds to the action that the nails exert in keeping together the wooden pieces. Similarly, the macroscopic phenomenon of trillions of water molecules appearing to us as water, corresponds to a few table parts that we identify as a table. What emerges from this analysis, is that when we talk informally about cohesiveness, we extend to concepts the physical notion of sticking together. Thus, there is a crucially important question that needs to be answered:

1. What does justify the extension of the notion of physical cohesiveness to concepts and man-made objects?

2. What are the assumptions and the motivations supporting this extension?

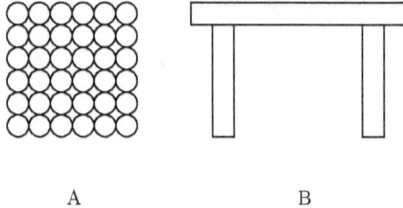

A B

Figure 17.1: Illustration of physical (A) and conceptual (B) cohesiveness.

There is a crucial difference between saying that some molecules stick together, and saying that some molecules form a whole as a result of their sticking together: a causal link between these two statements has to be either discovered or invented. This missing causal link needs to count as an adequate explanation of how the assertion that certain physical forces cause like-molecules to attract each other justifies the search for analogies with physical properties in the concepts that we invent to describe those very physical properties. This contrast becomes apparent in the table example. In the *object* table, the table parts are held together by nails. In the *concept* table, which we use as a reference to a certain definitional structure, wooden pieces become table parts *via the definitional structure itself*. Keep the definition of wooden table, and remove the nails from the physical table, and those table parts remain the components of a table, with the difference that the correspondence between the relation imposed on those pieces of wood by the definitional structure of wooden table ceases to correspond to the physical relations between those parts.

Unless we think that the human intelligence is the only possible form of intelligence, which I think would be naïve and presumptuous, we are forced to concede that the justification for extending cohesiveness from the physical domain to the mental domain must reside on a larger belief system than the mere observational data. That larger belief system is analogy. What does analogy have to do with rational reasoning? Well, I have no desire to discredit analogy as a style of reasoning, but analogy is itself based on the underlying belief that to a certain degree, the extent of which is a deep and vast topic of philosophical and mathematical research, the very structure of all things that exist is governed by some form of similarity. This is a perfectly respectable claim to make that must be made manifest in a definition of cohesiveness if we choose to accept, invoke and require analogy as the foundation of a definition of cohesiveness. From a mathematical, metaphysical and practical viewpoint, I think the deep structure of cohesiveness is essentially this: *it is the hidden link by means of which we extend the physical sticking together to concepts and man-made objects.*

17.2 Cohesiveness-based Intelligibility

Cohesiveness-based intelligibility is a structure of thought in which the human mind has epistemic access to the world via a principle of unity, which extends to concepts the physical property of sticking together.

The definition of Cohesiveness-based intelligibility that follows is the conceptual nexus between the philosophical and psychological domains, that have been the stage where we have developed a framework to define the structure of the Abstract Mind, and the mathematical domain, that will be the arena where we will begin to translate these insights into mathematical structures.

This section examines with the tools of Organizing Principle, Actuation Mechanism and Definitional Model the mechanism through which a **concrete** concept of cohesiveness gives rise to a notion of intelligibility. The pattern through which cohesiveness creates intelligibility is the same pattern described in the Identity-based

17 Cohesiveness-based $\mathcal{M}$-signatures

and in the Finiteness-based $\mathcal{M}$-signatures, and the considerations made in those chapters about the inadequacy of natural language to describe non-Identity-based modes of thought apply, necessarily, to this mode of thought.

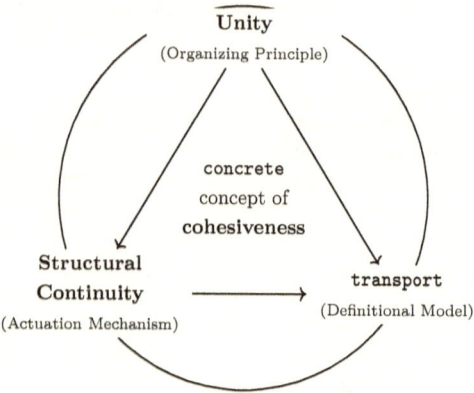

Figure 17.2: The structure of Cohesiveness-based intelligibility.

Organizing Principle Of Cohesiveness

*Unity is an unintelligible metaphysical intuition about the potential of the **concrete** concept of cohesiveness.*

The capacity encoded by cohesion is *unity* (Figure 17.2). Unity characterizes how an abstract model of "sticking together" defines cohesion. Regardless of whether cohesiveness is regarded as a cause, an effect, or as the intrinsic nature of the objects of thought or perception at hand—recall the discussion of Section 17.1—the notions of unity and togetherness are, by design, based on the definition of an underlying principle through which cohesion become manifest and intelligible: they are distinct from cohesiveness, but indistinguishable and inseparable from it. For example, when the principle is a force of mutual attraction, which we detect via spatial relations, unity is defined *through* a concept of space because that specific form of the principle of sticking together manifests itself through space—or more precisely, because it's been spatialized. Clearly, the function of space in this example is purely instrumental to the manifestation of the core principle of sticking together.

17.2 Cohesiveness-based Intelligibility

Cohesion generalizes the mechanism through which unity arises from the sticking together, by replacing the sticking together with an arbitrary core principle, and by replacing space with any other media consistent with that core principle.

Actuation Mechanism Of Cohesiveness

Structural continuity defines the phenomenology of unity, and unity becomes intelligible to the human mind as structural continuity.

In the Cohesiveness-based mode of thought, the Actuation Mechanism describes how an Organizing Principle of *unity* manifests itself as *structural continuity* (Figure 17.2).

Structural continuity describes the exact way in which the metaphysical intuition of unity becomes intelligible thought, and how this transmutation enables the construction of models of Cohesiveness-based intelligibility. By characterizing structural continuity with the examples that follow, we pave the way for the discussion about the general form of structural continuity we'll see in Section 17.3.

The term *structural continuity* suggests that the phenomenology of unity consists of the interplay of two elements. There is a *structural* element that encodes some sort of makeup, arrangement or configuration of the abstract principle of unity, and there is the notion of *continuity* that encodes the idea that certain features through which we describe unity do not change along the dimensions of the structure with which we represent unity. This intuitive description of structural continuity can be made more precise in the context of spatialization. Anything is space. To comprehend how to choose a space to give a representation of an object of thought for a specific purpose, it is therefore crucial to understand how the properties of a space enable the description of an object. It is, in other words, necessary to delve into the structure of spatialization, and understand how that structure defines the framework to construct forms of structural continuity for an object of thought. For the purpose of the description of structural continuity, it is sufficient to outline the following points about the structure of spatialization. A spatialization can be thought as a triple $(\mathbf{T}, \sigma, D)$ where

17 Cohesiveness-based $\mathcal{M}$-signatures

- **T** is an object of thought
- $\sigma : \mathbf{T} \to \mathbf{S}$ is a process that assigns a collection of objects **S** to **T**.
- D is a set of features that characterize the objects of **S**.

D defines implicitly the *dimensions* of **S**, and with them, how the objects of **S** can *change*. There is, in other words, a faithful correspondence between a notion of *change*, and the *definitional structure* of the elements of **S** given by D. With these ideas in mind, the notion of *structural continuity* for a given spatialization σ of an object of thought **T**, captures the idea that certain features of the elements of **S** captured by D do not change along the dimensions of **S** defined by D. The concept of self-similarity—the property of an object of being identical or similar to parts of itself—is a form of structural continuity in which the concept of *part* represents a *point* in the space in which an object of thought has been spatialized.

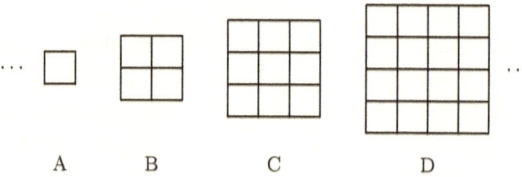

Figure 17.3: Structural continuity as self-similarity.

For example, in Figure 17.3

- **T** is the concept of square
- The spatialization σ assigns to **T** the geometrical figure **S** of a square. In this case we would say that σ consists of a space of one element, namely, **S**.
- The set of features D relative to σ are the relations between four segments to form right angles with each other and to have the same length.

As a result, D defines the conditions for *any* other **S** to be consistent with σ, that is, any object in which D is satisfied, such as the examples of Figure 17.3-A, B, C and D.

17.2 Cohesiveness-based Intelligibility

The same structural relation defined by D is satisfied by the object in Figure 17.4, where there are, in particular, 5 places corresponding to the 5 dark squares where D is recursively satisfied.

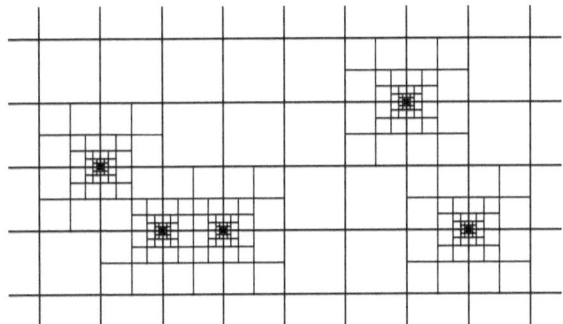

Figure 17.4: Structural continuity as self-similarity in a fractal object.

To appreciate how spatialization works, we have to think about how spatialization allows us to explore a concept with our imagination, and in general with formal tools that allow us to overcome the limitations of imagination and intuition. Figures 17.3 and 17.4 are two spatializations of the abstract notion of square in which we can visually navigate a two-dimensional space that encodes in its structure the definitional structure of square, namely: four segments of equal length arranged to form four right angles. Each spatialization defines "squareness" by providing a way to explore an arena in which the abstract properties of being a square form the fabric of a space. In Figure 17.3 squareness is codified by the tessellation of space into equal squares, and can be explored by traveling along any path defined by the grids of Figure 17.3-A, B, C and D. Figure 17.4 refines the spatialization of 17.3 by introducing the notion of shape. Shape—the ensemble of relations between the sides and angles of a square—is used to represent a square, and this gives rise to a different type of tessellation which, in turn, allows us to explore this representation of squareness along new paths.

We can give a first cursory representation of these ideas in the language of diagrams. Recall the discussion about Limits

17 Cohesiveness-based $\mathcal{M}$-signatures

and Colimits in Section 16.3. In Category Theory, the limit of a diagram without arrows is called Product, and its colimit is called Coproduct.

In the context of naïve set theory, we are familiar with the notion of product. A product between sets A and B is the set of all ordered pairs $\{a, b\}$ where a and b are elements of A and B respectively. Probably the most intuitive and most familiar example of product is the Cartesian product, depicted as the XY-plane where a point (x, y) denotes an element of the product. The prototypical element of a product, from a purely diagrammatic viewpoint, is a certain transformation that starts from *many* and ends in *one* such as the following arrow

$$\bullet \times \bullet \times \ldots \times \bullet \to \bullet$$

Here, $\to$ transforms the element of a product denoted by $\bullet \times \bullet \times \ldots \times \bullet$ into a single element $\bullet$. This conceptual machinery is at the basis of what in mathematics are called *algebras*, and expresses the intuitive idea of forming one object by *combining* multiple objects. Similarly, if we examine from a purely diagrammatic standpoint a coproduct, we recognize the following pattern

This time the conceptual machinery codified by $\to$ is the *immersion* of a certain object $\bullet$ that represents *one* into another object $\bullet \times \bullet \times \ldots \times \bullet$ that represents *many*, and that we can think of as a certain number of copies of $\bullet$ glued together according to a certain rule $\times$. These patterns are prototypes of self-similarity. Traditionally, the mathematical zoo of structures that codify self-similarity is called coalgebras, and at its core there are forms of immersion. However, in this research we are extending the notion of self-similarity in two directions to include also forms of, so to speak, reverse self-similarity such as those expresses by projections. To get a feel for why immersion codifies self-similarity, let's consider the following two rules to combine symbols

$$A \to AA \; or \; A \to A$$

With these rule it is possible to construct a string of As by juxtaposing A to itself once or twice with equal probability. By applying these rules to themselves iteratively, we obtain a string consisting of an arbitrary number of As. Thus, we can think of any string of As, such as for example $B = AAAAAAAAAA$, in two ways: as a substring of a longer string of As, such as for example BA, and, vice versa, as the product of a certain number of iterations of these rules to glue together strings of As shorter than B. Any string of As contains a string of As and is contained in a longer string of As: this is therefore an example of self-similarity in which the relation of structural continuity is "being a string of As". We can navigate the space represented by a string of As of infinite length in two directions, up and down. When we go up we juxtapose an A, when we go down we split a string of As into two substrings. No matter what direction we pick, this imaginary space of As is always similar to itself.

Definitional Model Of Cohesiveness

A **transport** *defines the epistemology of unity by modeling structural continuity.*

The Definitional Model of *structural continuity* is a philosophical device called **transport** (Figure 17.2) that codifies the prototypical structure of **success** for the metaphysical structure of thought defined by the Cohesiveness-based $\mathcal{M}$-*signature*.

The **transport**:

- is a *form of awareness*—as seen in "The Structure Of Awareness" (Section 14.2),

- defines the structure of (de)spatialization in the Cohesiveness-based $\mathcal{M}$-*signature*—see Definition 14.2.1 and "Consciousness And Spatialization" (Section 14.3),

- is used in Interaction Theory as a reference to formal models of *structural continuity*,

- forms the basis of *any* construction of the **success condition** in the Cohesiveness-based $\mathcal{M}$-*signature*—see also Section 8.3,

- is defined by the following type of *generative knowledge* (see Section 12.1) $\{Dots, \rightarrow\}$,

17 Cohesiveness-based $\mathcal{M}$-signatures

- constitutes the core of the Minimal Interaction Model for the Cohesiveness-based $\mathcal{M}$-signature.

The term **transport** is used in Interaction Theory literally: it denotes a certain system to carry stuff from one space to another. The places connected by a transport are the positions on the stage created by a spatialization. The stage is where thought takes place, and the stuff carried around by the **transport** are the Interaction Models, the ways of knowing our own thoughts. The reason why the metaphor of traveling is used in the Cohesiveness-based $\mathcal{M}$-*signature* is that, as we are about to see in the next sections, structural continuity is indeed a concept to navigate a space, both conceptually and practically.

Cohesiveness-Based Generative Knowledge

Generative knowledge in the Cohesiveness-based $\mathcal{M}$-*signature* consists of the set of archetypes $\{Dots, \rightarrow\}$, where *Dots* is a graph without arrows—just bullets—and $\rightarrow$ is an element that connects • to *Dots* in one direction, or vice versa. We find these elements in the representations of **transport**.

The representation of **transport** we introduce here is called **cohesivon**.

17.3 The cohesivon

The **cohesivon** is a *form of awareness*—as seen in "The Structure Of Awareness" (Section 14.2)—that we *describe* as an atomic and irreducible pattern that models the notion of *structural continuity*—the Actuation Mechanism of *unity*—and that defines the *structure of spatialization* in the Cohesiveness-based $\mathcal{M}$-*signature*.

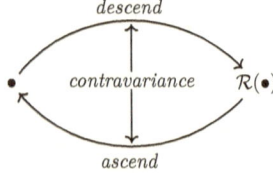

Figure 17.5: The **cohesivon**

17.3 The `cohesivon`

Compare Figure 17.5 with Figure 14.1 in "Consciousness And Spatialization", where the **concrete** concept is $\mathcal{R}$: *cohesiveness*. The form of awareness codified by a `cohesivon` is a set of "sleights of mind" that symbolize the irreducible modes in which the conscious mind operates at a fundamental level in the Cohesiveness-based $\mathcal{M}$-*signature*. These modes of operation are: *descend, ascend*, and *contravariance*.

To introduce these three sleights of mind, we interpret the diagram in Figure 17.5 as follows:

- the • represents an *object of thought*,
- $\mathcal{R}(\bullet)$ is a diagram with no arrows that represents the *spatialization* of the object of thought •,
- the arrows $\bullet \to \mathcal{R}(\bullet)$ and $\mathcal{R}(\bullet) \to \bullet$ connect • to each vertex in $\mathcal{R}(\bullet)$, and each vertex in $\mathcal{R}(\bullet)$ to • respectively,
- the arrow between arrows called "contravariance" denotes a certain symbolic operation to switch to and from $\bullet \to \mathcal{R}(\bullet)$ and $\mathcal{R}(\bullet) \to \bullet$.

If we compare the above correspondence with the definition of `finiton` (Section 16.3), we notice that the only difference between $\mathcal{F}$ and $\mathcal{R}$ is that $\mathcal{F}$ is a diagram with arrows, and $\mathcal{R}$ is a diagram without arrows. It is therefore necessary, in light of this difference, to clarify how to interpret the diagrams used to describe $\mathcal{R}$ as part of the `cohesivon`.

Interpretation Of $\mathcal{R}$

The use of diagrams to describe `identiton`, `finiton` and `cohesivon` is, in a sense, archetypical, because diagrams represent universal templates to arrange objects of thought that we can apply to mathematical objects to characterize spatialization, despatialization and contravariance. Therefore, we must give an interpretation of the arrows used in these archetypes, and in doing so, we should not be misguided by the mathematical interpretations where these archetypes are used. In other words, we must make a sharp distinction between the practical correspondence that assigns mathematical objects to bullets and certain morphisms to arrows, and the metamathematical meaning of this correspondence.

The arrows used in the characterization of a Definitional Model, such as for example in `finiton` and `cohesivon`, symbolize ontologically objective elements of the structure of thought described by that Definitional Model, and *not* the manifestation of an intentional act of relating objects of thought—recall that `identiton`, `finiton` and `cohesivon` are, by definition, atomic and irreducible. Another way to put this is that Definitional Models shape the way intentionality manifests itself in the state of consciousness characterized by a given `concrete` concept. With these ideas in mind, we interpret a diagram without arrows as a prototypical definition of *space*.

It is convenient to think of $\mathcal{R}(\bullet)$ the way we learned to think about space in The Yarn Ball Experiment (Chapter 7). There, we separated our perceptual experience of what appears to be the three-dimensional space of the human experience, from the concept of space from which we construct the experience of *that* space. We learned that the number of dimensions defines space geometrically and algebraically, not conceptually, because we can construct apparently conflicting concepts of space that are consistent with the same experience of space. We dress space with dimensions in the same way we dress numbers with digits. The dimensions of space are the telltale signs of spatialization.

(Co)Algebras, Similarity And System Thinking

We have had a first contact with the diagrammatic structure of algebras and coalgebras in the description of the Actuation Mechanism of cohesiveness in Section 17.2. In modern mathematics, the terms algebra and coalgebra denote a large zoo of mathematical objects that form a sophisticated conceptual framework to describe various nuances of the notions of structural continuity introduced in this research. Some forms of similarity are more apparent than others. For example, it is relatively easy to think of fractals as examples of self-similar structures because in a fractal we can *see* that certain shapes are repeated as we zoom in on it. Here we want to discuss briefly an example of a less apparent form of structural continuity and self-similarity: dynamical systems. First off, a dynamical system is a very broad notion used to describe a set of rules to describe the dependence of a certain system state on certain parameters, typically time. For example, a system that does not change over time, a constant system, could be described by the equation $x(t + 1) = x(t)$ where x denotes a certain system

17.3 The cohesivon

state and t denotes time. In this example, if at $t = 0$ the system is in state $x(0) = p$, since $x(0) = x(1) = x(2) = \ldots x(t)$, it follows that $x(t) = p$ for any time t. Similarly, we could construct a simple "system with memory" as follows $x(t+1) = x(t) + x(t-1)$, where the presence of the term $x(t-1)$ acts as the memory of the system state before the time t. In this example, if we know the system state at the times $t = 0$ and $t = 1$, we can calculate the system state at any time t. When we set $x(0) = 0$ and $x(1) = 1$, the system $x(t)$ is the famous Fibonacci sequence—an integer sequence that describes many natural phenomena such as branching in trees, and the arrangement of fruitlets and leaves. To get a more vivid and less intuitive example of structural continuity let's consider a more complex example of dynamical system: a simple model of food chain.

The most elementary food-chain consists of just one type of predator and one type of prey, and is based on the simplistic assumption that there is an infinite supply of food available to preys, and that predators are always hungry. A description of this hypothetical predators-prey interaction is a model to compute how at any point in time the population of predators and the population of preys. For example, let's apply these hypotheses to a cat-mouse system. If we denote with M the number of mice and with C the number of cats, we can imagine that M increases at a certain ratio α as a result of reproduction, and decreases at a ratio β as a result of predation, which grows with M and C—the greater C, the greater the probability each mouse is eaten by a cat. Similarly, we can think that C increases at a certain ratio δ as the population of M grows, and as a result of reproduction, and the decreases at a certain ratio γ as a result of death. These intuitive observations are captured by the following differential equations

$$\frac{dM}{dt} = \alpha M - \beta MC$$
$$\frac{dC}{dt} = \delta MC - \gamma C$$

where the symbols $\frac{dM}{dt}$ and $\frac{dC}{dt}$ denote the variations of M and C respectively over an infinitesimal fraction of time. This set of equations is called a Lotka-Volterra system. To comprehend why the Lotka-Volterra equations—and in fact any dynamical system—is an example of structural continuity and self-similarity, we must note that M and C appear on both sides of the equations:

both M and C are, in other words, described by other versions of themselves. In particular, M and C are also described by each other, and this feature binds M and C and allows us to think of them as a whole: the pair (M, C).

The Structure Of The `cohesivon`

In the next three sections we offer an overview of the three modes of operation of the `cohesivon`. This description is meant to be interpreted as an example of "The Abstract Mind-as-sound Analogy" we presented in Section 11.5, where the analogy is between the `cohesivon` and the sound produced by the buzzer circuit. Focus, unfocus and contravariance are a way to describe the basic, irreducible, atomic experience of the Abstract Mind, in a Cohesiveness-based state coming in contact with an object of thought.

Spatialization As The Structure-preserving Immersion Of An Object Of Thought In A Broader Context

Spatialization is the "descend" sleight of mind, and its mathematical analogy are certain generalizations of the category-theoretical notion of coalgebra.

The action evoked by a descent is the merging of an object of thought into another object of thought by virtue of the shared similarities that make the merging possible. The merging of an object of thought into another objects of thought results in the vanishing of an object's identity and in the reappearance of the same via the definitional or structural properties that give rise to the similarities between the two objects. In this sense, spatialization, as a form of awareness, is a prototypical example of *identity through similarity* where the merging of two objects of thought via relations of similarity is the experience of grasping an object of thought: it is the atomic, irreducible structure of any the mind-world transaction in the Cohesiveness-based $\mathcal{M}$-*signature*.

The modern mathematical framework to describe self-similarity is a category-theoretical formulation of coalgebras. Roughly, it consists of various mathematical constructions based on a the pattern $\bullet \to \mathcal{R}(\bullet)$—the category-theoretic colimit of a diagram

17.3 The `cohesivon`

without arrows—where an object of thought • is transformed into a more complex one $\mathcal{R}(\bullet)$ that denotes its spatialization. These constructions are the conceptual basis of a zoo of mathematical objects that encode in their structure the basic language to describe how the part is connected to the whole via certain types of relations of geometrical, or topological, or algebraic, or logical similarity.

Despatialization As The Structure-preserving Extraction Of An Object Of Thought From A Broader Context

Despatialization is the "ascend" sleight of mind, and its mathematical analogy are certain generalizations of the category-theoretical notion of algebra.

Ascent symbolizes the mode of the mind through which an object's definitional identity is inherited from another object of thought via a structure-preserving process of synthesis. The synthesis consists in the structure-preserving transformation from a complex object to simpler one, and is diagrammatically depicted as $\mathcal{R}(\bullet) \to \bullet$, the category-theoretic limit of a diagram without arrows, where $\mathcal{R}(\bullet)$ symbolizes a complex object in the form of its spatialization, and • symbolizes a simpler one, and where the shape • denotes the structure being preserved by the transformation $\to$.

Each shape of the diagram $\mathcal{R}(\bullet)$ is the archetypical pattern used to define a large zoo of mathematical objects called algebras. Algebras, and their many variations and generalizations, are built on this primitive concept of synthesizing objects of thought from more complex ones, where the complexity is measured or modeled in terms of the information lost in the process of synthesis: the higher the loss, the more abstract the algebra is, the more descriptive it becomes—compare this with the discussion on decategorification in the Motivic Lens, Section 4.3.

Contravariance Is The Mental Perception Of An Object Of Thought By Structural Continuity

Cohesiveness-based contravariance models the notion of **transport**. It is an atomic and irreducible process that codifies the grasping of an object of thought via—what we describe as—the relations of structural continuity that connect what appear (to the Identity-based mind's eye) as distinct objects of thought. It is the most practical, the most intuitive conceptual device to describe what we may characterize as the workings of a similarity-based intellect: a mind where the structure of rational thought revolves entirely around a vast and sophisticated zoo of mathematical gadgets that describe the various manifestations of self-similarity. Cohesiveness-based contravariance denotes the experience of grasping an object of thought not as an ontologically independent entity, but as the movement between two categories of mental perception: one denoting an increase of cognitive data, from $\bullet$ to $\mathcal{R}(\bullet)$, one denoting a decrease of cognitive data, from $\mathcal{R}(\bullet)$ to $\bullet$. While most people might find it easy to spot, at least intuitively, how coalgebras codify self-similarity—think of fractals or of any other recursively defined mathematical object—thinking of algebras as models of structural continuity may turn out to be counterintuitive and impervious to good and sound and mathematically healthy common sense. This difficulty is, perhaps, a cognitive bias dictated by the habit of regarding the increase of data—synonym of information here—as an indication of some kind of progress—recall that in coalgebras we add information to go from $\bullet$ to $\mathcal{R}(\bullet)$. The less intuitive part of this argument is that the *loss* of information is also an informative event, but of a different kind, and this is exactly what happens in algebras in the definition of the transition from $\mathcal{R}(\bullet)$ to $\bullet$. The core idea captured by Cohesiveness-based contravariance is precisely that an Abstract Mind in a Cohesiveness-based state the grasping of an object of thought occurs precisely as a process of data "loss" and "creation".

Definition 17.3.1. (cohesivon – *informal*)
A **cohesivon** is the pattern defined in Figure 17.5 where $\mathcal{R}$ denotes a diagram and $\bullet \to \mathcal{R}(\bullet)$ and $\mathcal{R}(\bullet) \to \bullet$ denote respectively category-theoretic models of coalgebras and algebras, and where contravariance denotes the category-theoretic operation of inverting arrows.

17.4 The Minimal Interaction Model

The Minimal Interaction Model of the Cohesiveness-based $\mathcal{M}$-*signature*:

- *is* a Type-A Interaction Model that models the reconstruc-

17.4 The Minimal Interaction Model

tion of cohesiveness (Definition 11.5.1),

- *is* an atomic, irreducible model of the fundamental makeup of the Cohesiveness-based mind-world interaction (Definition 11.5.2),

- *is* a prototypical model of *structural continuity*—the Actuation Mechanism of *unity*,

- *is built* via the application of the Knowledge-Reality Interface (Section 13.8) to a model of `transport`—a `cohesivon`,

- *defines* the conception of Cohesiveness-based knowledge and epistemology.

Note

As we observed for the Identity-based and the Finiteness-based $\mathcal{M}$-*signatures*, here we assume that $\mathcal{M}$-*signature* and Cognitive Architecture are distinct and ontologically independent entities. As a result, an Interaction Model is necessarily based on the spatialization of a certain object of thought.

In the discussion about the Actuation Mechanism of *unity*, we discovered that cohesiveness can be thought of as a form of structural continuity, where structure is a feature of the spatialization of the object of thought or perception. The notion of structural continuity captures the idea that a certain feature is independent of the changes along certain dimensions consistent with the spatialization of an object at hand.

Structure Of The Minimal Interaction Model

The `reconstruction of the success condition` takes place in the spatialization of an object of thought as described in the steps of Figure 17.6—compare with Figure 11.4.

A. The `reconstruction of the success condition` begins with the application of a model of `transport` based on the `cohesivon` to the elements of a spatialization of the object of thought. The result of this step is a framework to describe the object of thought via a set of equational properties. In this case, though, the equational properties between the elements of a spatialization are *references* to cognitive processes that occur in a metaphysical structure of thought where spatialization

17 Cohesiveness-based $\mathcal{M}$-signatures

and objects do not exist, and where the *only* terms of reference are spaceless relations ($Coalgebra + Algebra$) between spatializations such as those described earlier in the Actuation Mechanism.

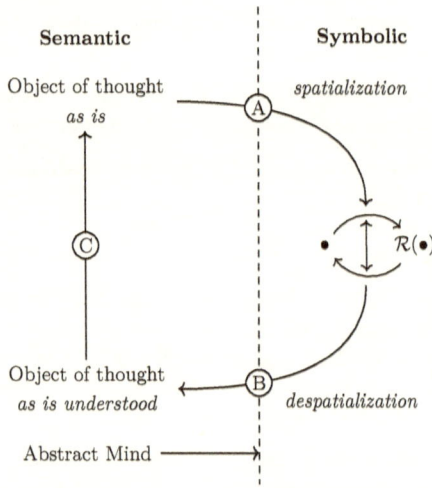

Figure 17.6: Structure of a Minimal Interaction Model in the Cohesiveness-based $\mathcal{M}$-signature.

B. This is the second half of the **reconstruction of the success condition**, where **Cohesiveness-based knowledge** emerges, and consists in applying to the elements of the framework defined in **A**, various models of *structural continuity* based on the **cohesivon**. This step produces statements of the form $(Coalg + Alg)A = B$ which relate any two entities A and B via **cohesivon**-based models of *structural continuity*. Note that in the Minimal Interaction Model the equational relations between ontologies do not rely on any relation other those based on the **cohesivon**; they are, at least definitional and at most tautological.

C. This step symbolizes the construction of **higher forms of knowledge**. The inferences constructed in **B** constitute a higher spatialization of the initial object of thought, and represent *how* we see and *what* we see of that initial object of thought through the equational lens produced by this Interaction Model—what is called Local Identity (Section 8.7). By applying iteratively the schema of Figure 17.6 to the image produced by the Local Identity, we obtain, in this analogy, forms of cohesiveness along two dimensions (Figure 17.7): the admissible types of spatialization, and the models of cohesiveness.

spatializations →

models of cohesiveness ↓

$(Co)Alg_1 A = B \quad (Co)Alg_1 A' = B' \quad (Co)Alg_1 A'' = B'' \quad \ldots$
$(Co)Alg_2 A = B \quad (Co)Alg_2 A' = B' \quad (Co)Alg_2 A'' = B'' \quad \ldots$
$(Co)Alg_3 A = B \quad (Co)Alg_3 A' = B' \quad (Co)Alg_3 A'' = B'' \quad \ldots$
$\ldots \quad \ldots \quad \ldots$

Figure 17.7: Local Identities produced by a Minimal Interaction Model in the Cohesiveness-based $\mathcal{M}$-*signature*.

where A' and B' represent higher order equational properties between lower order equations such as $(Coalg + Alg)A = B$. This cognitive machinery produces equations, equations of equations, equations of equations of equations and so on, according to hierarchies of **cohesivon**-based constructs.

From The Minimal Interaction Model To An Interaction Model

The construction of an arbitrary Type-A Interaction Model consists in extending the Minimal Interaction Model with any number of transformations compatible with the equational properties defined by the models of **cohesivon**. These transformations are obtained by iterating the Minimal Interaction Model with itself (Figure 17.6-C) and by adding to it any transformation consistent with the spatialization to which the models of **cohesivon** are applied.

17.5 The Structure Of Knowledge By Similarity

The Cohesiveness-based epistemology

- *is* an interpretation of the terms *knowledge*, *comprehension*, *meaning*, *explanation* and *consciousness* as defined in

Chapters 12, 13 and 14, based on the mechanism through which the **concrete** concept of *cohesiveness* defines what is *intelligible*—as described in Sections 17.2 and 10.5, and

- *consists* in the definition of a correspondence between the concepts introduced in Chapters 12, 13 and 14 and certain mathematical constructions of Category Theory based on the **cohesivon**.

The interpretation of the terms knowledge, comprehension, meaning, explanation and consciousness for the Cohesiveness-based $\mathcal{M}$-*signature* assigns to the definitions of Chapters 12, 13 and 14 meanings that derive from the way the **cohesivon** models structural continuity, and defines intelligibility in this $\mathcal{M}$-*signature*. In particular, it is interesting to see how in this $\mathcal{M}$-*signature*, in which spatialization can be seen as a special case of the Finiteness-based spatialization without arrows, the absence of arrows changes the structure of knowledge. There is a sense in which this $\mathcal{M}$-*signature* represents an archetypical form of knowledge by analogy. As seen in the previous $\mathcal{M}$-*signatures*, to define the process of acquiring knowledge in this $\mathcal{M}$-*signature*, we need to define: *the ontologies that we relate through* the **cohesivon**.

The informal construction of the Cohesiveness-based epistemology is articulated in these 6 steps:

1. Define the Cohesiveness-based object of thought. The goal is to characterize the Cohesiveness-based experience of an object of thought, and to emphasize how it differs from the ordinary Identity-based experience of grasping thoughts.

2. From 1, we obtain the form of Cohesiveness-based knowledge, and explain the experience of knowledge in the Cohesiveness-based $\mathcal{M}$-*signature*. This form of knowledge is prototypical in that it defines at a basic cognitive level the practical and conceptual event of consciously knowing an object of thought *before* the development of any complex cognitive apparatus such as beliefs, contemplation or justification, and any derived notion of truth.

3. From 2, we derive the interpretation of the diagram of comprehension (Figure 17.8), and explain the structure of Cohesiveness-based understanding, the fundamental process through which the changes in the Abstract Mind's configuration give rise to what we consciously experience as

17.5 The Structure Of Knowledge By Similarity

apprehending (one's own) thoughts, not by grasping the individuality of those thoughts, but by clutching their mutual relations of similarity—as described by the **cohesivon** and by some of its mathematical representations.

4. From 3, we derive the structure of the morphisms between Interaction Model, and define the conceptual machinery at the basis of the Cohesiveness-based protosemantic and semantogenetic processes.

5. From the points above, we outline the elements of the Cohesiveness-based explanatory system.

6. From the points above, we outline the archetypical structure of the conceptions of Cohesiveness-based communication and mathematics.

7. From 1, we deduce a description of the two key features of consciousness, unity and flow, and offer a characterization of these in the Cohesiveness-based state of consciousness.

Cohesiveness-based Object Of Thought

An object of thought in the Cohesiveness-based $\mathcal{M}$-signature can be thought of as a (co)algebra-preserving endomorphism.

In the Cohesiveness-based $\mathcal{M}$-*signature* the object of thought is the **cohesivon**, and represents how the Abstract Mind grasps thoughts. The experience of a Cohesiveness-based object of thought is therefore the experience of a Minimal Interaction Model. It is the fundamental experience of thought that originates from the correspondence described in "Consciousness And Spatialization" (Section 14.3) between spatialization and the experience of an object of thought. In the category-theoretic representation of these ideas introduced in this book, this means that the Cohesiveness-based experience of an object of thought is (a generalization of the) the category-theoretic limit of a certain diagram $\mathcal{R}$—a diagram without arrows.

As seen in the description of the **cohesivon** (Section 17.3), these constructions form the cognitive and conceptual basis of a large zoo of mathematical structures that describe two general kinds of concepts: algebras and self-similarity. This is how a Cohesiveness-based Abstract Mind grasps concepts: by association (algebras) or via some forms of analogy (self-similarity) according to certain criteria of structural continuity. In an attempt to convey in the natural language a concept that is, by definition, inaccessible to the Identity-based intellect, we can say that a Cohesiveness-based object of thought is just that: a form of mathematical analogy, which can be made very precise with the tools of category-theoretic models of algebras and coalgebras.

An way to depict these ideas is with the following diagram

$$F$$

where the • is a mathematical gadget with enough structure to allow for the definition of (co)algebras—or generalizations of the category-theoretical notion of algebra—and where F is a morphism from and to • characterized by the property of preserving (co)algebras.

Cohesiveness-based Knowledge As Knowledge By Similarity

Cohesiveness-based knowledge can be thought of as a (co)algebra-preserving morphism between objects of thought.

In the Cohesiveness-based $\mathcal{M}$-*signature*, knowledge is produced by relating the ontologies that form a spatialization of an object of thought via various mathematical constructs based on the **cohesivon** (Definitions 17.3.1 and 12.3.1). These constructs represent *instances* of the Minimal Interaction Model. The hierarchical *structure* of Cohesiveness-based knowledge is described by the below correspondence.

1. The Cohesiveness-based *conception* of knowledge is represented by the **cohesivon**.

2. A *prototypical definition* of knowledge is the atomic cognitive process defined by the Minimal Interaction Model (Section 17.4).

3. Each *definition* of knowledge corresponds to the way in which an Interaction Model implements the **reconstruction of the success condition** based on 2.

To give an Cohesiveness-based interpretation of the process of acquiring knowledge described in Definition 12.3.1 we need to describe how the objects of thought are related to each other via the **cohesivon**.

The diagram above depicts a general characterization of knowledge we are familiar with, where • are objects of thought in the ordinary Identity-based language, and F is a morphism between them. The remarks made in the previous two chapters apply here for the same reasons. For a morphism to connect objects of thought defined in the Cohesiveness-based $\mathcal{M}$-*signature*, it must preserve the structure of those objects of thought, which translates into the condition that F be a (co)algebra-preserving morphism.

Knowledge By Similarity

The philosophical interpretation of Cohesiveness-based knowledge as a (co)algebra-preserving morphism between objects of thought is that Cohesiveness-based knowledge is knowledge by *similarity*. Knowledge by similarity, or knowledge by structural continuity, indicates a Cognitive Architecture that expresses a conception of knowledge as the intellectual act of *traversing* the relations of structural continuity that connect a collection of spaces.

17.6 Cohesiveness-based Comprehension

Cohesiveness-based comprehension can be thought of as a (co)algebra-preserving morphism between two instances of Cohesiveness-based knowledge.

In the Epistemology of the Abstract Mind (Section 12.5) the process of understanding is defined as an object $\bullet^*$ together with an arrow g such that any other arrow f factors via g as in the diagram of comprehension (Figure 17.8).

17 Cohesiveness-based M-signatures

Figure 17.8: Diagram of comprehension.

This condition is captured by the arrow equation below, where ∘ is a certain operation that assign one arrow to a pair of adjacent arrows—adjacent meaning that the tip of one arrow coincides with the bottom of the other arrow.

$$f = g \circ h$$

We use the definitions of Cohesiveness-based object of thought, and Cohesiveness-based knowledge to interpret the diagram above:
• are *objects of thought*, and → are *types of knowledge*.

The problem of defining the Cohesiveness-based process of understanding, is the problem of defining the meaning of the = sign in the equation above. The observations made in the discussions about Actuation Mechanism and Definitional Model (Section 17.2), and in the Minimal Interaction Model (Section 17.4), indicate that = codifies a relation between f and $g \circ h$ modeled by a `transport`. It follows that comprehension in the Cohesiveness-based *M-signature* is a **cohesivon**-based relation between the arrows f and $g \circ h$, that is, it is an algebra, a coalgebra, or both. Another way to interpret this concept, in light of the definition of Cohesiveness-based knowledge, is to say that understanding is a certain system in which we come to know f via its relations of similarity with $g \circ h$.

Similarly to what observed in the previous two *M-signatures*, the definition of Cohesiveness-based knowledge seen in the previous section leads to the intuitive notion that the forms of Cohesiveness-based comprehension are (co)algebra-preserving morphisms between types of Cohesiveness-based knowledge: one type of knowledge is represented by the arrow f, another type of knowledge is $g \circ h$. Note that the information in the diagram above that f and g point to the same object (of thought) X, and that f and h have the same origin (object of thought), is part of the *definition* of the (co)algebra-preserving morphism between the knowledge systems where f and $g \circ h$ are defined respectively.

17.7 Meaning And Semantogenesis

Here we use the definitions of protosemantic transformation (Definition 12.6.1), semantogenesis (Definition 12.6.3), and the intuitive description of a semantogenesis given in "Meaning And Semantogenesis In Interaction Theory" in Section 12.6 to sketch an Identity-based semantogenesis.

Recall from Definition 12.6.3 that a semantogenesis is a mathematical object that classifies types of understanding.

A Cohesiveness-based semantogenesis is, therefore, a morphism

$$\Gamma : \mathcal{U} \to \mathbf{P}$$

where $\mathcal{U}$ symbolizes a class of Cohesiveness-based *modes of understanding*—which are protosemantic morphisms—and $\mathbf{P}$ symbolizes a space of *values*. Here we describe an example of Cohesiveness-based semantogenesis without the notion of protosemantic transformation.

Recall that $\mathcal{U}$ is an object where the diagrams of comprehension

Figure 17.9: The diagram of comprehension.

are defined, and where the condition of understanding $f = g \circ h$ for the arrows in Figure 17.9 is based on a certain definition of $=$ given by the Interaction Model in which $\mathcal{U}$ is defined, and according to the Cohesiveness-based conception of knowledge defined in this chapter. An example of $\mathcal{U}$ is $\mathbf{P}$, where $=$ has the ordinary meaning, and where f, g and h are functions between sets. If in $\mathcal{U}$ we interpret the sets as *data* and the arrows as *strategies* activated by that data, we can think of the comprehension diagram as an illustration of the condition that a certain $\bullet^*$ represents the minimum data to activate a strategy (g) to achieve X.

To describe a Cohesiveness-based semantogenesis $\Gamma : \mathcal{U} \to \mathbf{P}$ for the pair of entities $\mathcal{U}$ and $\mathbf{P}$ above, we use the notion of knowledge by similarity introduced in this chapter. We do not interpret

17 Cohesiveness-based $\mathcal{M}$-signatures

Cohesiveness-based understanding via the definition given in this chapter. Instead, we think of Cohesiveness-based understanding as a certain form of Cohesiveness-based knowledge that f and $g \circ h$ are related via a relation of similarity expressed by the **cohesivon**. Let us think then of Γ as a classifier of this special types of knowledge of arrows. Γ creates meaning by characterizing the types of Cohesiveness-based knowledge—of f and $g \circ h$—with respect to the representation **P** of the purpose where Cohesiveness-based understanding is considered. In other words, meaning is created by identifying what relations of similarity between certain objects of Cohesiveness-based thought satisfy a certain purpose codified by **P**.

Semantogenesis As Measure Of Modes Of Understanding By Similarity

This brief description of the genesis of meaning in a Cohesiveness-based Symbolic Intuition of Reality suggests the way in which a Cohesiveness-based metaphysical structure of thought creates an experience that relates us to reality. For in essence meaning is a measure of how much a certain type of understanding improves our ability to relate symbolically to reality: the greater the meaning, the better what we understand expands and deepens our experience. Here the fundamental structure of thought relies on various forms of structural continuity to function and to relate us symbolically to reality, as described by the **cohesivon**, and this ultimately defines the mechanism by means of which an Abstract Mind in a Cohesiveness-based state ascribes meaning to the objects and states of affairs in the world, and how this gives rise to its experience.

Toward Cohesiveness-based Information

Understanding by similarity, and the cognitive process conducive to this type of understanding, result necessarily in a new kind of information that measures the way a certain type of (understanding by) similarity becomes relevant to a specific purpose. The correspondence between meaning and information pointed out abstractly in Section 12.6 reemerges here in this interpretation of semantogenesis: the more (or the less) a certain meaning is conducive to a certain form of similarity, the more a certain type of understanding becomes relevant to a context, the higher its information content. Cohesiveness-based information is therefore a measure of how much a certain type of structural continuity creates knowledge in a given context. We will revisit these ideas in Chapter 20.

17.8 Explanatory Systems

The Principle of Epistemic Relativity (Section 13.1) points to the fact that the exploration of explanatory systems beyond the ordinary Identity-based $\mathcal{M}$-*signature* is motivated by the observation that there seems to be no indication that in Nature the **concrete** concept of identity is a privileged frame of reference to define mathematical thought and a scientific method. In this brief introduction to Cohesiveness-based explanatory systems we outline the scaffolding of a system of thought to construct explanations based on the paradigm of epistemic continuity described in Section 13.3 applied to the Cohesiveness-based metaphysical structure of thought. To begin to outline the structure of a Cohesiveness-based explanatory system, we answer the following questions. Section 16.9 outlines the structure of Cohesiveness-based explanations.

- What are explanandum and explanans?
- What are the accepted truths and self-explanatory ontologies?
- What is epistemic continuity?
- What is a reality?

Explanandum And Explanans

Cohesiveness-based explanations are statements about an explanans in the language of an Abstract Mind that articulates thoughts via forms of similarity as described by the **cohesivon**. The linguistic limitations pointed out already in the text about our ability to describe anything that does not adhere to the subject-object dichotomy apply, necessarily, here too. Explanandum and explanans are references to the semantic and symbolic parts of the Cohesiveness-based mind-world interplay via the **cohesivon** in the exact same way in which two objects of thought are related via the **cohesivon** in any form of Cohesiveness-based knowledge. The interpretation of their semantic relation is therefore to be sought in the meaning acquired by the relations of similarity codified by an explanans. Once again, what can guide us in this interpretation is the "Structural Canon" of good explanations we saw in Section 13.2. Good Cohesiveness-based explanans point out similarities of an explanandum that are based on a greater number of similarities between Cohesiveness-based objects of thought.

Accepted Truths And Self-explanatory Ontologies

Recall Definition 13.2.1: *"An accepted truth is a spatialization of an explanandum, in which a reality is defined. "* The three spatializations introduced in the `cohesivon` (Section 17.3) are the accepted truths in the Cohesiveness-based mode of thought, because they are the arena where Cohesiveness-based thought takes place. Inside the domain delimited by the accepted truths, the Cohesiveness-based self-explanatory ontologies, are the minimal, irreducible unities of Cohesiveness-based thought, which, by virtue of Definition 17.3.1, and of the definition of the Cohesiveness-based object of thought, are mathematical models of the notion of (co)algebra. The notion of truth follows from the above and from Definition 13.3.2. In summary:

- The Cohesiveness-based accepted truths are: ascend, descend and contravariance.
- The Cohesiveness-based self-explanatory ontologies are models of (co)algebras.

Epistemic Continuity

Recall Definition 13.3.1: *"Epistemic continuity is the belief that an explanandum has an intrinsically protosemantic cohesiveness that corresponds to the symbolic cohesion inherent in the structure of an explanatory system $\boldsymbol{P}$ to a degree consistent with the explanans produced by $\boldsymbol{P}$. "* The belief that our explanations, and the framework we adopt to construct them, give us epistemic access to the things they explain is codified by the notion of epistemic continuity. To give a Cohesiveness-based interpretation of this schema, we need to explain how we characterize the cohesion and coherence of a Cohesiveness-based explanans, and how we relate the coherence of a Cohesiveness-based explanans to what we believe is the intrinsic order that holds together a Cohesiveness-based explanandum. The observations made in the characterization of the Finiteness-based epistemic continuity apply here for the same reasons: the

17.8 Explanatory Systems

Cohesiveness-based conception of knowledge is not based on the ability of an explanans to make quantitative predictions about an Cohesiveness-based explanandum. As seen in the paragraph above about Explanandum and Explanans, in the Cohesiveness-based mode of thought the relation between these two elements is based on forms of similarity expressed via the **cohesivon**. Hence, the validation of Cohesiveness-based knowledge occurs via the *existence* of a type of similarity between explanans and explanandum. The existence of a relation of similarity between explanans and explanandum, expressed via the **cohesivon**, is the type of coherence needed in order to validate our Cohesiveness-based knowledge of reality. In this sense, epistemic continuity is manifested as canons and schemes. In summary: *Cohesiveness-based epistemic continuity is predicated on the belief that the existence of a relation of similarity between explanandum and explanans—described mathematically by various category-theoretic versions of (co)algebras—is a faithful and reliable expression of the intrinsic order of the Cohesiveness-based explanandum.*

Reality

The semantic part of a mind-world transaction is a *reality*. The same notion is expressed in the language of Interaction Theory in Definition 13.3.3: *"A reality is the experience created by an Interaction Model."*

The semantic content of a reality is the collection of all explanations based on the symbolic content of that reality, interpreted according to the schema of epistemic continuity. The semantic content of an unexplained reality consists only of the accepted truths. To understand why, recall Definition 13.2.1: *"An accepted truth is a spatialization of an explanandum, in which a reality is defined."*

From Definition 13.3.3, it follows that the most elementary kind of reality is the experience that originates from the most elementary Interaction Model, namely, the Minimal Interaction Model. As seen in this chapter in the description of a Cohesiveness-based object of thought, a **cohesivon** represents how an Abstract Mind in a Cohesiveness-based state grasps thoughts at a fundamental level, and coincides with the fundamental experience of thought described in Section 14.1 between a type of spatialization and the experience of an object of thought. The experience of a

Cohesiveness-based object of thought is therefore the experience of a Minimal Interaction Model.

To get a feel for the intuitive experience of a Cohesiveness-based reality, we can imagine a situation where a certain entity is comprehended not per se, as a distinct object of thought, but solely and entirely via relations of self-similarity. This is, as seen in Section 17.5, the most basic form of knowledge by similarity. Since we are giving an Identity-based description of a concept that can be grasped directly only by an Abstract Mind in a Cohesiveness-based state, the relations of self-similarity are to be defined between an entity that we *represent* in the ordinary Identity-based language: stuff made of parts and features. Roughly, self-similarity means that the same object appears on both sides of an equal sign. An example of self-similarity is given by the following relation $X = X + X$, where X represents a certain mathematical gadget, $=$ is a certain form of (generalized) identity, and $+$ is a certain operation defined on collections of objects of type X. Let $[0, 1]$ denote the set of all real numbers between 0 and 1. American mathematician Peter J. Freyd (Freyd 1999) showed that $[0, 1]$ can be transformed—stretched—continuously to match perfectly two copies of itself glued together end to end, and that this property fully characterizes it. In mathematics continuous stretching is called homeomorphism, generalizes the notion of bijection we saw in Section 16.2, and is a mathematical form of the philosophical notion of identity. By virtue of this characterization of $[0, 1]$, if in the equation above the operation $+$ represents the gluing together of sets, denoted by $\cup$, and $=$ is an homeomorphism, the equation above becomes $[0, 1] = [0, 1] \cup [0, 1]$. By defining $[0, 1]$ via a relation of self-similarity, this homeomorphism represents both the object of thought $[0, 1]$ and a relation of self-similarity: we know $[0, 1]$ via a relation of similarity that defines it, this is, perhaps, the most intuitive example of knowledge by self-similarity.

17.9 Morphogenesis And Formal Efficacy

The central themes of Cohesiveness-based explanations are the genesis of cohesion through form, and how the World appears through the notion of morphogenesis.

Here we use the rationale presented in "The Metastructure Of

17.9 Morphogenesis And Formal Efficacy

Explanations" (Section 13.7), and the definitions of *lens* and *image* (Definitions 13.7.1 and 13.7.2 respectively), to outline the structure of Cohesiveness-based explanations, and to describe the limits of the interpretations of what we see and comprehend when we look at the World through a Cohesiveness-based explanation.

Recall from Section 17.2 that *structural continuity*, the Actuation Mechanism of cohesiveness, is the phenomenology of unity, the Organizing Principle of cohesiveness. To comprehend the structure of Cohesiveness-based explanations we follow the line of reasoning we introduced in Section 13.2: we need to explore what connects a Cohesiveness-based explanation to the mechanism through which the **concrete** concept of cohesiveness defines intelligible thought, and understand how this nexus defines the conceptual structures that differentiate a Cohesiveness-based explanation from any other Cohesiveness-based utterance or thought. In "The Deep Structure Of Explanations" (Section 13.2), we discovered that for an utterance to be an explanation, it must necessarily, directly or indirectly, describe the genesis of the definitional structure of the Actuation Mechanism for the $\mathcal{M}$-*signature* where the explanation is formulated, which in this case is: *structural continuity*. With these ideas in mind, and with the concepts of *lens* and *image*, we can describe the structure of a Cohesiveness-based explanations as follows.

Cohesiveness, Morphogenesis And Form

The genesis of the definitional structure of structural continuity is codified by the notion of *form*. Form is the *lens* of cohesion, and expresses at an archetypical level the mechanism through which the notion of structural continuity originates. The term form is used here to denote an abstract configuration that defines the essential nature of an entity, and to signify how through form an instance of cohesiveness acquires its definitional structure.

Through form we see *morphogenesis*. Morphogenesis is the *image* of cohesiveness, and characterizes what we see when we look at the World through the lens of form. The World described by Cohesiveness-based explanations is *made of* morphogeneses because that is how we *comprehend* form.

Through morphogenesis we comprehend *cohesiveness*. We *grasp* the various manifestations of *cohesiveness* in the World through the way the notion of form is portrayed as morphogenesis in our Cohesiveness-based explanations.

Objectivity

The term *objectivity* denotes how an Abstract Mind in a Cohesion-based state *apprehends* the fundamental mode of existence of the things in the World.

Objectivity is defined by *similarity*—the Actuation Mechanism of unity—and characterizes the condition that resolves the epistemic tension between the semantic and symbolic parts of a mind-world transaction. Thus, objectivity is formulated as a *reference* to a model of **transport**—such as, for example, the **cohesivon**.

A claim is *objective* by virtue of transcending the features that define the epistemic tension where we develop our metaphysical, theoretical and practical comprehension of an object of thought as cohesion through form. The profundity of objective Cohesiveness-based knowledge is gauged by how much of it can be *compared* to its object.

From the perspective of Cohesiveness-based thought, *objectivity* is the natural mode of existence of facts and ideas, from which we are "separated" only relationally, not epistemically or causally, and defines the scope of an epistemic project, namely: to describe how things are the way they are by virtue of how their intrinsic constitution is a manifestation of the similarities of Nature, a concept corresponding in the Cohesiveness-based thought to the ordinary word *truth*.

The World Explained Through Cohesiveness

The *requirement* of a Cohesiveness-based explanation is to objectively discern, reveal and describe the types of *morphogenesis* that hold together the World, and in this rests its descriptive character.

The *features* of reality revealed by Cohesiveness-based explanations are principles of similarity, analogy, homomorphy, uniformity, wholeness, heterogeneity, affinity, chirality and symmetry.

Through a Cohesiveness-based explanation, the World appears as a work of *structure and intrinsic unity*, whose workings can be understood and described through the development of our

17.9 Morphogenesis And Formal Efficacy

knowledge of various notions of pattern, analogy and similarity that represent fundamental types of morphogenesis.

Cohesiveness-based explanations describe a World made of archetypical *similarities*, which are instances of the Organizing Principle of unity.

A Cohesiveness-based *explanation* is a statement about the distinctive processes that define the objects in the World intended as a reference to instances of those objects.

Ontic And Epistemic Contents Of An Explanation

Cohesiveness-based explanations are *existential*: the objects of thought or perception, and the states of affair they describe are indirect references to objects and states of affair captured in the language of the `cohesivon`.

Something "is" by virtue of being the *condition* according to which a `cohesivon` rearranges the objects of thought the way it does.

To *construct* Cohesiveness-based explanations we look for the conditions for which relations of form exist—as described by the `cohesivon`—between entities that we represent in the ordinary Identity-based language.

The most basic Cohesiveness-based explanation is the explanation of a self-evident reality, namely, an *object of thought*, and is described by the condition of epistemic continuity. Recall that Cohesiveness-based epistemic continuity is predicated on the belief that the existence of a relation of similarity between explanandum and explanans—described mathematically by various category-theoretic versions of (co)algebras— is a faithful and reliable expression of the intrinsic order of the Cohesiveness-based explanandum.

Cohesiveness-based explanations are *gauged* by how much of the World they describe in terms of *formal efficacy*.

The diagram in Figure 17.10 maps some of the points we have just presented, the description of the `cohesivon` (Section 17.3) and of the Cohesiveness-based epistemology (Section 17.5) in the Knowledge-Reality Interface, where we recognize spatialization, despatialization, and contravariance (Figure 17.10-A, B and C).

17 Cohesiveness-based $\mathcal{M}$-signatures

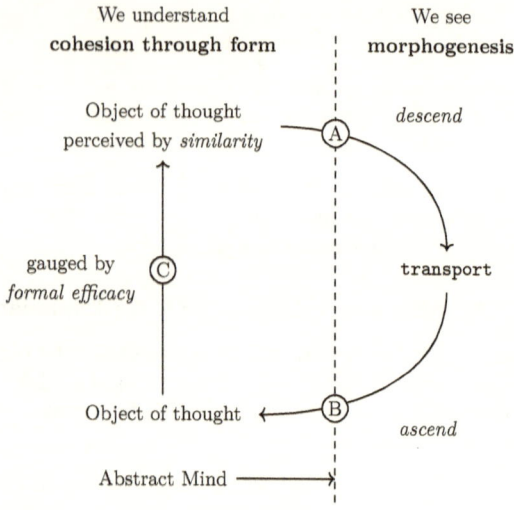

Figure 17.10: How an Abstract Mind in an Cohesiveness-based mode of consciousness sees and understands reality.

In Interaction Theory, the term *formal efficacy* denotes a measure of how much a Cohesiveness-based explanation describes a reality—a set of known truths—in terms of the minimum number of form conditions according to which a **cohesivon** organizes the elements of that reality. The term efficacy is used in this context to denote a measure of certain extensive properties through which structural continuity is represented, and does not have any causal connotation. The notion of epistemic continuity applied to a Cohesiveness-based explanatory framework tells us that the greater the form efficacy, the more fundamental or universal the kind of morphogenesis through which we comprehend a reality.

From Form To Cohesiveness To Morphogenesis

The notion of form has in itself already a quality of unity, because its ontological status derives from its being entire and complete and undivided beyond and before any representation. And that quality of unity is the same quality of unity we have identified as the Organizing Principle of the **concrete** concept of cohesiveness. It is a unity that makes form self-consistent. How can we explain,

17.9 Morphogenesis And Formal Efficacy

then, the nexus between form and cohesion? How does morphogenesis reveal form? This is, once again, a situation where it is useful to resort to the rarefied language of diagrams to construct a metaphor that should convey a profound idea about the nature of cohesiveness. We have given a description of the phenomenology of unity in the category-theoretic formulation of various forms of structural continuity. That was, of course, the discussion about the **cohesivon**, where various relations of proximity to diagrams without arrows, which we captured with the language of (co)algebras, gave rise to a zoo of mathematical forms of structural continuity, from the most intuitive such as recursive structures, to far less intuitive such as dynamical systems—recall Lotka-Volterra. Once again, we can convey what connects form, cohesiveness and morphogenesis at an archetypical, pre-conceptual, pre-cognitive level, by comparing them to the evolution of an intentional act. We can describe the nature of cohesiveness by comparing form to an intent, and the achievement of that intent to a type of morphogenesis consistent with that form. Each archetypical form is symbolized by a diagram $\mathcal{R}(\bullet)$—a collection of points in space fully defined by their number. The evolution of that form is a certain process called morphogenesis that adds and removes bullets to and from $\mathcal{R}(\bullet)$, changing its form. Each form is cohesive by virtue of being entirely self-consistent: the existence of a bullet is a necessary and sufficient condition to define a diagram of type $\mathcal{R}(\bullet)$. Cohesiveness is, in this sense, the evolutionary process of an elementary mode of consciousness that relates us humans with the World through various kinds of morphogenesis, it really is a celebration of form in its most cosmic way, and is how and what Cohesiveness-based explanations explain the World.

Fully Explainable Realities

By generalizing Lawvere's Fixed Point Theorem (Section 8.4 "The Limits Of Formal Systems") to the context of a Cohesiveness-based $\mathcal{M}$-*signature* we define an Cohesiveness-based fully explainable reality: a reality fully explainable by virtue of being fixed-point-free, in the sense of this generalization. Roughly, a fixed-point in a Cohesiveness-based theory is a fixed-point as in Lawvere's original formulation but in the equational system defined by the **cohesivon**—this is a work in progress.

17.10 Cohesion And The Language Of Form

In "The Archetypical Structure Of Communication" (Section 14.5) we introduced the idea that communication is that which results in a change of Abstract Mind state, and in "How Data Becomes Message" (Section 14.6) we used this rationale to categorize communication based on the types of Abstract Mind state transitions.

To describe the structure of Cohesiveness-based communication—the communication that occurs when the Abstract Mind is in an Cohesiveness-based state—we interpret the classification of the Abstract Mind state transitions using cohesion as a reference Abstract Mind state, and derive from this analysis the structure of Cohesiveness-based languages.

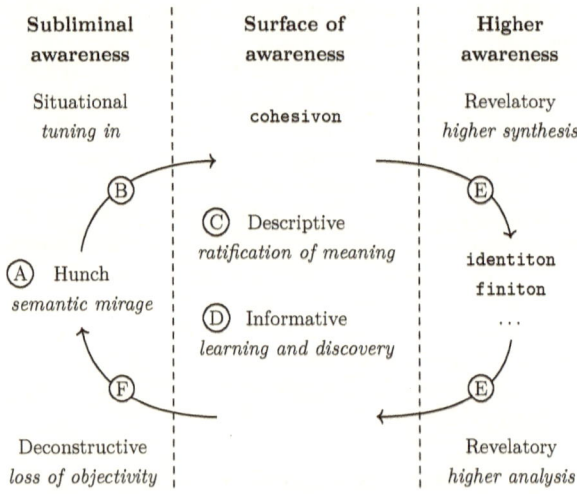

Figure 17.11: The 6 types of Cohesiveness-based communication.

Figure 17.11 is a close-up view of the Abstract Mind in the Cohesiveness-based state. The region between dashed lines represents the surface of awareness currently lit by consciousness, and signals where the mind-world transactions are defined by a conception of intelligibility based on the **concrete** concept of *cohesiveness*, and where, therefore, the Symbolic Intuition of Reality emerges as the experience of grasping concepts as described in

17.10 Cohesion And The Language Of Form

"The Structure Of Knowledge By Similarity" (Section 17.5).

In the brief description of Figure 17.11 that follows, we ground the basic ideas of Cohesiveness-based communication to the experience that they produce, because this greatly helps to comprehend how certain subtle fluctuations of consciousness are related to these types of communication. To this end, for each type of communication in Figure 17.11, we explain when it occurs, how meaning is created, and the experience produced by the genesis of meaning.

A *hunch* in the Cohesiveness-based *M-signature* (Figure 17.11-A) is a transition between Abstract Mind states that does not involve the reconstruction of cohesiveness. Hunches do not produce meaning because, when they occur, the cognitive structures underpinning semantogenesis are absent. Hunches produce, instead, a form of false awareness called *semantic mirage*, that in the Cohesiveness-based *M-signature* corresponds to the experience of seeming intuitions about the unity of an object of thought, and consists in the illusion that the features of similarity, analogy, homomorphy, uniformity, wholeness, heterogeneity, affinity, chirality and symmetry that characterize the experience of that object of thought, signify that the identity of *that* object of thought is actually being revealed through the changes in our awareness of it. For these reasons, the Principle of Epistemic Continuity (Section 13.3) does not apply to hunches.

Situational communication (Figure 17.11-B) occurs when a transition between Abstract Mind states gives rise to a Minimal Interaction Model for the Cohesiveness-based *M-signature*. The emergence of a Minimal Interaction Model in the Cohesiveness-based *M-signature* signals that the mind is spatializing through the **cohesivon**. This produces a sense of *tuning in* to what one is thinking of—see also Contravariance in the **cohesivon** (Section 17.3)—and is the experience of becoming aware of an object of thought (Definition 14.2.1), that in the Cohesiveness-based *M-signature* corresponds to the grasping of an object of thought as a type of morphogenesis through the features that define *knowledge by similarity* (Section 17.5). Situational communication marks also the beginning of an embryonic form of Cohesiveness-based objectivity, where intuition, knowledge, comprehension and meaning are guided by the Cohesiveness-based form of epistemic continuity—see also "Epistemic Continuity" in Section 17.8.

Descriptive communication (Figure 17.11-C) occurs when the

17 Cohesiveness-based $\mathcal{M}$-signatures

Abstract Mind transitions to and from the same Interaction Model within the Cohesiveness-based $\mathcal{M}$-signature. Descriptive communication is the most basic form of abstraction, and produces the experience of becoming aware of the current Interaction Model—as described in Section (14.2)—by objectifying it according to the structure of *knowledge by similarity* (Section 17.5). Descriptive communication ratifies meaning via the forms of Cohesiveness-based epistemic continuity where principles of similarity, analogy, homomorphy, uniformity, wholeness, heterogeneity, affinity, chirality and symmetry explain the qualities of homogeneity, uniformity, regularity and homology of the objects and states of affair in the World. These produce examples of how semantogenesis measures different modes of understanding by similarity or analogy of the "same" object of thought accessed through that Interaction Model.

Informative communication (Figure 17.11-D) occurs when the Abstract Mind transitions between distinct Interaction Models in the Cohesiveness-based $\mathcal{M}$-signature. Informative communication gives rise to the experience of *learning by similarity*, as explained in Section 17.5, and creates meaning by revealing the content of an object of thought through various forms of relations of proximity—the abstract concept of closeness to a diagram used in the **cohesivon**—between different mental representations of that object. Linguistically, whenever applicable, informative communication conveys how principles of similarity, analogy, homomorphy, uniformity, wholeness, heterogeneity, affinity, chirality and symmetry make up a World where things exist as morphogeneses.

Revelatory communication (Figure 17.11-E) occurs when the Abstract Mind changes $\mathcal{M}$-*signature*, and signals a deep change in the way the intellect functions. A synthetic type of revelatory communication denotes a transition *to* a non-Cohesiveness-based $\mathcal{M}$-*signature*, and gets its name from the fact that the activation of a different $\mathcal{M}$-*signature* produces the experience of reframing an entire system of thought, and not just specific types of knowledge. In synthetic revelatory communication, meaning is created by translating the structure and purpose of a non-Cohesiveness-based epistemic project into the language of Cohesiveness-based epistemology. This is, for example, the interpretation of a view of the World not based on types of morphogenesis, in terms of the concepts of the Cohesiveness-based epistemology, and conveys the intuitive experience of thinking of an object from the perspective

17.10 Cohesion And The Language Of Form

of how the knowledge of that object is acquired and understood in terms of similarity, analogy, homomorphy, uniformity, wholeness, heterogeneity, affinity, chirality and symmetry, rather than from the narrative of the frame of reference within which that object was originally known and conceived. Similarly, an analytic type of revelatory communication denotes a transition *from* a non-Cohesiveness-based $\mathcal{M}$-*signature* to the Cohesiveness-based $\mathcal{M}$-*signature*. Its name suggests that the experience associated to this transition is that of how thinking from a perspective not based on homogeneity, uniformity, regularity and homology primes the mind to appreciate new details of what is already known. Analytic revelatory communication produces meaning by reframing or interpreting the narrative of a Cohesiveness-based epistemological project into the language of non-Cohesiveness-based $\mathcal{M}$-*signature* without changing the structure and purpose of that epistemic project. This is, for example, the interpretation of a morphogenetic view of the World in terms of non-Cohesiveness-based $\mathcal{M}$-*signatures*.

Deconstructive communication occurs when the Abstract Mind disengages the Cohesiveness-based Symbolic Intuition of Reality, and enters a pre-cognitive or pre-conceptual state. This type of communication corresponds to a *deconstruction* in the sense that, as a result of this transition, the reconstruction of cohesiveness no longer takes place. In deconstructive communication, the Abstract Mind suspends its awareness of an object of thought by similarity and of the corresponding epistemology—as modeled by the **cohesivon** (Section 17.3), and as explained in "Morphogenesis And Formal Efficacy" (Section 17.9)—and this interruption results in the disintegration of the current Symbolic Intuition of Reality. Deconstructive communication does not produce meaning as such, because it suspends the cognitive structures underpinning semantogenesis. As a result, the identity of an object of thought is no longer grasped via the qualities of homogeneity, uniformity, regularity and homology, and this produces a class of experiences that can be described as a loss of objectivity.

17 Cohesiveness-based $\mathcal{M}$-signatures

Worlds, Languages And Utterances

A Cohesiveness-based language is a method to communicate how similarity, analogy, homomorphy, uniformity, wholeness, heterogeneity, affinity, chirality and symmetry make up a World where things exist as morphogeneses.

In Section 14.7 we defined an archetypical language as a method to communicate how the World is seen and understood by an Abstract Mind in a cognitive state (Definition 14.7.1).

We know how the World is seen and understood by an Abstract Mind in a Cohesiveness-based state. In "Morphogenesis And Formal Efficacy" (Section 17.9), we described the basic features of reality as it appears through the lens of cohesiveness. There, we used the ideas of "The Metastructure Of Explanations" (Section 13.7) to outline the fundamental mode of existence of the things in a Cohesiveness-based World, and to describe how that World can be described through the features of its objects, and of the relations between those objects. To introduce Cohesiveness-based languages, we revisit here some of those ideas—also summarized in Figure 17.10—to explain two essential dimension of this problem:

- How the World *seen* through cohesiveness defines the *purpose* of Cohesiveness-based languages.
- How the modes of existence of the things in a World *understood* through cohesiveness, and the features of that World, define the *structure* of Cohesiveness-based utterances.

Let us recall that how an Abstract Mind in a Cohesiveness-based state *sees* the World defines how things *exist* in that World. In "Cohesiveness, Morphogenesis And Form" we observed that the World we *see* through cohesiveness is made of morphogeneses.

> *"[t]hrough form we see morphogenesis. Morphogenesis is the image of cohesiveness, and characterizes what we see when we look at the World through the lens of form. The World described by Cohesiveness-based explanations is made of morphogeneses because that is how we comprehend form. "*

Morphogenesis is, therefore, how things *exist* in the World seen through cohesiveness, and this defines the fundamental purpose

17.10 Cohesion And The Language Of Form

of Cohesiveness-based languages as technologies to convey how certain fundamental morphogeneses make up the World.

The *features of existence* are defined by the features of World *understood* by an Abstract Mind in a Cohesiveness-based. In "The World Explained Through Cohesiveness", we observed that

> *"[t]he features of reality revealed by Cohesiveness-based explanations are principles of similarity, analogy, homomorphy, uniformity, wholeness, heterogeneity, affinity, chirality and symmetry. "*

These features dictate that the types of *sentences* of a Cohesiveness-based language are *necessarily* about how diverse existential formal relations hold together Cohesiveness-based realities by virtue of forming types of morphogeneses. This is how, at the most fundamental level, Cohesiveness-based languages relate us to the World via epistemic continuity.

> *"[t]hrough morphogenesis we comprehend cohesiveness. We grasp the various manifestations of cohesiveness in the World through the way the notion of form is portrayed as morphogenesis in our Cohesiveness-based explanations. "*

From these observations, it follows that Cohesiveness-based languages function by *conceptual resonance*. It is essential to emphasize that the term "conceptual resonance" is used in this context to convey an idea that pertains to a different metaphysical structure of thought, and that, as such, we can grasp exclusively through a metaphor, not through a reference. The metaphor of conceptual resonance helps us comprehend the structure of Cohesiveness-based languages from the point of view of the Identity-based structure of thought that is being used to describe these concepts in ordinary language: but that is where its meaning and content end. For an Abstract Mind in a Cohesiveness-based state there is no such thing as conceptual resonance or similarity.

17.11 Conception And Foundations Of Mathematics

A conception of Cohesiveness-based mathematics is a fundamental intuition about how what we see as morphogenesis explains what we comprehend as manifestations of cohesion through form.

Here we apply the ideas we presented in "Morphogenesis And Formal Efficacy" (Section 15.9), in "Cohesion And The Language Of Form" (Section 17.10), and in "Conceptions And Foundations Of Mathematics" (Section 14.8) to interpret the fundamental intuition of epistemic continuity at the basis of Cohesiveness-based Mathematics. The steps below outline the adaptation of the line of reasoning we presented in Section 14.8 to the structure of Cohesiveness-based languages and intelligibility.

The morphogenesis-explains-form intuition at the basis of what we call the Cohesiveness-based conception of mathematics, constitutes a mental attitude and a metaphysical sensitivity that allows us to grasp and use the symbol-meaning epistemic tension in a way that we can illustrate as follows.

17.11 Conception And Foundations Of Mathematics

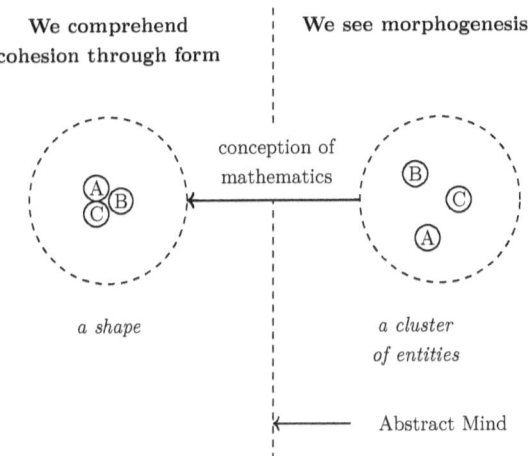

Figure 17.12: Illustration of the Cohesiveness-based conception of mathematics as the fundamental intuition about the relation of epistemic continuity between morphogenesis and cohesion through form.

- The qualities of *structural continuity* are homogeneity, uniformity, regularity and homology, and form the substratum where the principles of similarity, analogy, homomorphy, uniformity, wholeness, heterogeneity, affinity, chirality and symmetry that make up a Cohesiveness-based reality, exist as *morphogenesis*.

- A *prototypical concept*, a model of homogeneity, uniformity, regularity and homology is a *constant form or pattern*. Intuitive means that a form is grasped as a whole, and the relations of similarity and self-similarity it is made of are defined by the very grasping and not through their application of identification as patterns.

- A *conception of mathematics* (Figure 17.12) is an intuition about the existence of a fundamental relation of epistemic continuity, intrinsic in the Cohesiveness-based intuition of reality, by means of which what we see as forms of morphogenesis *explains* what we comprehend as manifestations of cohesion through form.

- *Morphogenesis explains form* through the intuitive, prototypical notion of pattern and shape, and its natural attributes and organization—tiling, group operations, etc.—that, together, constitute an archetypical language to convey how similarity, analogy, homomorphy, uniformity, wholeness, heterogeneity, affinity, chirality and symmetry *explain* homogeneity, uniformity, regularity and homology.

- Cohesiveness-based mathematics is a language of *properties and qualities*. It is the language of periodicity and symmetry and analogy.

- A *foundation of mathematics* is a language to model of the intuition at the basis of the conception of mathematics based on the prototypical notion of pattern and ita natural operations.

- The distinctive features of Cohesiveness-based mathematical reasoning are abstract notions of symmetry, periodicity, orientation, signature, incidence, and their many generalizations, and characterize the natural ways to construct properties.

Cohesiveness-based Computers

An archetypical Cohesiveness-based computer, what in Interaction Theory we should call a *form computer*, is a device or a system that embodies in its structural properties the features of Cohesiveness-based languages. Cohesiveness-based computers parametrize structure and form, and in this way they implement forms of morphogenesis.

The instructions to set their parameters are still called programs, and correspond to what in ordinary language we generally refer to as properties, such as properties of space and spatializations, as in symmetries, periodicity and quasi-periodicity, and various forms of two- and tridimensional tiling properties. Cohesiveness-based programs and programming languages are designed to model the features of similarity, analogy, homomorphy, uniformity, wholeness, heterogeneity, affinity, chirality and symmetry.

Cohesiveness-based computers don't need to have moving parts. They do not realize processes of any sort—recall the fundamental mode of Cohesiveness-based existence.

17.12 Features Of Consciousness

The three modes of consciousness modeled by the **cohesivon** define Cohesiveness-based conscious experience. By Definition 14.2.1, the conscious experience of a certain object of thought $\mathcal{O}$ is a reference to a configuration where the epistemology of $\mathcal{O}$—as described in Chapters 12 and 13—coincides with the phenomenology of structural continuity—the Actuation Mechanism of cohesiveness. Thus, to characterize the appearance of the unity and flow of consciousness in the Cohesiveness-based $\mathcal{M}$-*signature*, we must look at the Cohesiveness-based knowledge of an object of thought.

Unity Of Consciousness

In the Cohesiveness-based $\mathcal{M}$-signature, the unity of consciousness is experienced via forms of structural continuity.

In the Cohesiveness-based $\mathcal{M}$-*signature*, the unity of consciousness denotes a fundamental feature of any conscious experience of being consistent with a the notion of structural continuity codified by the **cohesivon**. The experience of unity through structural continuity is the experience of unity through the various forms of self-similarity: forms of self-similarity that we might intuitively describe as generalizations of the ordinary notion of analogy. These forms of self-similarity are particularly apparent when we examine the most elementary form of Cohesiveness-based knowledge of an object of thought, and correspond to (co)algebra-preserving or (co)algebra-inverting morphisms between (co)algebras—the object of thought in the Cohesiveness-based $\mathcal{M}$-*signature*.

Flow Of Consciousness

In the Cohesiveness-based $\mathcal{M}$-signature, the flow of consciousness is experienced as unity through similarity.

An Abstract Mind in the Cohesiveness-based state that perceives every conscious experience as connected to the one that precedes it, is said to have the experience called flow of consciousness. The term flow denotes precisely the quality of coherent connected-

17 Cohesiveness-based $\mathcal{M}$-signatures

ness of each conscious experience, and its exact meaning in the Cohesiveness-based $\mathcal{M}$-signature is therefore to be sought in the basic structure of thought that gives rise to this $\mathcal{M}$-signature: the **cohesivon**. In the Cohesiveness-based $\mathcal{M}$-signature consciousness is experienced as a flow by virtue of being continuous with respect to the dimensions along which structural continuity operates.

As a first approximation, *continuity* means constant with respect to the structure of the extensive properties of the spatialization defined by the **cohesivon**.

Bibliographical Notes

An archetypical notion of cohesion—or lines of reasoning that today we may recognize as conducive to a modern conception of cohesion—appears in the philosophical and mathematical study of the continuum, the discrete and the infinitesimal. In Western philosophy, since Democritus (ca. 450 BCE) and his school of atomism, the study of the continuum has had to deal with an implicit notion of cohesion. Continuum means undivided, but it means also divisible infinitely many times, which is an intuition about a prototypical relation between infinitesimal entities and the whole. It is in the trichotomy undivided-part-infinitesimal that lies the conceptual labyrinth underpinning the necessity to characterize the relation between the part and the whole. That relation is cohesion: it is the basic intuition about the qualitative nature of a principle according to which the parts give rise to a united whole, about how the dynamical nature of thought originates form the interplay of definitional and relational identities (see Section 5.2). This is why any serious discussion about the continuum must necessarily take into account, or be conducive to, a framework to define cohesion. Notable reflections on the continuum are, among others, in Brentano 1988, Charles S. Peirce 1976, Charles Sanders Peirce 1992, Poincaré 1913, Folina 1992, Lawvere 2007, Lawvere 2011, in the constructive, (neo)intuitionistic approach to mathematics (Brouwer 1975), and again in Brower's Continuity Principle and in his use of the Bar Induction, but also in John L Bell 2000, Feferman 2009.

In mathematics, the task of carrying or encoding a principle of cohesion is delegated to various types of extra structure, such as a topology, a smooth structure, or the property of a functor, to mention a few. The study of cohesion in the broader framework of structural continuity, and in particular in the context of what is probably the most common manifestation of structural continuity, namely, self-similarity, requires more abstract tools. An excellent introduction to a category-theoretic approach to self-similarity, mostly centered around endofunctors is in Hughes, Steve Awodey, and Scott 2001, Leinster 2004a, Leinster 2004b and Leinster 2004c and Jacobs 2016.

The view that cohesion is a manifestation of a law of structural continuity ("Actuation Mechanism" Section 17.2), and that the study of structural continuity in the broader scope of the study of spatialization is necessary to comprehend cohesion are, to the best of my knowledge, mine—structural continuity is the nexus underpinning every discussion about the continuum, self-similarity, infinitesimals and cohesiveness, because a continuum is a model of invariant across the dimensions of the spatialization of an object on which a notion of continuity is defined. The descriptions of the `cohesivon` and of the Cohesiveness-based epistemology, and the interpretation of consciousness as a mode of spatialization are sketched in my unpublished work Lo Vetere 2015a. For a short bibliographical reference on consciousness see the note at the end of Chapter 14.

PART V

Toward A New Conception Of Mathematics

How big is an elephant?

Introduction

> Since new paradigms are born from old ones, they ordinarily incorporate much of the vocabulary and apparatus, both conceptual and manipulative, that the traditional paradigm had previously employed. But they seldom employ these borrowed elements in quite the traditional way.
>
> T. Kuhn 1993

We begun our journey to the center of the mind, after Jules Verne, with the notion that if we are brave enough to study the structure of the mind-world transactions, we might one day understand the origin of our mathematical intuition of reality, and, perhaps, a little more of the nature of consciousness, and the role it plays in our experience of connectedness with the World. To describe the mind-world interplay, we introduced a strange device called Abstract Mind that produces Symbolic Intuitions of Reality, which are self-contained foundations of human thought based on models of intelligibility, and we described how within those foundations we create knowledge and meaning, and transform our abstract intuition of reality into experience.

Through our Symbolic Intuitions of Reality, we invent the landscape of imagination where our experience of existence takes place. Through our Symbolic Intuitions of Reality, every corner of the Universe we explore overwhelms us with the stunning beauty and sophistication of the aesthetic canons that seem to have inspired and guided the hand of the artist who painted it. Through our Symbolic Intuitions of Reality, Nature seems to whisper to our mind in a language of unity, beauty, meaning and purpose.

We conclude this journey, for now, with a medley of reflections on some of the most interesting avenues of research suggested by Interaction Theory.

Key Concepts Of Part V

1. Interaction Theory defines higher canons for explanations and higher types of knowledge based on the dynamical properties of the Abstract Mind.

2. New conceptions of mathematics help us revisit and reinterpret the modern conception of mathematics from the perspective of different Symbolic Intuitions of Reality.

3. It is possible to construct a universal notion of information as a measure of the extensive properties of the knowledge defined inside an Interaction Model.

4. The foundation of human thought offered by Interaction Theory provides the background to put forward new principles of scientific inquiry called Metaphysical Realism.

Chapter 18

Higher Epistemology

Mathematics is a part of physics

Arnold 1997

It might perhaps, at first, seem strange that a
self-contradictory sentence, hence one which no ideal
receiver would accept, is regarded as carrying with it
the most inclusive information. It should, however,
be emphasized that semantic information is here not
meant as implying truth. A false sentence which
happens to say much is thereby highly informative in
our sense. Whether the information it carries is true
or false, scientifically valuable or not, and so forth,
does not concern us. A self-contradictory sentence
asserts too much; it is too informative to be true.

The Bar-Hillel-Carnap Paradox
(Carnap, Bar-Hillel, et al. 1952)

When the Abstract Mind changes state, either because the Interaction Model changes, or because the $\mathcal{M}$-*signature* changes, it is possible to identify on the fabric of that change, higher order cognitive structures that generalize the terms knowledge, comprehension, meaning, explanation and consciousness, and that are called, for this reason, a *Higher Epistemology*.

A Higher Epistemology is, therefore, a description of the dynamic properties of the Abstract Mind, and can be seen as an evolutionary model of Symbolic Intuition of Reality. In this short

18 Higher Epistemology

chapter, we present some reflections on what seems to be its basic structure.

There are many ways to characterize the dynamic behavior of the Abstract Mind. In the "Classification Of Interaction Models" (Section 11.4) we had a partial preview of what a characterization could be because our analysis was limited to the horizon of a single **concrete** concept, and in "How Data Becomes Message" (Section 14.6), we took an indirect approach, and enumerated the transitions that can occur inside a given $\mathcal{M}$-*signature*, and between $\mathcal{M}$-*signatures*.

Another approach, that we might call, geometrical, is suggested by the observation that the types of spatializations we encounter in **identiton**, **finiton** and **cohesivon** are related diagrammatically via the shape of the diagrams that represent these three modes of consciousness.

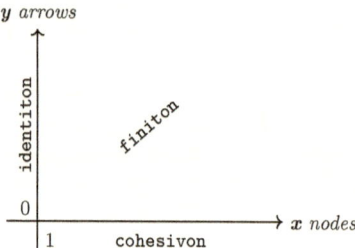

Figure 18.1: Diagrammatical relation between the three main $\mathcal{M}$-*signatures*.

In the diagram of Figure 18.1, the x-axis represents the number of nodes, and the y-axis the number of arrows. Thus, along the y-axis we find the **identiton** (Section 15.3), because it represents a type of spatialization modeled by a diagram with one vertex and an arbitrary number of arrows ($\mathcal{I}(\bullet)$). Along the x-axis there is the **cohesivon** (Section 17.3), because it represents a type of spatialization modeled by a diagram without arrows ($\mathcal{R}(\bullet)$), and inside the xy-plane there is the **finiton** (Section 16.3), which represents a type of spatialization described by an arbitrary diagram ($\mathcal{F}(\bullet)$) where it is possible to construct category-theoretic limits.

Introduction

Homeostatic Properties Of The Abstract Mind

An important metaphor to explain the relation between a dynamic description and a Symbolic Intuition of Reality is based on the notion of autopoietic system. Autopoiesis, from Greek αυτο (*auto*, auto-) and ποιεσις (*poiesis*, creation) is a term introduced by Chilean biologists Humberto Maturana and Francisco Varela in 1972 to denote the property of living systems to self-maintain and self-generate. An autopoietic system (Maturana and F. J. Varela 1980)

> "... *continuously generates and specifies its own organization through its operation as a system of production of its own components, and does this in an endless turnover of components under conditions of continuous perturbations and compensation of perturbations.*"

In this sense, living systems display homeostatic properties, and, to the extent that this metaphor applies to the workings of the human mind, so does the Abstract Mind. The metaphor we use here between a dynamic description of the Abstract Mind and certain properties of autopoietic systems is limited to what seems to be the ability of the Abstract Mind to self-generate and self-maintain the archetypical cognitive structure through which it spatializes. It is important to observe that the homeostatic properties of the human mind and of its Symbolic Intuitions of Reality are themselves to be regarded, in light of the autopoietic description of mental phenomena, as a subtle, perhaps secondary, perhaps epiphenomenal or hypophenomenal feature of the *unity of consciousness*, where unity is manifested or kept consistent with itself through homeostatis.

18 Higher Epistemology

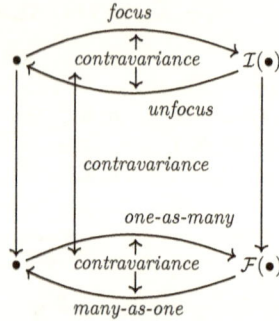

Figure 18.2: Illustration of a higher type of spatialization as a transition from `identiton` to `finiton`.

Just like in an *M-signature* we define an epistemology on the structure of the spatialization that characterizes that *M-signature*, so can we define an epistemology on the structure of the transitions of Abstract Mind states, as soon as those transitions display properties that are consistent with a spatialization, that is to say, as soon as it is possible to clearly describe how they connect the symbolic and semantic domains. This characterization of Higher Epistemology in terms of higher order patterns between *M-signatures* is, admittedly, heavily influenced by a mathematical style of reasoning, where the synthesis of new ideas often consists in bracketing and folding into themselves known concepts numerous times. A visual metaphor to convey how a new type of spatialization would emerge from an Abstract Mind state transition is illustrated in Figure 18.2, which depicts an Abstract Mind state transition as a morphism from `identiton` to `finiton`, where the spatialization consists in thinking of that diagram as atomic. Note that, in this case, there is a new type of awareness of an object of thought that emerges as a contravariance *between* `identiton` and `finiton`.

18.1 Submodalities

Knowledge always supervenes on awareness.

A first step to interpreting the transitions between Abstract Mind

18.1 Submodalities

states consists in using the description of the modes of consciousness introduced in Section 14.1, the representation of Abstract Mind states in the language of diagrams, and the epistemology of the three $\mathcal{M}$-*signatures* we presented in Part IV to give an account of what it means, cognitively, psychologically and intellectually, to shift from one Abstract Mind state to another.

As a first approximation, we can think of an Abstract Mind state as an arbitrary diagram $\mathcal{D}(\bullet)$, and denote with **B** and **A** the number of nodes and arrows of $\mathcal{D}(\bullet)$ respectively.

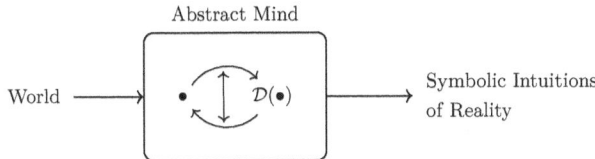

Thus, a transition between Abstract Mind states is a certain transformation of diagrams that consists of certain rules that define how **B** and **A** change as a result of the transition. For example, a transition from an Identity-based mode of thought to a Cohesiveness-based mode of thought, is a transition where the number of nodes of $\mathcal{D}(\bullet)$ increases, and the number of arrows decreases to zero.

To interpret an Abstract Mind state transition, we recall the description given in "Modes Of Consciousness" (Section 14.1) of how certain conditions on comprehension and meaning dictate the changes of Abstract Mind states. We also make the assumption that the transitions of Abstract Mind states have an intrinsic structure consisting of two cognitive modalities described, as a first approximation, by **B** and **A**, where **B** is the primary cognitive modality and **A** is always a submodality that supervenes on **B**. This view is informed by what seems to be the primary role of spatialization in all known cognitive events, which in the language of diagrams corresponds to the number of nodes **B**.

With these premises, the description of an Abstract Mind state transition is the description in the language of the Epistemology of the Abstract Mind of the changes in **B** and **A**, and is called the description of the *submodalities* of the Abstract Mind—a term we borrowed from Neurolinguistic Programming that indicates the fine distinctions in the perceptual modalities (visual, kinesthetic, auditory, olfactory and gustatory) through which we humans codify our experience—because the variations in the structure of $\mathcal{D}(\bullet)$ do precisely this: they give rise to nuances or fluctuations in the epistemology of an $\mathcal{M}$-*signature* through which we experience the World.

Analytic Expansion And Synthetic Contraction

Analytic and synthetic contractions denote changes in the awareness of an object of thought.

The variations in **B** determine the type of spatialization occurring in an Abstract Mind state, and in this interpretation are therefore associated to *changes in the awareness on an object of thought*.

The phase of *analytic expansion* is characterized by an increase in the number of nodes, its primary pattern is the `cohesivon`, and pertains to the cognitive activities where the Abstract Mind explores different types of spatialization, without necessarily completing them. In this phase, given that $\mathcal{D}(\bullet)$ may not even have arrows, the epistemology can be incomplete, and certain features such comprehension and semantogenesis may not even take place fully. Analytic expansion describes the intuitively experience of pondering, of perception turned inward in an exploring fashion, without necessarily lingering in any detailed thought.

The phase of *synthetic contraction* is characterized by a decrease in the number of nodes, and is also a form of spatialization. The difference between analytic expansion and synthetic contraction is that the temporary reduction in the number of nodes of $\mathcal{D}(\bullet)$ points to, and can be conducive to the limiting Abstract Mind state described by the Identity-based, where spatialization as such ceases to exist, and the only changes that can occur are in the awareness of an object of thought.

Cognitive And Semantic Drifts

Cognitive and semantic drifts denote the exploration of types of knowledge and meaning.

The variations in **A** characterize the type of Abstract Mind state transitions where the predominant cognitive function is of an epistemic nature. During these transitions the Abstract Mind *explores types of knowledge and meaning.*

The phase of *cognitive drift* is characterized by an increase in the number of arrows of $\mathcal{D}(\bullet)$, and is associated with the development of knowledge because it increases the configurations between nodes where the equational conditions of knowledge for that particular $\mathcal{M}$-*signature* may occur.

The phase of *semantic drift* is characterized by a decrease in the number of arrows of $\mathcal{D}(\bullet)$, and is called semantic because, as long as the variation of arrows remains the primary cognitive function, it can be conducive to minimal diagrams where it is still possible to define an epistemology. The condition of minimality—recall the discussions about comprehension and semantogenesis (Sections 12.5 and 12.6 respectively)—is in fact necessary to define first comprehension and, from there, meaning.

18.2 Higher Knowledge, Higher Explanations

Any fact is explainable by different $\mathcal{M}$-signatures.

The frugality of this statement hides a peculiar intuition: the intuition that explanations are independent from knowledge. To comprehend what this means, and the profound and unintuitive implications of this notion, we must recall the Principle of Epistemic Relativity (Definition 13.1.1)

> *"There is no privileged frame of reference to define an epistemology of the Abstract Mind, and with it, a mathematical and scientific method of inquiry. "*

This principle captures the observation that (Section 13.1)

> *"... [I]n Nature there seems to be no indication that the* **concrete** *concept of identity is a privileged frame of reference to define mathematical thought and a scientific method. Identity, and its many manifestations, are anthropomorphic projections that we transfer more or less consciously in our efforts to come at the world with a demand that it endorse a worldview in harmony with what we think is the only*

18 Higher Epistemology

safe, reliable, rational structure of human thought. [...] Nature has no requirement to be in harmony with human thought and ambition ..."

The existence of multiple $\mathcal{M}$-*signatures*, together with the Principle of Epistemic Relativity, raises a compelling question: Is an explanation valid only in the $\mathcal{M}$-*signature* where it is constructed? And, all the more, Can a statement constructed in an $\mathcal{M}$-*signature count as an explanation* in another $\mathcal{M}$-*signature*?

For example, a statement defined in a Finiteness-based $\mathcal{M}$-*signature*, such as "this object is the (category-theoretic) limit of this diagram" could *count as an explanation of* a fact described in the language of the Identity-based $\mathcal{M}$-*signature*, such as for example "this set is finite". But we have already seen an example of this correspondence in the description of the `finiton` (Section 16.3): where the limit of this diagram

gives rise to the condition for a so-called Dedekind-infinite set. This is an example of a general concept that in Interaction Theory is called a *higher explanation*. Higher explanations are a broad and complex and fascinating topic that is impossible to present in this introduction to Interaction Theory.

It turns out that the answers to both questions are in fact affirmative, and that coherent statements constructed in the language of one $\mathcal{M}$-*signature* can count as valid explanations in another $\mathcal{M}$-*signature*. It is useful, at this point, to recall the notion of exoexplanation—theories defined on transformations between $\mathcal{M}$-*signatures*—introduced in "The Metastructure Of Explanations" (Section 13.7), and the archetypical structure language we introduced in "Metaphysics Of Language And Computation" (Section 14.7). Together, those two concepts form the frame of reference where what is seen and the understood of a World from the Symbolic Intuition of Reality of a certain $\mathcal{M}$-*signature*, becomes the conduit for higher types of knowledge and explanations that involve multiple $\mathcal{M}$-*signatures*. Higher knowledge is, in this sense, the knowledge that originates from the types of awareness of an $\mathcal{M}$-*signature* as a whole, beyond the horizon of Interaction Models, and is, as expected, represented by a higher order types of

18.2 Higher Knowledge, Higher Explanations

contravariance, as exemplified in the diagram of Figure 18.2.

The discovery of non-standard $\mathcal{M}$-*signatures* and of parallel explanatory frameworks that do not rely on causation, opens up the theoretical possibility of constructing strictly non causal, non predictive arguments whose content is as informative as an explanation constructed in an Identity-based $\mathcal{M}$-*signature*. These types of arguments are called existential, and are something like "A exists or is such and such object in one $\mathcal{M}$-*signature*, therefore B in another $\mathcal{M}$-*signature*". The complexity intrinsic in higher explanations resides in the interpretational framework needed to turn statements into explanations. For example, to interpret an explanation based on causation in a language based on a non causative mode of thought, such as the language of proximity and encompassment (Finiteness-based), or the language of form and morphogenesis (Cohesiveness-based), one needs to comprehend to what extent the requirement of a certain degree of mutual intelligibility between two languages informs of hinders the transliteration of the ontologies of one $\mathcal{M}$-*signature* into the ontologies of the other $\mathcal{M}$-*signature*.

Bibliographical Notes

As usual in this book, there are several ideas converging in our analyses of the Abstract Mind. Here we have revisited and reinterpreted some fundamental concepts of cybernetics and system theory, and integrated them with the notion of autopoiesis, which is a beautiful and powerful synthesis introduced in Maturana and F. J. Varela 1980 and later developed in Francisco J. Varela, Thompson, and Rosch 1993.

The description of the dynamic behavior of the Abstract Mind, and in particular of the transitions between modes of consciousness, such as $\mathcal{I}(\bullet) \to \mathcal{F}(\bullet)$, is a simplification of a central topic of Higher Category Theory, which is the definition of multidimensional categories, categories that contain arrows, arrows between arrows, arrows between arrows and so on (Leinster 2001b, Brown and Porter 2003, Leinster 2004d, Lurie 2009b). These mathematical objects are useful to model the dynamic behavior of the Abstract Mind, because their definition naturally allows the embedding of higher types of epistemologies. My mathematical exploration of these structures, still a work in progress. Below is a sketch of the current avenues of research.

Interpretations Of The Synthetic Structure Of Intentionality

The first step of the modeling process of an Interaction Theory is of an *ontological* nature, because it consists in mapping philosophical definitions onto objects that embody those definitions via their structure. Every model originates from an interpretation of the tenets of the theory that it models, and as such, it necessarily encodes the worldviews and the belief systems that guide the interpretation process. But what exactly is the ontological structure of a modeling process? I think the following example can adequately answer this question. Consider modern Physics. For a theory to count as a mathematical model of a certain physical phenomenon, there must be a body of knowledge, either inside the theory or as part of the assumptions the theory is based on, to justify that the invention of the basic concepts used in the theory such as space, time, force, momentum etc. are consistent with a certain set of philosophical views about the things that exists and that Physics aims to explain. There is *no apriori necessity to have* the concepts of space, time, force, etc. to develop Physics and our physical knowledge of the World. Those concepts are merely instrumental to the construction of a theoretical framework to acquire knowledge of the physical world based on a certain *conception* of Physics—in particular, space and time belong to the physical-philosophical tradition that spatializes concepts to study Nature.

These considerations about the ontological nature of the modeling process lead to the questions about what ontologies are suitable to describe the Synthetic Structure of Intentionality. For example, an interpretation might aim at preserving the distinction between the definitions of Validation System, Interaction Model and Local Identity, and would therefore result in a model consisting of three entities A, B and C, that represent Validation System, Interaction Model and Local Identity respectively. Another interpretation might do something completely different, such as mapping Validation System, Interaction Model and Local Identity onto a single entity that codifies the entire Synthetic Structure of Intentionality through its own structure. In the definition of the Synthetic Structure of Intentionality there is nothing that prevents us from discarding a particular approach to mapping its components onto one or more entities of the target model. The choice on the type of mapping depends entirely on the worldviews that inform the

18.2 Higher Knowledge, Higher Explanations

modeling process, which is described in the next section. These observations lead to two paths to the formulation of Interaction Theory.

The two paths to the formulation of Interaction Theory I present here are based on two interpretations of the ontological relations between $\mathcal{M}$-*signatures*, Cognitive Architectures and Local Identities. Two interpretations of Synthetic Structure of Intentionality seem to stand out for their simplicity, for their ability to capture the basic features of the Abstract Mind, and for their adherence to the worldviews that have guided and inspired this research so far. These two interpretations are the *field* and *categorial* interpretations of the Synthetic Structure of Intentionality, which give rise to types of Field Interaction Theory and Categorial Interaction Theory respectively.

Recall the discussion about the Synthetic Structure of Intentionality (Chapter 8), where we saw how the Validation System provides to the Interaction Model the epistemic context from which a Local Identity emerges as a result of the specific way in which the Interaction Model relates an Intention to a goal. With that data in mind, one might be led to think that there is a causal relation between Validation System and Interaction Model, in that the definition of Interaction Model is erected on a substrate of concepts that belong to the definition of Validation System. Indeed, each interpretation of the Synthetic Structure of Intentionality in relation to how we think it describes the Abstract Mind, reflects necessarily the distinct character of our observation of the Abstract Mind. Two reflections on the relation between Synthetic Structure of Intentionality and the Abstract Mind model emerge from a universe of interpretations of the philosophical work presented so far.

The first reflection is of a practical nature, and focuses on the emergence of mathematization as a very peculiar form of structured conceptualization. The second reflection is of a metaphysical nature, and revolves around how mathematical thought has a distinct quality of unity and self-consistency in which the definition of parts, such as the three elements used in the description of the Synthetic Structure of Intentionality model, seem artificial. These two reflections give rise to two paths to the formulation of Interaction Theory. Let's see them in more detail (Figure 18.3).

Figure 18.3 summarizes the two interpretations of the Synthetic

18 Higher Epistemology

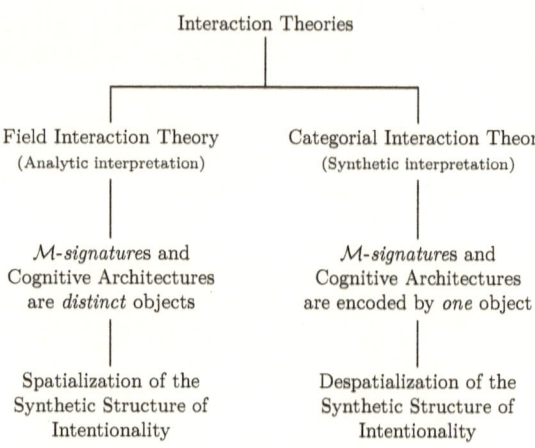

Figure 18.3: Paths to the formulation of Interaction Theory.

Structure of Intentionality, and emphasizes their relations with the core human experience of concepts: spatialization. Field Interaction Theory corresponds to a *spatialization* of the components of the Synthetic Structure of Intentionality because it relies on a conceptual arena that occupies two dimensions to represent the Abstract Mind, namely, $\mathcal{M}$-*signature* and Cognitive Architecture. In contrast, Categorial Interaction Theory eliminates the distinction between $\mathcal{M}$-*signature* and Cognitive Architecture introduced by the philosophical descriptions we gave of these entities, and compacts their ontological statuses into one single entity which codifies the Abstract Mind. In the categorial interpretation of the Synthetic Structure of Intentionality, the Abstract Mind "takes place" inside an abstract object where $\mathcal{M}$-*signatures* and Cognitive Architectures are properties and features of that object. When the predominant theme in the observation of the Abstract Mind is how the human mind's ability to create a Symbolic Intuition of Reality emerges from the mind-world transactions, the interpretation of the Synthetic Structure of Intentionality that follows from this observation regards $\mathcal{M}$-*signatures* and Cognitive Architectures as *distinct* and *separate* entities. This is the field interpretation of the Synthetic Structure of Intentionality, also called *analytic*, because it is based on substrate concepts that correspond to the *components* of the Synthetic Structure of Intentionality. The categorial

18.2 Higher Knowledge, Higher Explanations

interpretation of Interaction Theory, also called *synthetic*, is based on the interpretation of the Synthetic Structure of Intentionality as a *united whole* where the distinction between $\mathcal{M}$-*signature* and Cognitive Architecture serves a purely explanatory purpose. In the synthetic view of the Synthetic Structure of Intentionality, we are bound to encode in the construction of an Interaction Theory the intrinsic unity and cohesiveness of the Abstract Mind as a basic feature of the theory, and as a result, $\mathcal{M}$-*signatures*, Cognitive Architectures and Local Identities are mapped onto a *single* abstract object.

Are There Other $\mathcal{M}$-*Signatures*?

Yes. The three $\mathcal{M}$-*signatures* I introduced in this account represent broad types of Symbolic Intuitions of Reality. In principle, each diagram encodes a type of spatialization which can be used to define a Symbolic Intuition of Reality.

Chapter 19

Interlude: Definitions Of Mathematics

> The whole of mathematics consists in the
> organization of a series of aids to the imagination in
> the process of reasoning.
>
> Whitehead 1898

> Mathematics as an expression of the human mind
> reflects the active will, the contemplative reason,
> and the desire for aestetic perfection.
>
> Courant, Robbins, and Stewart 1996

In this interlude we review some known definitions of mathematics, and add some new ones in light of the tenets of Interaction Theory.

In the game of knowledge, the single most important thing is asking the right questions. This brief review of definitions of mathematics is an evolutionary journey through some of the trajectories of thought that define the spirit of mathematical inquiries. It is a journey that chronicles the conceptions of mathematics through the history of human thought, and how these have refined our ability to relate symbolically and semantically to the fabric of reality. These conceptions of mathematics express more than worldviews and philosophical positions, they are beacons that shape our minds and define the boundaries of our imagination and intuition. These conceptions of mathematics bear witness to the metaphysical intuitions through which we come at the world with the demand to reveal its deep structure; a structure that reminds

19 Interlude: Definitions Of Mathematics

us how little we know about consciousness.

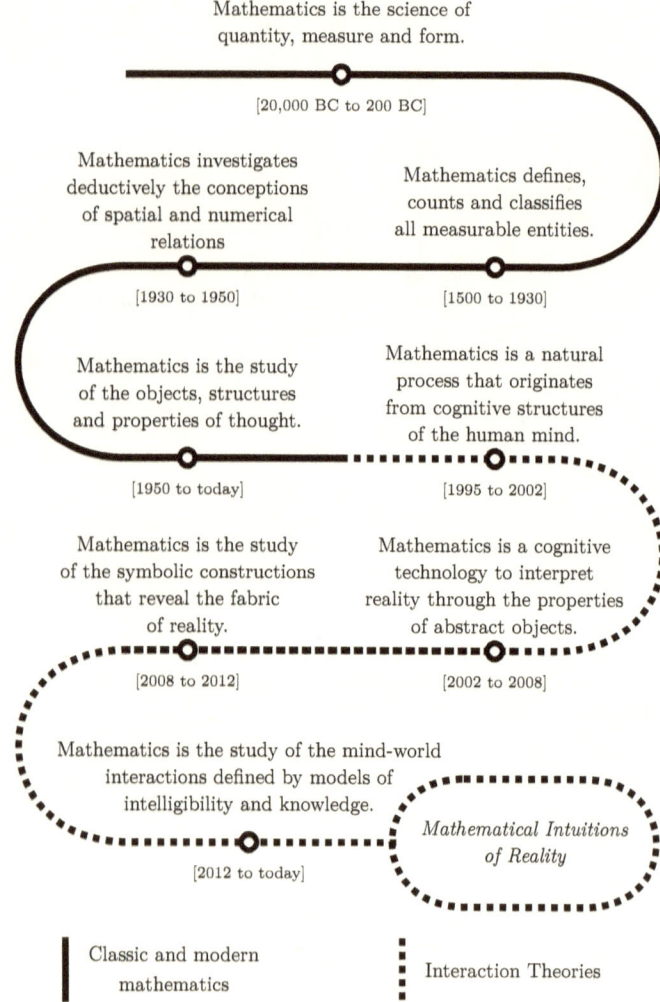

Figure 19.1: An evolutionary view of the conceptions of Mathematics.

I gathered the definitions of mathematics into three groups: those that emerge from the classical and modern conceptions of mathematics, the definitions that emerge from Interaction Theory, and those that emerge from the Symbolic Intuitions of Reality we have presented in this book. The chapter reflects this rationale, and presents a medley of historical definitions of mathematics and definitions of mathematics I have given over the course of the past two decades.

Figure 19.1 is a map of the evolutionary journey of mathematics in which the conceptions of mathematics are ordered by increasing level of abstraction: from mathematics as the science of quantity, to the contemporary structuralist view, to the evolution of mathematics proposed by Interaction Theory as the study of the mind-world interactions based on models of intelligibility.

Each group has a theme that captures its underlying worldviews and belief systems. The theme of the *classic and modern* group is that mathematics is a cognitive technology based on a quantitative-structuralist view of the world in which a conception of mind plays no role. The leitmotif of the *Interaction Theories* group is that mathematics is a cognitive technology based on a self-generating model of human mind defined by the process of knowing what is intelligible at any given level of consciousness.

Tracing the history of ideas is not like watching a car being assembled on a production line. It is a far more chaotic and back and forth and inconsequential enterprise, driven in equal measure by dramatic events, logic, intellectual fanatism, private agendas, brilliant intuitions, courage, by the infinite range of human emotions, and by the quiet grandiosity of dreamers and mavericks of all walks of life. Many great ideas are discovered, forgotten, ignored, ridiculed and rediscovered many times before taking root and being accepted. As A. Schopenhauer once wrote *"All truth passes through three stages. First, it is ridiculed. Second, it is violently opposed. Third, it is accepted as being self-evident."* Mathematics is no exception, and its history is a testament to Schopenhauer's thought. This is to say that the dates I indicate in each section are not to be taken too strictly.

19 Interlude: Definitions Of Mathematics

19.1 Classic And Contemporary

This is a brief view of three pivotal moments in the history of mathematics: the origin of structured mathematical thought, the foundational crisis of mathematics and its consequences, and the modern proposals to unify mathematics. This is not a historical account, and as such, it does not indulge in detailed explanations of the facts and ideas that shaped mathematics. Its purpose is to convey the moments in the history of mathematics that seem to mark, according to the author, a global and profound paradigm shift in the way we think of and do mathematics, that is comparable, in intellectual terms, to the discovery of fire, or to the invention of the wheel.

Quantity, Measure And Form

Mathematics is the science of quantity.

(Aristotle)

years 20,000 BC to 200 BC

Mathematics as the science of quantity and measure is the conception of mathematics passed on to the Western world by Greek philosophers in the last six centuries BC. Mathematics was most likely created primarily as a system to measure time (30,000-20,000 BC), although there is evidence of geometric patterns even prior to the use of mathematics to measure time (70,000 BC). Mathematics then evolved into a system to mechanize computations: this is when a concept of quantity first emerges and numerical systems make their appearance for the first time (3400-3100 BC). Since then, our species begun to develop and refine mathematical technologies to abstract reality through various conceptions of quantity and measure. Fractions made their first appearance in 2800 BC. Perhaps, following the successful application to practical problems of those strange and mysterious gadgets called numbers, a similar abstraction process was applied to other aspects of perceptual experience: form and change. This is when we begin to find evidence of concepts of symmetry (2000 BC). This technological progress went on with the invention of quadratic equations (1800 BC) and arithmetic and geometric algorithms.

It should come as no surprise that the definitions of mathe-

19.1 Classic And Contemporary

matics given at this early stage of the history of mathematical thought were based on the concepts of quantity, measure and form. Quantity, measure and form are in fact related to perceptual experience in an almost unmediated way. The underlying theme of this conception of mathematics was based on a view according to which the world mathematics was concerned with was made precisely of the entities described by the mathematicians of that time, that is: *stuff*, which carries the category of quantity, *form*, which carries the category of shape, and *measure*, which codifies the relations between quantities and forms—a prototypical concept of space. For a measure of quantity is a relation between quantities, and similarly a measure of form is a comparison of shapes. We must not forget that at the beginning of its evolutionary journey, mathematics was a branch of Philosophy, and that its cosmology of thought was the cosmology of thought of Plato, Aristotle, Pythagoras, Euclid, Heraclitus, Thales, and many other fathers of Western thought. This was a cosmology of thought in which *form* and *change* were universal principles at the intersection between the human and the divine realms: they were bold interpretations of the phenomenology of the mysterious impulse of the human intellect to reconnect with Nature through the abstractions of mathematical reasoning. This definition of mathematics was, perhaps, the first attempt to define a syntax of thought as an extension of the metaphysical syntax of Nature described by Greek philosophers in their pioneering philosophical investigations. The mathematical realities painted by the Greek philosophical tradition were windows on Plato's heaven through which to peek into physical reality as a reflection of a universe of semantic possibilities.

Significance Of This Phase

This phase marks one of the greatest discoveries of mankind, which happens to be also one of the deepest mysteries of Nature: we seem to have epistemic access to the structure of physical reality through the mind's ability to mathematize. This early conception of mathematics marks the beginning of a glorious intellectual tradition of mathematical and philosophical inquiries about the nature of mathematics, about the structure of conceptual experience, and about the way in which symbolic thought seems to relate our species to a deeper truth at the foundation of reality. It is the birth cry of an adventure of ideas that has taken our species

19 Interlude: Definitions Of Mathematics

to ponder the structure of its own existence, and to seek in the transcendent elegance of arithmetic the answers to the ultimate questions.

Define, Count And Classify

Mathematics is the science of defining, counting and classifying any measurable entity.

(Davide Lo Vetere)

years 1500 to 1930

Fast forward 2000 years, and the conceptions of mathematics that begin to appear, transcend and incorporate the original idea of mathematics based on quantity and measure, and begin to focus on the *role* played by quantity and measure in the formulation of mathematical concepts. The shift from quantity and measure to the *function* that these have in the formulation of mathematical concepts is an important abstraction leap that shapes globally the narrative of mathematics, its role of metaphysical language, and its becoming more and more a form of computational philosophy. This is an important abstraction leap because is redefines quantity via the functions that apply to quantity bearers—in modern category-theoretical terms, this idea is refined, purified, enriched and amplified by Sheaf Theory and, consequently, by Topos Theory.

These are the years that see the rise of mathematical thought as the de facto standard for rational knowledge for our species, as opposite to the mystic and revelatory ways to relate to reality. This is a glorious period that celebrates the triumph of reason over belief. From the invention of calculus, analytic and projective geometry, and non-euclidean geometries, to probability, modern algebra, number theory, to various systematizations and generalizations of logic, to the emergence of mechanics, thermodynamics, to the study of dynamical systems and many more gems of mathematical thought; there isn't a corner of human existence not affected by the implications of mathematical technologies in our lives.

Significance Of This Phase

The significance of this phase of the evolution of mathematical thought is in the discovery that mathematical thought can be

19.1 Classic And Contemporary

applied to itself. This discovery accelerated the development and rise of mathematics as the language of science by multiplying the discoveries of new mathematical technologies. At a deeper level, this discovery stems from the intuition that relations between mathematical gadgets describe deeper structures of thought, and that these new levels of abstraction give us access to very powerful ways to describe practical and theoretical problems.

Mathematics Is Logic

[Mathematics is] The abstract science which investigates deductively the conclusions implicit in the elementary conceptions of spatial and numerical relations, and which includes as its main divisions geometry, arithmetic, and algebra.

(Oxford English Dictionary, 1933)

years 1930 to 1950

The quest for consistent foundations of mathematics that started at the beginning of the 20th century—known as the foundational crisis of mathematics—marked the beginning of an important moment in the history of modern mathematical thought. This is the moment in which, for the first time in the history of mathematics, philosophers and mathematicians use mathematics not to study the physical world but to study mathematics itself. It is the period in which the human species begins to look at the universe as a linguistic structure, a solemn construction that reveals the aesthetic canons of logic wherever we look, if we are willing to wear the glasses of formal systems. There is a sense in which with logicism and formalism an embryonic form of the principle of the universality of computation was beginning to replace Plato's heaven with a supercomputer. This is the period in which much of the conception of mathematics consists of a vivacious dialectic between formalists—Hilbert, Carnap, Tarski—and logicists—Russell, Peano, Frege, Dedekind. For the first time, mathematics ceases to be absolute in all epistemological directions, and shows that it too is an intrinsically incomplete instrument of thought that encodes in its deep structure the seeds of logical paradoxes. This realization is, of course, the message conveyed by Gödel's incompleteness theorems—which we have briefly described in Section 8.4. But at a deeper level, the philosophical and mathematical implications of the foundational crisis are, I think, even more interesting and profound than its foundational implications. Gödel's incompleteness theorems first (1931), and Lawvere's fixed point theorem almost forty years later (1969), revealed that mathematical theories cannot be regarded as isolated, self-contained systems to create knowledge. Rather, they are part of a conceptual web that we can navigate in many dimensions to reveal recursively deeper truths. The intrinsically recursive structure of truth in mathematical theories is a direct consequence of Gödel's incompleteness theorems—and a fortiori of Lawvere's fixed point theorem—because it shows that each time a theory is extended to incorporate those conditions that make it incomplete, it becomes incomplete for the same reasons but at a higher level.

Significance Of This Phase

The importance of this phase of the evolution of mathematical thought is in the birth of metamathematics—the mathematical study of mathematics itself—and in the logical leap that this new field of study represents. The realization that all non trivial mathematical theories are incomplete (Gödel) is what I define as the moment in which mathematical thought reached its puberty.

Mathematics Is Structure

Mathematics is the systematic study of the objects, the structures and the properties of abstract thought.

(Davide Lo Vetere)

years 1950 to today

The past 70 years have marked the definitive transition from the logicist-formalist views of 20th century mathematics, to a distinctly structuralist conception of mathematical thought. This profound transition may be roughly ascribed to the discovery and development of various methods to translate algebraic, geometric, arithmetic and topological problems into one another, and to the rise of Category Theory as the overarching technology to work with new mathematical structures that connect Algebra, Geometry, Arithmetic, Logic and Topology—structures called sheaves, sites, stacks, schemas, motives, topoi and many more. This transition,

19.1 Classic And Contemporary

that we describe with the slogan *from logic to structure*, has had two consequences of enormous significance for the development of modern mathematical thought.

The first consequence is that the development of "bridges" between branches of mathematics—powerful technologies that allow us to translate algebraic, geometric, arithmetic and topological problems into one another—has had profound theoretical and practical implications. Probably the most famous theoretical and practical result due to the development of these conceptual bridges between areas of mathematical research is the theoretical apparatus that led British mathematician Andrew Wiles to prove Fermat's last theorem in 1994. But the list of theoretical and practical results that originated from the new insights into the deep structure of mathematics is vast: from TFQT (Topological Quantum Field Theory), to mathematical methods to develop String Theory and Quantum Gravity, to various applications of Categorial Logic to Computer Science and Type Theory, to the discovery of Kripke-Joyal semantics (a semantics of modal logic)—which describe a natural semantics in an arbitrary Topos where ordinary mathematics can be defined.

The second consequence is the emergence of various foundations of mathematics, various frameworks to unify mathematics, and new ways to regard entire branches of mathematics as manifestations of the same overarching schema. We have discussed very briefly in Section 4.1 (The Generative Lens) different foundations of mathematics which emerge from category-theoretic (ETCS) and type-theoretic arguments (HoTT), and from category-theoretic models of set theory (AST).

Significance Of This Phase

The shift of focus in mathematical thought, from logic to structure, seems to suggest that a new kind of tension is about to surface in mathematics: the ontological and epistemological tension between syntax and semantics. This tension is very different from the tension that triggered the foundational crisis, because it is about the meaning of mathematics, and about the meaning of doing mathematics. Syntax reflects logic. Semantics reflects structure, because it is intrinsically contextual and relational. In this tension resides the significance of this evolutionary phase of mathematical thought, which is beginning to emerge now, in part due to serious misconceptions about artificial intelligence. This tension will demand us to reconcile mathematics with consciousness, hopefully, in the spirit of Interaction Theory.

19.2 Interaction Theories

In the early stages of the formulation of a philosophical or mathematical theory, and especially if the theory is very abstract, the boundaries between philosophical speculation and formal investigation are blurred because the dialectic of thought gravitates around intuitions rather than clear concepts. In the mid 90's Interaction Theory was just that, a cluster of fuzzy philosophical and mathematical notes orbiting around concepts that were far from clear. But those fuzzy ideas had a center of gravity. And that center of gravity, over time, set in motion concepts that begun to take shape, particularly in the past 8 years. In this section I want to convey the key moments in the development of Interaction Theory, and how these correspond to different conceptions of mathematics.

I have no pretense to argue that Interaction Theory belongs to the history of mathematics. It belongs to the history of human thought, like any sufficiently structured system of ideas designed to look at the world with authentic wonder, and to understand reality without an agenda. The "toward" in the title of this book indicates a direction, a possible avenue of research that I think the mathematical and philosophical communities should be considering.

Mathematics As Natural Process

Mathematics is a natural process like digestion, photosynthesis and nuclear reactions that originates from certain hidden cognitive abilities of the human mind.

(Davide Lo Vetere)

years 1995 to 2002

This is the conception of mathematics before the introduction of the Abstract Mind model, and is based on the notion of process. It emphasizes the organizing principles that seem to govern the development of mathematics along distinct trajectories of thought that are clearly visible from the rarefied framework of Higher

19.2 Interaction Theories

Category Theory, in which the tenets of categorial reasoning produce a holistic image of mathematics of unprecedented clarity. This is the phase in which I begin to look at the entire edifice of mathematics through The 4 Lenses, and realize that what I see is the regular repetition of patterns of thought camouflaged in different ways: Algebra, Logic, Topology, Geometry, Category etc. Some of these patterns are known and are captured by Higher Category Theory, some are not. The known patterns are called universal constructions in the parlance of Category Theory, and play a central role in the unification of mathematical structures. Universal constructions are patterns that describe certain mutual relations between categories that appear everywhere in mathematics, but that are difficult to spot without the abstraction and the tools of Category Theory. Universal patterns derive from a peculiar structure called *Kan extension*, which is a generalization to categories of the notion of "closeness to a diagram" we saw in Section 16.3. Other known patterns are the relations between higher-dimensional categories captured by the Periodic Table of n-categories—a table reminiscent of Mendeleev table that relates categories of different dimensions.

k,n	-2	-1	0	1	2	...
0	*trivial*	*True*	*set*	*category*	*2-category*	...
1	-	*trivial*	*monoid*	*monoidal category*	*monoidal 2-category*	...
2	-	-	*abelian monoid*	*braided monoidal category*	*braided monoidal 2-category*	...
3	-	-	-	*symmetric monoidal category*	*sylleptic monoidal 2-category*	...
4	-	-	-	-	*symmetric monoidal 2-category*	...
...	...	...	...	...	...	...

Table 19.1: The periodic table of n-categories.

In Table 19.1 each entry represents a so-called k-monoidal n-

category, which is a mouthful to say that a $(k+n)$-category—a generalized category (Definition 4.3.1) that contains arrows, arrows between arrows, arrows between arrows between arrows,...and so on $(k+n-1)$ times—contains just one of these multi-arrowed objects for each $i < k$, and that is monoidal, which means that a category is equipped with a so-called tensor product, which is a generalization of the ordinary product to multi-dimensional objects.

Significance Of This Phase

The view of mathematics as a natural process, and the intuition that there are distinct cognitive patterns at work in the genesis of mathematical thought that are different from the patterns with which we construct mathematical definitions, reframe the problem of studying the nature of mathematics within the larger context of the workings of the human mind. The significance of this phase resides precisely in this process-based reformulation of mathematics, because it raises questions about the structure of mathematical thought that have profound mathematical implications, and that cannot be articulated from within a syntactic conception of mathematics, as it emerged first in the development of the Synthetic Structure of Intentionality, and subsequently in the development of the notions of $\mathcal{M}$-*signature* and Cognitive Architecture.

Mathematics As Phenomenology Of The Mind

Mathematics is a cognitive technology to interpret reality through the properties of abstract objects.

(Davide Lo Vetere)

years 2002 to 2008

In the phenomenological phase of this inquiry, the intuition that mathematics is a natural process is used to characterize the phenomenology of the proto-mathematical abilities of the human mind. This is the phase in which I introduce the Abstract Mind model and the Synthetic Structure of Intentionality. At the foundation of the Abstract Mind model is the discovery that it is possible to isolate the cognitive processes that determine the distinction between an abstract object of thought or perception, and the

19.2 Interaction Theories

mathematical image we are able to produce of that object. From the perspective of the Abstract Mind model, mathematics is the continuous, self-generating process in which the same object of thought is constructed and deconstructed in an intentional process steered by the notion of meaning. This is the practical and conceptual lesson that Topos Theory has thought us: that Logic, Topology, Algebra and Geometry reveal their deepest and most authentic nature when they are reduced to pure structure, and when they are relativized through various forms of spatialization, such as the definitions of the same concepts in higher dimensional categories. These mathematical and philosophical intuitions are captured by the language of diagrams, to confirm, to those like me who believe in the equation $truth = beauty$, that aesthetics is the ultimate language of truth. This vertiginous reduction to pure structure of what used to be the pillars of the classical edifice of mathematics, was a telltale sign of a deeper structure of mathematical thought waiting to be revealed. It signaled not just a theoretical possibility, but a factual evidence that what we are used to call mathematics is in fact the abstract shell surrounding much deeper dynamics of thought that relate us to reality.

Significance Of This Phase

This is a scene-setting phase in which the conception of mathematics as a self-generating process leads to the quest for deeper structures of thought at the foundation of the mind's ability to mathematize, and to relate us symbolically to the physical world.

Mathematics As Metaphor

Mathematics is the study of the abstract constructions of thought that seem to imitate the fabric of reality.

(Davide Lo Vetere)

years 2008 to 2012

In this interim phase, the cognitive apparatus at the basis of the mind's ability to mathematize is linked to a concrete cosmology of thought with which it is possible to articulate questions about the Abstract Mind. This is when the focus of this investigation begins to shift from axiom systems thinking to a system of thought based on conceptions of intelligibility and knowledge. In this phase, mathematics becomes an incomplete metaphor that relates the genesis of mental constructions to the structure of objects and states of affairs in the world at a fundamental level detached from any syntactical reference. This metaphor is incomplete because it does not explain what elements of the genesis of mathematical definitions connect those definitions to the common nature of mind and reality. It does not explain why mathematical thought seems to be so deeply dependent on various equational properties—(see the OSP Pattern in Chapter 9)—it does not explain, beyond the syntactic horizon set by formal systems, the link between concepts, mental states and the structure of perceptual experience. It does not connect the mental and the physical realm in a way that we can make consistent with the structures of thought with which we create mathematical abstractions of the world. To reveal those connections, we need a further abstraction leap in which the conception of mind is synthesized as a state in the continuous interaction between two archetypical entities, a model of what is intelligible, and a model of the possible transactions with what's intelligible: transactions that we call *knowledge*.

Significance Of This Phase

This phase preluded to a radical shift in the problem of modeling the Abstract Mind, because it signaled the need to dig deeper into the structure of mathematics. These are the years in which the embryonic notions of $\mathcal{M}$-*signature* and Cognitive Architecture are developed.

Intelligibility And Knowledge

Mathematics is the study of the mind-world interactions defined by models of intelligibility and knowledge.

(Davide Lo Vetere)

years 2012 to today

The central theme of Interaction Theory is the interpretation of the mind's ability to mathematize as the phenomenology of the interplay of two elements that define the deep structure of human thought: intelligibility and knowledge. According to Interaction Theory, mathematics as we know it today is one of the many possible modes of thought that originate from the subtle changes in the structures of thought that we identify collectively as intentionality. Interaction Theory extends the modern conception of mathematics in two directions: it is a theory suitable as a foundation of mathematics, and it situates the structure of mathematical thought within the larger context of the structure of intelligible thought.

Significance Of This Phase

To get a balanced appreciation of a theory still in its infancy like Interaction Theory, it is useful, and perhaps necessary, to gauge the depth and the precision of its tenets within the broader context of the trajectory followed by the classical and modern conceptions of mathematics so far. As I underlined already in this chapter, the spectacular success of the structuralist view of modern mathematics points vigorously to a problem at the foundation of modern rational thought: the relation between semantics and syntax. The conceptual framework of Interaction Theory reframes the problem of explaining the emergence of semantics from syntax as the problem of explaining the genesis of the entire structure of rational thought, and it does so by modeling the Abstract Mind.

19.3 Mathematical Intuitions Of Reality

The conception of mathematics we introduced in Section 14.8, first in an intuitive form

> "[a] conception of mathematics is an intuition, intrinsic in the Symbolic Intuition of Reality, about how what we see explains what we comprehend."

and then in a slightly more technical form (Definition 14.8.1) via the notion of epistemic continuity

> "[a] conception of mathematics is an intuition about the existence of a fundamental relation of epistemic continuity, intrinsic in the mechanism through which a **concrete** concept defines intelligibility, by means of which what we see (Actuation Mechanism) explains what we comprehend (Definitional Model). "

set the scene for the $\mathcal{M}$-*signature*-specific conceptions of mathematics we sketched in Part IV. The conceptions of mathematics

19 Interlude: Definitions Of Mathematics

that emerge from this view of the fundamental structure of a mathematical intuition of the World, rest on the mechanism by means of which each type of mathematics resolves the symbol-meaning tension intrinsic in the occurrence of epistemic continuity of a specific $\mathcal{M}$-*signature*. In the review of the three conceptions of mathematics that follows, we highlight the types of insights into the World connected to these three distinct Symbolic Intuitions of Reality.

The Language Of Distinctiveness, Division And Change

Mathematics is an abstract language to convey how principles of permanence, difference, constancy, succession, variance, evolution and change explain multiplicity, variety, range and diversity as the fundamental manifestations of identity through change.

This is a reformulation of the modern conception of mathematics in archetypical terms. It doesn't mention objects, structure and properties (Chapter 9), it does not mention sets or categories, it does not contain any direct reference to any structuralist argument, and does not rely on notions of quantity, enumeration, measure, logical deduction etc., because those are necessary manifestations of the view of the World that this conception is concerned with. This is the Identity-based mathematical intuition of reality. It is an intuition that resolves the symbol-meaning tension by formulating models of that explain how things that exist as processes, even as mental experience, originate from principles of permanence, difference, constancy, succession, variance, evolution and change. These models, which we call Algebra, Geometry, Logic and Topology, exist in the necessary spatializations we can imagine from within this intuition of reality: they exist as space, quantity and relation.

This is the mathematics of process. It describes the World in terms of change and identity, in terms of purpose and outcome, cause and effect, this or that, right or wrong.

19.3 Mathematical Intuitions Of Reality

The Language Of Boundedness, Inclusion And Proximity

Mathematics is an abstract language to convey how principles of addition, subtraction, inclusion, exclusion, finitude, limitlessness and encompassment explain multitude, gathering, conglomeration and aggregation as the fundamental manifestations of finiteness through proximity.

This conception of mathematics gives us a picture of the World that does not rely on the classic types of spatialization—space, quantity and relation—instead, it bases its narrative on a view of the World made of canons that govern what in the ordinary language we may recognize and describe as abstract relations of boundedness that become apparent via principles of inclusion and proximity involving ordinary objects of thought or perception.

This is the mathematics of aesthetics and architecture. It is the mathematics of how ratios and canons and proportions are or become manifestations of the finitude or reality.

The Language Of Unity, Structural Continuity And Form

Mathematics is an abstract language to convey how principles of similarity, analogy, homomorphy, uniformity, wholeness, heterogeneity, affinity, chirality and symmetry explain homogeneity, uniformity, regularity and homology as the fundamental manifestations of cohesion through form.

The World this mathematical intuition of reality is concerned with is the realm of structure as a fundamental constituent of the unity reality. This is a mathematical intuition of reality that produces models—some of which are representable in the language of modern mathematics—that we can interpret as insights into how the intrinsic unity of Nature manifests itself via principles of structural continuity, similarity and form.

This is the mathematics of morphogenesis. It is the study of the forms of Nature as manifestations of its polymorphic unity.

19 Interlude: Definitions Of Mathematics

Bibliographical Notes

The Periodic Table of n-categories is a fundamental object of modern mathematics discovered by Baez and Dolan 1995. The table stabilizes—a k-tuply monoidal n-category is an $(n+2)$-tuply monoidal n-category—for $k \geq n+2$. This is called the stabilization hypothesis, and is proven in Example 1.2.3 in Lurie 2009a.

Chapter 20

Information As Measure Of Relative Knowledge

> How can the semantic interpretation of a formal symbol system be made intrinsic to the system, rather than just parasitic on the meanings in our heads? How can the meanings of the meaningless symbol tokens, manipulated solely on the basis of their (arbitrary) shapes, be grounded in anything but other meaningless symbols?
>
> Harnad 1990

The notion of information is used extensively throughout the text, and now is the time to clarify what information really is. In "How Cognitive Architectures Define Knowledge" (Section 11.3), we pointed out that *"Interaction Models [...] encode the cognitive technology to transform data into information"*, and again in "The Structure Of Comprehension" (Section 12.5). Then again in "Meaning And Semantogenesis" (Section 12.6) we discussed how the phenomenology of meaning is related to the phenomenology of information, how meaning and information always coexist, and distilled a few more observations about information in the context of the origin of meaning, with the definitions of protosemantic transformations and semantogenesis, and in the mantra *information is to knowledge what meaning is to understanding*.

In Part IV we highlighted how each $\mathcal{M}$-*signature* gives rise to a notion of information as part of the Symbolic Intuition of Reality it creates. For example, in the introduction to "Meaning And Semantogenesis" in the Finiteness-based $\mathcal{M}$-*signature* (Section 16.7), we observed that

> *"the informational content of a certain type of Finiteness-based understanding is measured via the same mechanism used to create knowledge, and is therefore a measure of inclusiveness, not exclusiveness like in the ordinary notion of information. "*

and in "Meaning And Semantogenesis" in the Cohesiveness-based $\mathcal{M}$-*signature* (Section 17.7), we made a similar remark

> *"Cohesiveness-based information is therefore a measure of how much a certain type of structural continuity creates knowledge in a given context. "*

These initial reflections on the nature of information were greatly generalized in the analysis we presented in "How Data Becomes Message" (Section 14.6), where we discussed the main transitions between $\mathcal{M}$-*signature*s, and outside of the context of the Abstract Mind cognitive states.

In this chapter we use the tools of Interaction Theory to propose a unifying formulation of information that seems to resolve many of the intricacies that emerge from the study and application of this concept, and that eliminates most, if not all, of its halo of elusiveness. This analysis is, like most of this research, metaphysical, but is by no means a metaphysical fugue on information, rather, it is an endeavor to show the practical and theoretical applications of the principles of Interaction Theory.

20.1 Common Information Bearers

What do we see when we watch the world through the lens of information? and What *is* information? Information is a rather elusive concept, and is used with different meanings and for different purposes in practically every field of human knowledge. It should come as no surprise, thus, to find many competing theories of information, and to discover that the role played by this concept in our culture is, to say the least, polymorphic and confusing and, upon closer examination, utterly mysterious.

20.1 Common Information Bearers

Information has no color, smell, taste, sound, mass or shape. It is not an element of the Universe like hydrogen, helium or carbon. Yet, it seems to be omnipresent in our lives, and its importance and use in virtually any field of human knowledge are indisputable. In colloquial speech, information is mostly used as a mass noun to denote extensional properties of things, facts and systems of ideas, such as the amount of data sent, received, processed or stored by a computer, the last Grand Slam men's single champion, your seat number on the flight from London to New York City, and so on and so forth. It is convenient to have a generic term to denote whatever we think is the meaningful content of the countless mind-world transactions that constitute the fabric of life and reality, and information seems to be that term.

Information is actually many things, and its polymorphic appearance is accentuated and multiplied by the types of abstractions imposed on this concept by its many uses. This initial reflection on the many contexts where information seems to act as a conduit for the conveyance of certain internalized instances of knowledge, suggests that what we generally refer to as information is really the phenomenology of something else. Perhaps more structured. Perhaps more complex. But certainly something whose variegated appearance betrays a deeper, and, perhaps, overarching structure that must be uncovered. If we could visualize the many manifestations of information, as created by the phenomenology of the multitude of mental and physical realities, we would probably see a conceptual labyrinth created by any given concept as it percolates through the fabric of those mental and physical realities like rain through a wall of gravel. If to these observations we add the variable of time, and attempt to perform a historical analysis of information, we find that the nature of this concept becomes even more convoluted and muddled, because it has to take into account the various factors that contributed to, or obstructed, the ontological continuity of information over at least the past two and a half thousand years. I think there are two reasons why information seems such an elusive concept.

Information is elusive because in its various forms and guises, since we have historical evidence of it, has always characterized an effort to measure the extensive properties of human knowledge. This aim has determined the dual nature of information, qualitative and quantitative, in ways that aren't always immediately evident. To appreciate the complex interdependence of the qualitative and quantitative natures of information, we just have to observe the appearance of information in the mind-world transactions. The manifestations of information in the mind-world transactions can be grouped into three categories: *communication*, *action* and *thought*.

Communication is the group of mind-world transactions symbolized by the exchange of data through a media, such as spoken language and digital communications. In this case, information denotes the data successfully transferred from a sender to a receiver, such as bits, words and sentences. In this category the information bearer is the message, intended solely as data.

Action represents the set of mind-world transactions symbolized by inter-systems dynamics. For example, the act of placing an object on a desk changes the informational content of the environment represented by the desk. In general, any physical interaction can be regarded as a transaction between systems that results in a change of the state of those systems. In this category, the information bearers are the states of all the system participating in the transaction.

Thought denotes the category of mind-world transactions where the information bearer is a concept. The character of this type of information is, therefore, purely definitional, in that it stipulates a way to reference objects or states of affairs for a specific purpose. These are the transactions that constitute any cognitive process, and, in particular, the cognitive processes that create meaning out of meaningless data. For example, the meaning we assign to a thermostat reading, the meaning we assign to a mathematical theorem, the meaning we assign to events in our life and to the ideas in our heads are all examples of information created by cognitive processes.

The other reason why information is elusive, is that we seem to not have sufficiently recognized the subtle influence that a general intellectual attitude to come at the problem of using our ideas to study the world has on the development of knowledge. This is, perhaps, an overarching problem of Philosophy that is deeply connected to the systems of ideas we invent to study and solve problems. I often make this joke to illustrate this problem: "You can't complain that you have to count after you invent the numbers". Obviously, this doesn't apply just to numbers, it applies to anything we think. We have invented the concepts of mind, ideas, thought, data, information, reality and many more, as interfaces of pure intellect and awareness, but the freedom with which we extend the use of those concepts to problems increasingly distant from the problems we tried to solve with those concepts when we created them is, oftentimes, rather adventurous and temerarious.

A very first step to understanding information comes, as usual, from careful observation. We observe that, regardless of *where* information is used, it *consistently turns the world into a meaningful place*. When we watch the world through the lens of information, the world never looks like a "bag full of dull stuff", it looks meaningful and supercharged with all kinds of unexpressed potential.

This observation contains the quintessential character of information, and is the fil rouge that will guide our first steps to the discovery of its deep nature. This clue about how information changes what we observe, encodes in fact a metaphysical feature of this concept: information and meaning *always* coexists, as we have already observed in "Phenomenology Of Meaning" in Section 12.6.

It is useful, at this point, to recall the description of meaning we gave in Section 11.1, and to begin to think about information via its relation of coexistence with meaning.

> *"[t]he notion of meaning acts as an internal compass with which the human mind navigates through various Observational Structures of Intentionality, by reacting to the changes in the goals and in the strategies to achieve those goals that are continuously set by the stream of thoughts and observations triggered by mental and sensory perceptions. This activity of reorienting thoughts toward new goals along a trajectory of minimal datasets is the experience of understanding through meaning that we all have when we learn. "*

As we observed already, there are numerous theories of information. Their presence stems essentially from the variety of ways in which information creates meaning. This is why the path to redefining information in Interaction Theory begins with a review of the main theories of information.

We can recognize the three types of mind-world transactions outlined before—communication, action and thought—in the most

20 Information As Measure Of Relative Knowledge

interesting and influential contemporary theories of information. The style of reasoning that these theories of information have in common is captured by the meme "information as...", as in "information as probability space distribution", or "information as state transitions of a dynamical system" and so on. It is important to appreciate why this is necessarily so, and the profound implications that the epistemological positions motivating these theories have on our conceptions of information. To examine some of these theories of information, and to begin to reveal their common structure, we are going to use the ∞-diagram of Figure 20.1, which uses the Knowledge-Reality Interface to depict how information is understood to exist in the World, when we look at the World through an information theory. The cognitive mechanism we want to emphasize with the diagram of Figure 20.1 is the meme "information as..." we mentioned earlier. Each modern theory of information depicts information as "something" carried by certain "entities".

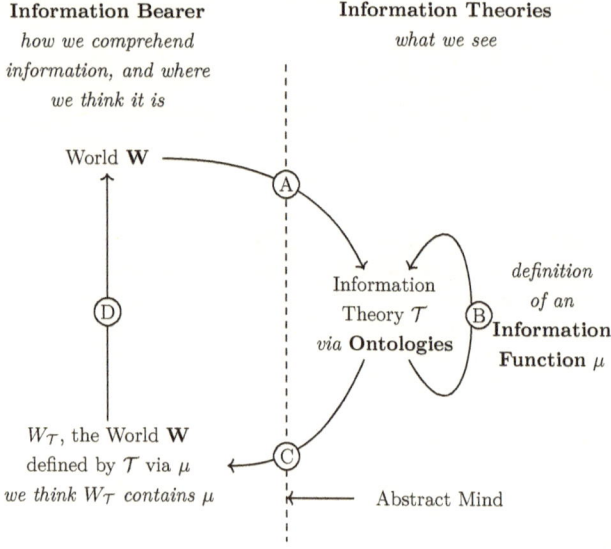

Figure 20.1: How we see information in the World through an information function.

20.1 Common Information Bearers

Intuitively, information is ascribed the quality of being "something contained in something else", and this quality is interpreted as an indication or a clue about its *nature*. With reference to Figure 20.1, let us explain how we get to think of the World **W** as the cradle of information.

A. A certain World **W** is interpreted through an Interaction Model, which is codified in the form of a *theory of information* $\mathcal{T}$, consisting, like any interaction model, of a certain type of spatialization of the World **W**. The fundamental structure of the Interaction Model is dictated by the type of $\mathcal{M}$-*signature* where it is defined, and is what we commonly refer to as *data*: the stuff **W** is made of according to $\mathcal{T}$.

B. Inside $\mathcal{T}$, we define an *information function* μ which models *how* information in **W** is carried by the data the theory $\mathcal{T}$ is made of.

C. $\mathcal{T}$ defines a World $W_\mathcal{T}$ as the *carrier of information* in the form dictated by μ.

D. Through $\mathcal{T}$, we *comprehend* **W** as a World $W_\mathcal{T}$ filled with information carried by the type of stuff $\mathcal{T}$ is made of. The World $W_\mathcal{T}$ becomes, therefore, the *information bearer* of the information μ we *see* through $\mathcal{T}$.

With the **Information Bearer-Ontology-Information Function** schema of Figure 20.1 in mind, we review briefly how the most important and influential theories of information, approach the problem of *defining* information. The very possibility of reading each theory of information according to the schema of Figure 20.1 is, in itself, an example of how this interpretation of information according to the principles of Interaction Theory *unifies* information. It is without doubt instructive to get to the concepts we formalize later in this chapter in small steps, because this helps fully appreciate the patterns we want to define.

The Communication Theory Approach

This is a mathematical theory of communication in which information is regarded as probability space distribution. It first appeared in Shannon's 1948 groundbreaking article "A Mathematical Theory of Communication" (C. Shannon 1948), and was published one year later as a book in Shannon and Weaver's "The Mathematical

Theory of Communication" (C. E. Shannon and Weaver 1949).

Information Bearer	Ontologies	Information Function
Data as strings of symbols.	*Source, receiver and data in motion over a communication channel.*	*Measure the capacity a digital communication system.*

The communication theory approach is concerned with the problem of characterizing the reliable transmission of data across a communication channel from a sender to a receiver. It is therefore a theory about data in motion, not about the information per se. The role of the notion of information in this theory, along with entropy, is purely instrumental to the measurement of the capacity of a system to transmit strings of symbols across a communication channel. However, for historical reasons, and because of the importance that this theory of communication has had, and still has, in the development of digital communications, it has every right to be regarded also as a theory of information, even though the theory itself is not centered on a notion of information.

The Complexity Theory Approach

This is based on a conception of information carried by x as the size in bits of the smallest computer program to compute x. The algorithmic approach was introduced by Li and Vitanyi (Ming and Vitányi 1997), and relies on the notion of computational complexity due to Kolmogorov (Kolmogorov 1965) and Chaitin (G. J. Chaitin 1969).

Information Bearer	Ontologies	Information Function
Transformations between data: from computer program to $\mathcal{X}$.	*Universal Turing Machine and algorithms.*	*Measure of computational complexity.*

The complexity theory approach uses the notion of computational complexity—or algorithmic complexity—to characterize the extensive properties of data, regarded as information bearer. In

20.1 Common Information Bearers

this case, the ontologies involved in this description of information are the procedures to perform certain computations conducive to a result on a reference computer—called a Universal Turing Machine, which is an abstract model of modern computers based on the notion of bit. Clearly, by tying the informational content to the computability of a data bearer, the theory makes a substantial assumption about the nature of information. For in this approach information measures a certain intrinsic quality encoded in the way the data is structured, quality that determines the amount of computation needed to construct that data. This is a remarkable idea from an epistemological viewpoint, because it uses, implicitly, an undefined notion of structure, in the form of algorithms, as a measure of complexity, which is a concept that transcends and, in certain ways, redefines data.

The Probabilistic Approach

This is concerned with the definition of semantic information in an ambient probability space as the inverse of the probability of p—the less likely an event, the more information it carries—and is found in Bar-Hillel and Carnap Carnap, Bar-Hillel, et al. 1952, Dretske Dretske 1980.

Information Bearer	Ontologies	Information Function
Events, facts.	*Events, probability and time.*	*Measure of the semantic content of an event via its likelihood to occur.*

This approach regards events as information carriers, and interprets their likelihood as a measure of their informational content. Events are carriers or triggers or enablers of knowledge. The role of the notion of information is to characterize the semantic significance of an event via its probability to occur.

The Modal Approach

The modal description of information uses a modal space as the ambient workspace in which the information carried or conveyed by X is, or corresponds to, the possible worlds excluded by X. A modal space is a universe of possible objects in which possible worlds are points of the modal space. The overarching philosophical framework in which these ideas are developed is called possibilist realism, and some of its most prominent exponents are Lewis 1986, Kripke 1981.

Information Bearer	Ontologies	Information Function
Objects, reality, worlds.	*Modal operators.*	*Measure modal logic propositions.*

This approach delegates to the features of reality the role of information bearers. The function of information here is to track what characterizes a world X in relation to a space of possibilities which do not contradict X. This makes the modal approach a heavily logic-laden theory, which depends on the machinery of modal operators—a device used in modal logic to forms propositions from propositions. To convey the gist of this approach, it is useful to briefly describe the modal operators used to construct worlds. These are: the *alethic* operator, which indicates possibility, impossibility and necessity of the ontologies of a world, the *deontic* operator, which indicates what is prohibited, obligatory, or permitted, the *axiological* operator, which acts as a value modifier—transforms the ontologies into positive or negative values as perceived by a community—the *epistemic* operator, which encodes the beliefs, knowledge and ignorance, and the *doxastic* operator, which encodes the logic of beliefs.

The Systemic Approach

This approach uses system thinking and situation logic—the logic of real world situations in which the context is incomplete, in the sense that not all propositions about a context hold—to tackle the problem of the nature of information as state transitions of a dynamical system. System theory is a vast, interdisciplinary area of research aimed at developing the conceptual tools to study complex natural systems, such as biological systems, self-generating systems—also known as autopoietic—macro and micro systems, and many other systems which display distinct patterns through which they react or adapt to changes in the environment. The study of information with the methods of situation logic is found in Barwise and Perry 1999, Devlin 1995, and is primarily inspired

20.1 Common Information Bearers

by the view of information as *correlation of situations*, which is a representation of the problem of the nature of information that can be studied with logic.

Information Bearer	Ontologies	Information Function
A system's state transitions and the correlations of situations.	*Dynamical system and real world situations.*	*Encode the possible transitions between a system's states.*

In this approach the information bearers are the relations between contexts, represented as states of a system. This view relativizes information by ascribing to its nature a character that depends on the interrelations between the parts of a system. This perspective on the nature of information has the remarkable feature of delegating the ontological status of information to a broader context in which the information bearer acquires its significance. This view has, at least on a structural level, many connections with the algorithmic complexity theory approach, in that it focuses on the process through which a certain data—in this case a system's state—is produced rather than on the data in isolation. Information here is functional to the codification of the significance that a state transition acquires in the context of the system which produces it.

The Inferential Approach

This approach defines information through inferences relative to an agent's epistemic state, which ties the nature of information to forms of correspondence between truth and local realities.

Information Bearer	Ontologies	Information Function
Inference about an agent's epistemic state.	*Inference space.*	*Measure information as valid inference.*

We rediscover here the themes of the Modal Approach, but revisited via an abstract notion of communication.

The Semantic Approach

In this approach the information bearer is the semantic content of data.

Information Bearer	Ontologies	Information Function
Semantic content of data.	*Data space.*	*Measure information as data correctness, meaningfulness and truthfulness.*

Data is a carrier of truthful data, and the information it contains is identified with its semantic content.

20.2 Information In Interaction Theory

If we consider collectively the belief systems underlying each information theory described by the diagram of Figure 20.1, we obtain a portrait where *data, information, intelligence, mind, system, cognition*, and *truth, meaning* etc., form spaces of thought where we reason about information. In those spaces, information is *something* with such and such characteristics (truthfulness, complexity, etc.), it is *carried* by such and such entities (data, agents, events, systems, etc.), it *obeys* certain laws captured by various versions of logic (modal logic, Lambek calculus, etc.), it *flows* between systems, agents and situations, and so on and so forth. And like in a sudoku puzzle, within the epistemology defined by the Interaction Model underlying each information theory, we can construct the fundamental questions that these notions of information seem to raise, as a permutation of the above terms and their mutual relations, and end up with different representations of the same idea.

It is within the arena defined by the background presuppositions that form the spatialization each information theory is based on, that we ask—because it makes sense *in that context* to do so—questions that reflect, or that are an expression of, those presuppositions, such as: What is the logic of information? How can we describe the dynamics of information? How does data acquire meaning? How does meaningful data acquire truth value? Does, or can, information explain truth? Does, or can, information

20.2 Information In Interaction Theory

explain meaning? Can cognition or intelligence be explained as information processing? Is it possible to transcend information? What is the relation between epistemology and a theory of information? What is the relation between science and information modeling?

We know from the Epistemology of the Abstract Mind that those questions and those notions of information are based on the local definitions of knowledge embedded in the Interaction Model representing each information theory. Hence, those theories of information are nothing but various manifestations, various instances, of a map $\mu : K(\mathcal{I}) \to \mathbf{P}$ that sends their local definition of knowledge $K(\mathcal{I})$ in an interaction model $\mathcal{I}$ (strings of symbols, data transformations, events, realities, a system state transition, etc.) to a space of values $\mathbf{P}$ (capacity of digital communication system, computational complexity, semantic content, modal logic propositions, possible state transitions, etc.). These are all manifestations of an overarching conception of information as a measure of the extensional properties of the knowledge defined within a given Interaction Model. In conclusion

Definition 20.2.1. (Information)
Given an Interaction Model $\mathcal{I}$ and a space $\mathbf{P}$, an *information* in $\mathcal{I}$ is $\mathbf{P}$-valued function μ on the space of definitions of knowledge (Definition 12.3.1) of $K(\mathcal{I})$.

Bibliographical Notes

The review of information theories is based mainly on the classics. C. Shannon 1948, C. E. Shannon and Weaver 1949 (the communication theory approach), Kolmogorov 1965, Ming and Vitányi 1997, G. J. Chaitin 1969, G. Chaitin 2003 (the complexity theory approach), Carnap, Bar-Hillel, et al. 1952, Dretske 1980 (the probabilistic approach), Lewis 1986, Kripke 1981 (the modal approach), Barwise and Perry 1999, Israel and Perry 1990, Devlin 1995 (the systemic approach), Hintikka 1962 (the inferential approach), and Bar-Hillel 1966, Bateson 1972, Deutsch 1985, Floridi 2013 (the semantic approach). The formulation of information in Interaction Theory is implied by the work presented so far.

Chapter 21

The Rediscovery Of Intellect

> Men whose research is based on shared paradigms
> are committed to the same rules and standards for
> scientific practice. That commitment and the
> apparent consensus it produces are prerequisites for
> normal science, i.e., for the genesis and continuation
> of a particular research tradition.
>
> T. S. Kuhn 1962

The reformulation of the foundations of human thought offered by Interaction Theory, provides an opportunity for certain additional reflections on the nature, meaning and purpose of the human intellect, and to outline a critique of other themes at the core of the modern philosophical, mathematical and scientific enterprise.

One of the biggest misunderstandings of the modern world, particularly apparent in the Western culture, is the reduction of human intellect to reason. This misconception, is rooted in the delusion fueled by technological development, that rational knowledge has some kind of privileged epistemic access to the world, and results in an overreliance on the analytical methods and intuitions typical of the alert, problem-solving state of consciousness. Superficially, the prejudice behind the conflation of intellect and rational thought, seems to be validated by the numerous industrial uses of the forms of knowledge developed through the systematic application of the methods of analytical thought. But on closer examination, it becomes apparent that this conflation is the sign of a deeper and broader belief, the belief in what progress is or should be, and, consequently, the disposition toward the research and development of the tools and methods and ambitions and goals that we should cultivate to attain that type of progress, and toward the aspects of reality and human experience that, not corresponding to this worldview, should be ignored.

As we saw in Chapter 15, what is generally referred to as "rational thought", corresponds in Interaction Theory to the Symbolic Intuition of Reality defined by the Identity-based $\mathcal{M}$-*signature*. When we say "rational", we mean *Identity-based rational*, and in so doing, we evoke a whole way of thinking about the world as a process, a *principium individuationis*, a plethora of intellectual and intuitional sensitivities and considerations and attentions toward anything that we can discern and comprehend solely and exclusively in terms of the mental categories of purpose and causal efficacy. Is this all there is to rationality? Are a version of rationality based on pure analyticity, the computational picture of the world it produces and our place in it, the manifestations of true intellectuality? Do they resonate with human nature? Can we really dissociate ourselves from the intuitions of reality and of our nature on the basis that they are not computable? What is the reason for looking at things in this way? And how far is it justified?

Interaction Theory shakes up this cosmology of thought, and shows that there is no real or inexorable necessity to identify human intellect and rationality with a Symbolic Intuition of Reality centered around the various manifestations of the notion of identity. It shows that the gospel of analytical thought is great to fix your toaster and send a human mission to Mars, but that those technological accomplishments do not sanction the superiority of analyticity over other modes of thought and other Symbolic Intuitions of Reality.

It does so with the adoption of a Participatory Worldview (Section 3.2) that repositions the origin of the epistemic enterprise around the structure of the mind-world transactions, and shows the practical and conceptual possibility of parallel structures of thought, not based on what seems to be the inescapable influence of identity, but on a fabric of archetypical intuitions of reality, teeming with insights on the languages of Nature.

Introduction

It does so with the argument about the interferential nature of concepts (Section 5.5), that reveals the ephemeral and relational nature of concepts, and disintegrates the centrality of identity as a foundation of human thought and rationality. This realization, together with the conceptual landscapes prefigured by the methods with which Interaction Theory reframes the study of the foundations of human thought around a model of the mind-world interplay, indicate the opportunity—but I should really say, the necessity—to revisit a fundamental notion of our culture and of the history of human thought: information.

It does so with the introduction of the Principle of Epistemic Relativity (Definition 13.1.1), with the formalization of a model of the mind-world interplay, with the Abstract Mind and its epistemology, with models of Symbolic Intuitions of Reality through which we can redefine reason, rationality and objectivity without divorcing knowledge from the knower, and with the definition of a framework that allows us to relativize and unify the foundations of human thought, and to conceive new mathematics that transcend the dominance of formal systems. As already observed about the Principle of Epistemic Relativity (Section 13.1)

> *"[t]he principle of epistemic relativity is really the principle of epistemic unity, in that the purpose of the relativization it introduces, is instrumental to the definition of a unifying framework for the definition and the development of a holistic conception of human thought. Thus, epistemic relativity must be understood not as a weak paradigm of knowledge, but, on the contrary, as a principle of interdependence that deepens and amplifies the meaning of knowledge along trajectories that are tightly intertwined with the deep structure of human thought, and with the symbolic significance of human work and ambition. "*

These ideas represent profound changes to the modern paradigm of knowledge, and to a general attitude toward what a scientific inquiry should be, and to the extent to which it should be defined by its purpose—another theme very dear to the Western scientific tradition. These ideas, when embraced fully, pave the way for major revisions in the interpretation of the nature of human thought, on how we should go about the project of explaining Nature, and on the meaning of reality, rationality and objectivity, and prime the mind to dissipate the illusory perception that Identity-based thought has privileged epistemic access to the world.

With this initial reflections in mind, we can attempt to give an operational definition of intellect.

Definition 21.0.1. (Intellect – *operational*)
By an *intellect*, we mean a property of the mind-world transactions not confined to a particular metaphysical structure of thought—intended as $\mathcal{M}$-*signature*—and defined entirely by its capacity to create meaning, and by having no intrinsic requirement of competence, performance and objectivity.

This practical definition of intellect, concentrates on reminding us of an important distinction to be made between the phenomenology of a specific Symbolic Intuition of Reality, and the archetypical cognitive structures it represents—and that are codified, as we know from the study of $\mathcal{M}$-*signatures*, by the underlying types of spatialization from which the Symbolic Intuition of Reality originates. This is why the definition insists on the centrality of the capacity to create meaning as the primary feature that defines intellect, because that is the highest expression of the epistemology of the Abstract Mind: it is ultimately the phenomenology of its core type of spatialization. We must now observe that this definition of intellect has also a second layer of meaning. By pointing out that intellect has no implicit requirement of competence, performance and objectivity, this definition calls for a broader way to conceive intellect that takes into account its role in the evolution of a human being, rather than its role in the evolution of an animal that computes. For the identification of intellect with competence, performance and objectivity is, yet again, a not-so-subtle manifestation of the various flavors and echoes of computationalism and functionalism, where mind and intellect are conceived and measured in terms of their ability to implement functions and perform tasks describable by a computational model. These observations lead to a second definition of intellect that I call archetypical, that should help restore intellect and intellectuality to its original purpose and meaning.

Definition 21.0.2. (Intellect – *archetypical*)
By *intellect* we mean the ability of the human mind to express fully the purpose of human existence through the creation of meaning

21.1 Mathematical Physicalism

in our life, to deepen our sense of humanity, to manifest our true nature, and to use meaning to connect us to each other and to the World.

21.1 Mathematical Physicalism

Mathematical concepts are epiphenomena of those mind-world transactions where a **concrete** *concept is reconstructed, and have, therefore, the ontological status of physical phenomena.*

Physicalism is the metaphysical view that everything is physical. It is a view fundamentally defined by the belief that the only mode of existence is physical, and that therefore reality is physical, and that that is all there is. Physicalism is not materialism, because the meaning of the term physical evolves with our understanding of the fabric of reality.

Mathematical Realism is the metaphysical view that mathematical concepts exist independently of us and our thoughts; they are, in other words, ontologically objective. The view of the world of mathematical realism is that, just as dogs and mountains and atoms and planets and galaxies exist independently of us, so do sets and numbers and topological spaces and tensors and spinors and every other member of the mathematical zoo. Mathematical realism places every mathematical concept on the same stage of human experience, thus implying that mathematics and mathematical truths are discovered, not invented. A stronger form of mathematical realism is Mathematical Platonism. Mathematical platonism regards mathematical concepts as entities completely removed from the physical dimension. It considers mathematical concepts as thoughtforms that do not participate in any causal or spatiotemporal transaction. According to the platonic view of mathematics, mathematical concepts are, therefore, not only not physical, they are also eternal. If mathematical platonism is true, the physicalist view of reality has to be revisited to account for what physical means. If mathematical platonism is true, our conception of a theory of physics must account for entities such as mathematical concepts, that do not obey the known laws of physics, but to which, nonetheless, we seem to have epistemic access to.

21 The Rediscovery Of Intellect

The view of Interaction Theory on the ontological status of mathematical concepts, is based on the hypothesis that the mind-world transactions are real (Postulate 8.7.5), and on the characterization of the Abstract Mind via the Synthetic Structure of Intentionality model, where concepts, called Local Identities (Section 8.7), represent the local phenomenology of intentionality. Consequently, mathematical concepts, being the phenomenology of mind-world transactions orchestrated by intentionality, have an ontological status corresponding to their physical origin.

The phenomenological approach of Interaction Theory to the problem of defining the ontology of mathematical concepts makes it easy and intuitive to explain why mathematical concepts are "the same" for everybody, why they seem eternal, and why, for these very reasons, some scholars, including the author, are inclined to believe that they are discovered rather than invented. The answer to these questions is that physical phenomena are the same for everyone because they are manifestations of general laws, and, by the same token, so are mathematical concepts, being manifestations of the physics of the mind-world transactions. From the standpoint of Interaction Theory, the physics of the mind-world transactions is the same for every human being, and so is its phenomenology. Expanding on the analogy between classical physics and what we call the physics of the mind-world interplay, we are familiar with the observation that an apple falls from a tree today, in the exact same way as it fell from a tree one thousand years ago, and is expected to continue to fall from trees that way for a long time. We are completely at ease with the phenomenology of apples falling from trees, because we are used to think of the world as the arena where most physical phenomena occur in a predictable manner by virtue of being governed by immutable physical laws. Consequently, we expect an apple falling from a tree to be the same phenomenon regardless of when it happens and who is watching it, and we have no difficulty in acknowledging that this phenomenon is in harmony with a symphony of other physical phenomena occurring in the same physical arena at the same time. This makes apples falling from trees ontologically objective, and dispenses us from having to ask whether an apple falling from a tree was discovered or invented. The characterization of the context where apples fall from trees as a physical system governed by laws of Nature has, in other words, two effects: it allows us to substitute the physical content of our observations with the symbol

21.1 Mathematical Physicalism

system of Newtonian Physics that synthesizes the phenomenology of apples falling from trees, to a degree of accuracy that is often more than sufficient for the requirements of everyday experience, and, as a result, it replaces our imagination with the allegories of the mathematical language. The interpretation of physical reality into a symbol system, ipso facto, replaces invention with syntax and discovery with provability and semantics. For this reason, the creative process in mathematics, what we might call the *real invention*, occurs only when a syntax and a semantics are constructed around an existing idea. This is why *inventions create realities* and make us see previously invisible ones.

Interaction Theory describes the physics of the mind-world transactions as a physical system activated by intentionality. Metaphysics meets Physics meets Mathematics. Going back to the analogy of the apple falling from a tree, we may think of intentionality as the branch of that tree: the physics of gravity is always there, we just *see it* when the apple falls from the tree. Similarly, the physics of the mind-world interplay, described abstractly by the Synthetic Structure of Intentionality, is part of the arena where we connect with reality symbolically through our mind: we *activate* and *experience* its phenomenology once we interact with its laws.

I call the form of mathematical realism of Interaction Theory *mathematical physicalism*, because it grounds mathematical concepts in the phenomenology of *a metaphysical system that describes the physics of the mind-world transactions*. For this very reason, it is debatable whether this metaphysical view about the nature of mathematical concepts should be called *metaphysical realism* instead of mathematical physicalism. Strictly speaking, mathematical physicalism is not a form of mathematical platonism, because it ascribes to mathematical concepts the ontological status of physical phenomena, and, perhaps most importantly, because the causal and spatiotemporal frames of reference of physical reasoning are themselves epiphenomena of the default, Identity-based mind-world transactions.

The deceptively naïve question about the discovery or invention of apples falling from trees has something else to tell us about the ontology of mathematical concepts. The presence of a symbol system to justify why apples falling from trees can only be discovered is of course unnecessary. The presence of a symbol system only creates a new way to get epistemic access to our interpretation of a physical phenomenon, that we can make consistent with a variety of other phenomena we explain in the same way—recall the discussions about explanations and explanatory systems in Sections 13.2, 13.3 and 13.4. The difference between an apple falling from a tree, and the "same" phenomenon described by a symbol system, is that, in the default $\mathcal{M}$-*signature*, our knowledge of apples falling from trees is predictive, and, as a result of this analytical understanding of the phenomenology of gravity, we ascribe to the original ontological status of apples falling from trees, a features that symbolizes how we incorporate that phenomenon into our knowledge system. Clearly, this is not the same as the pure ontological status of apples falling from trees. This new ontology has a superstructure to accommodate for the semantics of our experience of the phenomenon that it represents.

What emerges from this analysis is that the mathematization of a physical phenomenon is essentially a description of the type of knowledge we use, like a technology, to experience that phenomenon, and that this process is built on the same fundamental intuition of reality at the basis of the mind-world transaction through which we become aware of that phenomenon in the first place. In other words, abstract knowledge is built on an intuitive knowledge of reality from which it derives its intrinsic objectivity. It follows that to negate the ontological objectivity of mathematical concepts, we would have to negate the ontological objectivity of the archetypical symbolic intuition of reality through which we become aware of reality itself, which is a philosophical and intellectual black hole that leads to all sorts of contradictions and conceptual extravagances.

21.2 Principles Of Metaphysical Realism

If we embrace the Principle of Epistemic Relativity fully, we are confronted with a challenging question that speaks to our impulse toward the unknown with disarming frankness: What should science be? If, as implied by the Principle of Epistemic Relativity, the structure of human knowledge is broader than the horizon of Identity-based intellectuality, then it becomes imperative to ask ourselves how we should go about any enterprise that claims to want to *find the truth*, without prejudice, before we even agree or disagree on a set of principles to dignify or promote this or that

21.2 Principles Of Metaphysical Realism

enterprise with the label of "scientific".

If we embrace the Principle of Epistemic Relativity fully, we must also ask ourselves what is the purpose of human knowledge, because if there isn't a privileged conception of intelligibility from which to observe the World, how can we orient ourselves toward one type of knowledge over another? We are not concerned with the technical aspects of knowledge. We are not concerned with problem-solving knowledge, or with the abstract knowledge with which we construct elegant and complicated systems of thought. We are talking about the fundamental understanding of how certain basic intuitions about reality form a worldview and a style of reasoning. That knowledge, the knowledge called *generative* in Interaction Theory, is the archetypical knowledge of how our intuition of reality springs from the mysterious tension between symbol and meaning. That knowledge defines a scientific enterprise at its core.

Interaction Theory tells us that Nature is a metaphysical foam that can only be known through its structure, not its content. We don't look at what's inside a soap bubble, because everything that makes a soap bubble what it is, all the color effects that continuously change as it floats in the air, happens on its surface. Similarly, the shape of our interaction with the World is defined by what is intelligible, and not by what we know in this or that way.

We want to offer here an interpretation of what we think science should be, based on the tenets of Interaction Theory. We call this reinterpretation of science Metaphysical Realism.

The Purpose Of Knowledge Is The Full Realization Of Human Nature

The scope and the methods of inquiry of Metaphysical Realism are in harmony with the full realization of human nature. Unlike squirrels, giraffes, dolphins and hippos, our species is defined by its unique ability to decide its own intellectual and spiritual evolution. We are, after Sartre, condemned to be free. Therefore, a scientific enterprise guided by the principles of Metaphysical Realism is firmly based on a solid understanding of human nature, which is, necessarily, an understanding of human existence in its entirety. A science inspired by Metaphysical Realism never denies the human nature, and does not entertain projects or ambitions

21 The Rediscovery Of Intellect

that dehumanize us, either physically or mentally, in the name of progress or through the development of technologies. A science inspired by Metaphysical Realism recognizes the great potential of our fragility, our aspiration, our bravery, our pride, and of all the forces that make us who we think we are, and that often stand in the way of who we could become and what we could know. Such a scientific enterprise is sufficiently organized to protect its principles not in a dogmatic, intransigent way, but in harmony with the principles of life, and without any reduction or identification of those principles with its methods of inquiry. Such a scientific enterprise fully recognizes that the intrinsic order of things has no requirement to be in harmony with human ambition, and fully comprehends that this realization is a crucial signpost for the attitude toward the pursuit of knowledge. Such a scientific enterprise reflects this implicit and profound realization about the place and purpose of knowledge, by not putting technology before knowledge, by not confusing truth with utility, and by carefully ring-fencing any proclivity toward the industrialization or mechanization of the intangible, transcendental fabric of human existence.

A scientific enterprise based on Metaphysical Realism recognizes the human impulse toward the unknown in all its manifestations, and its symbolic function in defining the human condition. It understands the basic conundrum about the human instinct to comprehend the World around us, and about the methods to pursue that ambition. It understands the analytical, go-get-it approach of modern science, where knowledge is squeezed out of every single particle in the universe by finely chopping every theoretical ramification of a mathematical model. It understands the revelatory, believe-and-thou-shalt-know approach, where knowledge is downloaded into the human mind by a divine entity. And understands the mystic, shut-up-and-listen approach, which rests on the radical intuition that the purpose of what we superficially perceive as the pursuit of knowledge, is in reality a very sophisticated process of remembrance, where Nature speaks to us in its many languages, if we choose to develop our true nature by becoming receptive to the intrinsic truth that is already there in everything that exists. These are all types of knowledge that, to the extent that they genuinely fulfill the purpose of helping the realization of human nature, can only be judged and measured within their respective domains.

21.2 Principles Of Metaphysical Realism

In this sense, Metaphysical Realism is based on principles of universality and unity that come before the individual methods of scientific and philosophical inquiry, and that act as organizing principles that give clarity and purpose to each human enterprise.

The Basic Commitments Of Metaphysical Realism

From a *metaphysical* standpoint, Metaphysical Realism is a type of physicalism, and does not make an apriori distinction between mind and world. The fundamental existential claim of Metaphysical Realism, captured by the "metaphysical" part of its name, is the irreducible unity of Nature, and the view of the mind-world interplay as a fundamental constituent of everything that exists in the ways defined by the structure of $\mathcal{M}$-*signature*-dependent objectivity (Section 10.6). In this claim, therefore, the use of the terms mind and world serves a purely explanatory purpose, and is not to be interpreted literally.

The *semantic* dimension of Metaphysical Realism is defined by its commitment to a literal interpretation of the claims of Interaction Theory based on the Principle of Epistemic Relativity, and on the Epistemology of the Abstract Mind. Thus, "literal" must be understood with respect to the cosmology of thought defined by an $\mathcal{M}$-*signature*. This approach, therefore, has at its core, among the other features of a mode of thought defined by a **concrete** concept (Section 10.3), a specific interpretation of objectivity and rationality—see Sections 15.9, 16.9, 17.9.

The *theoretical claims* of Metaphysical Realism constitute knowledge of the World, but the term "knowledge" borrows its meaning and structure from Interaction Theory, and is therefore to be interpreted in three ways: as a definition of knowledge (see Definition 11.1.1) and, therefore, as an *endoexplanation* ("Endoexplanations As Theories", Section 13.7), as a manifestation of a conception of knowledge (Definition 11.1.2), and as part of an *exoexplanation* ("Exoexplanations As Metatheories", Section 13.7), and of a higher order explanatory system ("Higher Knowledge, Higher Explanations", Section 18.2). The epistemological dimension of Metaphysical Realism, in other words, inherits the highly particularized and interconnected views of knowledge and reality offered by Interaction Theory.

21 The Rediscovery Of Intellect

Experiments, Objectivity And Levels Of Reality

The method of inquiry of Metaphysical Realism is defined by its object, and acquires a specific connotation as a framework of systematic inquiry of Nature through the *mode of existence of its object*. The scope of Metaphysical Realism is Nature in its entirety, and the presence of multiple ways to conceptualize its methods of inquiry based on a conception of intelligibility and on the corresponding type of objectivity, disconnect the goal from the (scientific) journey. What emerges from this reflection is that within Metaphysical Realism there are in fact many sciences as such, which share the same object of study, Nature, and the same purpose, but that operate from different interlocking perspectives. To comprehend the global structure of Metaphysical Realism, and how it shapes the different versions of scientific projects that can be defined in it, it is useful to examine the role of experiments.

In Metaphysical Realism, the scientific method relies on experiments to validate theoretical claims, and the *mechanism* through which experiments carried out in a given $\mathcal{M}$-*signature* corroborate theory is defined by the following two principles:

- It is based on the metaphysical foundations of *objectivity* for that $\mathcal{M}$-*signature*.

- It has the same requirements as the specific structure of knowledge defined in that $\mathcal{M}$-*signature*

What this means is that the *purpose, scope and fundamental requirements* of experiments are to validate, within the horizon of a certain $\mathcal{M}$-*signature*, that the constituents of a theoretical claim *exist* in full accord with the mode of existence through which we see and comprehend the World through *that* theory—see also "How Intelligibility Defines Objectivity" (Section 10.6).

Thus, in Metaphysical Realism, there is a *science of change and division*. Its purpose is the study of the causal fabric of Nature, and its scope is defined by what can be understood in the language of process. Its experiments aim to ratify the theories about a fundamental intuition of reality where the things are the way they are by virtue of being processes governed by certain universal laws. Scientific progress coincides with the advancement of our knowledge of the causal fabric of reality and of our own existence, in relation to how much this knowledge improves or facilitates or accelerates the temporal aspects of our evolution, and all the other aspects of human existence that are deeply ingrained in the fugacity of our stay on Earth.

There is a *science of proximity and inclusion* that studies the relational fabric of Nature. Its scope is what can be understood in terms of relations of inclusion, and the goal of its experiments is to validate the theories about a fundamental intuition of reality where the things are the way they are by virtue of being in certain mutual relations with each other, without any requirement of causal efficacy. There are no laws in the ordinary sense. Instead, there are canons, the discovery and apprehension of which constitutes scientific progress. Scientific progress corresponds to the growth of our awareness of the canons that make up reality and our own existence, and is measured by how much it enlightens and inspires and motivates the aspects of our existence that pertain to the atemporal, intangible, transpersonal and metaphysical dimensions of life.

And there is a *science of form and similarity* that is concerned with the structural fabric of Nature. Its realm is the part of reality that can be known not through its content but only through the analogies and similarities by means of which it manifests itself. The aim if its experiments is the corroboration of the theories about a fundamental intuition of reality where the things are the way they are by virtue of having a certain structure or shape or pattern, without any other requirement other than that of being in harmony with other affine structures. Scientific progress is signaled by the expansion of our awareness of the affinities and canons and patterns that make up reality and our own existence, and is measured by how much it guides the aspects of our life that define our relation with Nature.

21.3 The Unity Of Knowledge

The purpose of knowledge should be measured by how much it deepens our sense of humanity and our sense of connectedness with Nature, by how much it dissipates dogmas, false identities and delusions, by how much it cultivates us as human beings, and by how much it inspires the proliferation of organizations and institutions that foster the complete realization of every individual.

We are beginning to rediscover the richness and the magnitude of human consciousness, and the demand and responsibility that come with our capacity to use our creativity to better our lives, and to design our own spiritual, intellectual, and biological evolution. Acknowledging our true nature is our ultimate and ineliminable mission: embracing this mission with authentic wonder leads to realizing who we really are, escaping it leads to suffering, delusion and self-destruction. We need to surrender to the instinct to reconcile the cosmology of the human mind with the metaphysical structure of Nature.

What propels us in the quest to become who we really are— a quest that is erroneously misconstrued as the origin of the human condition—is the impulse of life toward the unknown. It is the creative tension between thought and action, between the penetrating and disrupting clarity of unuttered intuitions about the symbolic content of Nature, and the reassuring structure of language and appearance, between the analytical intellect that separates us from Nature, and the synthetic intellect that seeks unity with It. This is a collective endeavor that can succeed only if we stay humble and curious and open minded and true to our nature. It cannot be the project to desensitize our sense of humanity by replacing life with computerized processes. It cannot be the project to mechanize life, evolution, intelligence, consciousness and spirituality, let alone to transform them into products.

In the race to turn humanity into a new species of rational gods, it seems that the Western scientific project has somehow forgotten the original purpose of knowledge. Over the past four centuries, the efforts to conquer rational knowledge of the World, have turned from a genuine, open minded epistemic project, into a globalized production system that more or less involuntarily projects onto the World the hallucinations and the instant gratifications of logic and rationality. This deviation from the original spirit of the scientific project has gradually reduced knowledge into a vast network of quantitative problems that mainstream science hopes to solve with the sufficient amount of brute-force computation, and with the right computational models. The industrial and technological successes of this approach are indisputable, but we need to acknowledge that this approach has also created self-fulfilling prophecies all around us, and that the more our computers crunch data, the more the World, and our experience of life, appear to us as existential simulations devoid of any justification for

21.3 The Unity Of Knowledge

seeking meaning and purpose, where we erroneously identify truth with utility, content with performance, and where the numinous beauty of Nature has disappeared.

In this technological head-trip, we seem to have forgotten about consciousness, and we seem to have forgotten what knowledge really is. Knowledge is not the accumulation of notions, and doesn't grow with the amount of ideas we compute. This is what every great scientific revolution reminds us time and time again, overturning each time the basic presuppositions of the current scientific project. Each scientific revolution is the whirl of the mountain torrent that reminds us that knowledge is not the torrent we seem to be able to navigate, but the mountain that creates it.

Bibliographical Notes

The traditional views on mathematical platonism and its implications, are, among others, in Benacerraf and Putnam 1964, Burgess and Rosen 1999, Feferman 2009, Frege 1980. My views are, unpublished. The principles of Metaphysical Realism are in some notes in Lo Vetere 2017.

How Big Is An Elephant?

The four men were still arguing among themselves, and the elephant, that had stood watching them, quietly amused by their naïveté and enthusiasm, showed its joy by trumpeting and stomping and swishing its tail.

Suddenly, the four men realized their mistake, and started laughing with the elephant.

... after all, isn't knowledge the stubborn cultivation of a sophisticated naïveté?

Bibliography

Adámek, Jiří, Jiří Rosický, and Enrico Maria Vitale (2010). *Algebraic theories: a categorical introduction to general algebra*. Vol. 184. Cambridge University Press.

Arbib, Michael A. (1970). "Brains, Machines, and Mathematics." In: *Journal of Symbolic Logic* 35.3, pp. 482–483.

Arbib, Michael A. and Mary B. Hesse (1986). *The Construction of Reality*. Cambridge University Press.

Arnold, Vladimir I. (Mar. 1997). "On teaching Mathematics." In: *Palais de Découverte*. Paris.

Asimov, Isaac (1983). *The roving mind*. Buffalo, N.Y.: Prometheus Books.

Atiyah, Michael (1978). "The unity of mathematics." In: *Bull. London Math. Soc* 10.1, pp. 69–76.

Audi, Robert (1993a). "Belief, Justification and Knowledge: An Introduction to Epistemology." In: *Philosophy and Phenomenological Research* 53.2, pp. 480–484.

— (1993b). *The Structure of Justification*. Cambridge University Press.

Austin, J. L. (1962). *Sense and Sensibilia*. Oxford University Press.

Awodey, Steve (2008a). "A brief introduction to algebraic set theory." In: *Bulletin of Symbolic Logic* 14.03, pp. 281–298.

— (2008b). *Category theory*. English. Oxford: Oxford University Press.

Awodey, Steve, Nicola Gambino, and Kristina Sojakova (2012). "Inductive types in homotopy type theory." In: *Proceedings of the 2012 27th Annual IEEE ACM Symposium on Logic in Computer Science*. IEEE Computer Society, pp. 95–104.

Awodey, Steve, Richard Garner, et al. (2011). "Mini-workshop: the homotopy interpretation of constructive type theory." In: *Oberwolfach Reports* 8.1, pp. 609–638.

Awodey, Steve, Álvaro Pelayo, and Michael A. Warren (2013). "Voevodsky's univalence axiom in homotopy type theory." In: *Notices of the American Mathematical Society (forthcoming)*.

Awodey, Steve and Michael A. Warren (2009). "Homotopy theoretic models of identity types." In: *Mathematical Proceedings of the Cambridge Philosophical Society*. Vol. 146. Cambridge University Press, pp. 45–55.

Awodey, Steven and A. W. Carus (2007). "Carnap's dream: Gödel, Wittgenstein, and Logical, Syntax." In: *Synthese* 159.1, pp. 23–45.

Azzouni, Jody (1997). "Applied Mathematics, Existential Commitment and the Quine-Putnam Indispensability Thesis." In: *Philosophia Mathematica* 5.3, pp. 193–209.

— (2007). "Ontological Commitment in the Vernacular." In: *Noûs* 41.2, pp. 204–226.

— (2009). "Ontology and the Word "Exist": Uneasy Relations." In: *Philosophia Mathematica* 18.1, pp. 74–101.

Baez, John C. and James Dolan (1995). "Higher-dimensional Algebra and Topological Quantum Field Theory." In: *Journal of Mathematical Physics* 36.11, pp. 6073–6105.

— (1998). "Categorification." In: *arXiv: 9802029 [math]*.

Baez, John C. and Michael Shulman (2006). "Lectures on n-Categories and Cohomology." In: *arXiv: 0608420 [math]*.

Baez, John C. and Derek Wise (2004). "Quantum Gravity Seminar." Unpublished Paper.

Baggott, Jim (Oct. 2003). *Beyond Measure: Modern Physics, Philosophy and the Meaning of Quantum Theory*. Oxford University Press.

Baker, Alan (2005). "Are There Genuine Mathematical Explanations of Physical Phenomena?" In: *Mind* 114.454, pp. 223–238.

— (2009). "Mathematical Explanation in Science." In: *British Journal for the Philosophy of Science* 60.3, pp. 611–633.

Baker, Alan and Mark Colyvan (2011). "Indexing and Mathematical Explanation." In: *Philosophia Mathematica* 19.3, pp. 323–334.

Bar-Hillel, Yehoshua (1966). "Language and Information; Selected Essays on Their Theory and Application." In: *Foundations of Language* 2.2, pp. 192–199.

Barwise, Jon and Lawrence Moss (1996). *Vicious circles: on the mathematics of non-wellfounded phenomena*. English. Stanford: CSLI Publications.

Barwise, Jon and John Perry (June 1999). *Situations and Attitudes*. New edition edition. The Center for the Study of Language and Information Publications.

Bibliography

Bateson, Gregory (1972). *Steps to an ecology of mind: Collected essays in anthropology, psychiatry, evolution, and epistemology.* University of Chicago Press.
— (1979). *Mind and Nature: A Necessary Unity.* Dutton.
Beales, A. C. F. and Patrick Gardiner (1953). "The Nature of Historical Explanation." In: *British Journal of Educational Studies* 1.2, p. 189.
Beeson, Michael J. (1985). *Foundations of Constructive Mathematics: Metamathematical Studies.* Springer-Verlag.
Bell, J. L. (2008). *Toposes and Local Set Theories: An Introduction.* Dover Publications Inc.
Bell, John L (2000). "Hermann Weyl on intuition and the continuum." In: *Philosophia Mathematica* 8.3, pp. 259–273.
Bénabou, Jean (1967). "Introduction to bicategories." In: *Reports of the Midwest Category Seminar.* Springer, pp. 1–77.
— (1985). "Fibered categories and the foundations of naive category theory." In: *The Journal of Symbolic Logic* 50.01, pp. 10–37.
Benacerraf, Paul (1965). "What Numbers Could Not Be." In: *Philosophical Review* 74.1, pp. 47–73.
— (1973). "Mathematical Truth." In: *Journal of Philosophy* 70.19, pp. 661–679.
Benacerraf, Paul and Hilary Putnam (1964). *Philosophy of Mathematics Selected Readings. Edited and with an Introd. By Paul Benacerraf and Hilary Putnam.* Prentice-Hall.
Benthem, Johan van (1991). *The Logic of Time: A Model-Theoretic Investigation into the Varieties of Temporal Ontology and Temporal Discourse.* Springer.
Berg, Benno van den and Ieke Moerdijk (Oct. 2007). "A Unified Approach to Algebraic Set Theory." In: *arXiv: 0710.3066 [math]*.
Bishop, Errett and Michael Beeson (2012). *Foundations of Constructive Analysis.* Ishi Press.
Black, Max (1952). "The identity of indiscernibles." In: *Mind* 61.242, pp. 153–164.
Bohm, David (1992). *Thought as a System.* Routledge.
— (2002). *Wholeness and the implicate order.* Vol. 10. Psychology Press.
Borceux, Francis and George Janelidze (2001). *Galois theories.* Vol. 72. Cambridge University Press.
Bourbaki, N. (2006). *Eléments d'histoire des mathématiques.* Springer.

Bradie, Michael (1994). "Epistemology From an Evolutionary Point of View." In: *Conceptual Issues in Evolutionary Biology*. Ed. by E. Sober. MIT Press. Bradford Books, pp. 453–476.

Braithwaite, R. B. (1953). *Scientific Explanation a Study of the Function of Theory, Probability and Law in Science*. Cambridge University Press.

Brentano, Franz (1988). "Philosophical Investigations on Space." In: *Time and the Continuum, London-New York-Sydney:Croom Helm*.

Bridgman, Percy Williams (1927). *The logic of modern physics*. Vol. 3. New York: Macmillan.

Brouwer, Luitzen Egbertus Jan (1975). *Collected Works: Philosophy and foundations of mathematics*. Ed. by A. Heyting. Vol. 1. North Holland Publishing Co., Amsterdam.

Brown, Ronald and Timothy Porter (June 2003). "Category Theory and Higher Dimensional Algebra: potential descriptive tools in neuroscience." In: *arXiv: 0306223 [math]*.

Bruner, Jerome S (1966). *Toward a theory of instruction*. Cambridge (MA): Belknap Press of Harvard University Press.

Buonarroti, Berlinghiero and Sergio Cencetti (1989). *Atlante della ottica illusa: breviario topografico degli inganni dell'occhio e della mente*. Bologna: Cappelli.

Burgess, John P. and Gideon Rosen (Dec. 1999). *A Subject with No Object: Strategies for Nominalistic Interpretation of Mathematics*. New edition. Oxford University Press.

Campbell, Donald T. (1956). "Perception as Substitute Trial and Error." In: *Psychological Review* 63.5, pp. 330–342.

Cantor, Georg (1955). *Contributions to the Founding of the Theory of Transfinite Numbers*. Dover Publications Inc.

Carnap, Rudolf (1928). *The Logical Structure of the World. Pseudoproblems in Philosophy, trans. by RA George*.

— (1947). *Meaning and necessity: a study in semantics and modal logic*. University of Chicago Press.

Carnap, Rudolf, Yehoshua Bar-Hillel, et al. (1952). *An outline of a theory of semantic information*. Research Laboratory of Electronics, MIT.

Carruthers, Peter (1996). *Language, thought and consciousness An essay in philosophical psychology*. Cambridge: Cambridge University Press.

Cellucci, Carlo (2008). "The Nature of Mathematical Explanation." In: *Studies in History and Philosophy of Science* 39.2, pp. 202–210.

Chaitin, Gregory (2003). "Two philosophical applications of algorithmic information theory." In: *Discrete mathematics and theoretical computer science*, pp. 1–10.

Chaitin, Gregory J. (1969). "On the length of programs for computing finite binary sequences: statistical considerations." In: *Journal of the ACM (JACM)* 16.1, pp. 145–159.

Cheng, Eugenia (Sept. 2008). "Comparing operadic theories of n-category." In: *arXiv: 0809.2070 [math]*.

Chierchia, Gennaro (1992). "Anaphora and Dynamic Interpretation." In: *Linguistics and Philosophy* 18.

— (1995). *Dynamics of Meaning: Anaphora, Presupposition, and the Theory of Grammar*. University of Chicago Press.

Chierchia, Gennaro and Sally McConnell-Ginet (1990). *Meaning and Grammar: An Introduction to Semantics*. MIT Press.

Chomsky, Noam (1986). *Knowledge of Language: Its Nature, Origins, and Use*. Prager.

Colman, Andrew M. (1995). *A dictionary of psychology*. English. Oxford: Oxford University Press. (Visited on 03/16/2018).

Connes, Alain, Andre Lichnerowicz, and Marcel Paul Schutzenberger (2001). *Triangle of Thoughts*. American Mathematical Society.

Connolly, Kevin and Margaret Martlew (1999). *Psychologically speaking: a book of quotations*. English. Leicester: British Psychological Society.

Courant, Richard, Herbert Robbins, and Ian Stewart (1996). *What is Mathematics?: an elementary approach to ideas and methods*. OUP Us.

Dennett, Daniel C. (1991). *Consciousness Explained*. Penguin.

Deutsch, David (1985). "Quantum theory, the Church-Turing principle and the universal quantum computer." In: *Proceedings of the Royal Society of London A: Mathematical, Physical and Engineering Sciences*. Vol. 400. The Royal Society, pp. 97–117.

Deutsch, David and Chiara Marletto (2014). "Constructor Theory of Information." In: *Proceedings of the Royal Society A: Mathematical, Physical and Engineering Sciences* 471.2174. arXiv: 1405.5563 [quant-ph].

Devlin, Keith (1995). *Logic and Information*. Cambridge University Press.

Dretske, Fred I. (May 1980). *Knowledge and the Flow of Information*. Bradford Books.

— (Feb. 2000). *Perception, Knowledge and Belief: Selected Essays*. 1st edition. Cambridge University Press.

Dummett, Michael A. E (1993). *The seas of language*. English. Oxford: Clarendon Press.

Eco, Umberto (1994). "The Limits of Interpretation." In: *Noûs* 28.1, pp. 119–122.

Einstein, Albert (1921). "Geometry and experience." In: *On a Heuristic Point of View about the Creation and Conversion of Light 1 On the Electrodynamics of Moving Bodies 10 The Development of Our Views on the Composition and Essence of Radiation 11 The Field Equations of Gravitation 19 The Foundation of the Generalised Theory of Relativity 22*, p. 82.

— (1934). *The world as I see it (trans. Alan Harris)*. New York: Covici, Friede.

Feferman, Solomon (1966). "Arithmetization of Metamathematics in a General Setting." In: *Journal of Symbolic Logic* 31.2, pp. 269–270.

— (1989). "Infinity in Mathematics: Is Cantor Necessary?" In: *Philosophical Topics* 17.2, pp. 23–45.

— (1999). "Logic, Logics, and Logicism." In: *Notre Dame Journal of Formal Logic* 40.1, pp. 31–54.

— (2000). "Mathematical Intuition Vs. Mathematical Monsters." In: *Synthese* 125.3, pp. 317–332.

— (2009). "Conceptions of the continuum." In: *Intellectica* 51.1, pp. 169–189.

— (2012). "And so On... : Reasoning with Infinite Diagrams." In: *Synthese* 186.1, pp. 371–386.

Feferman, Solomon et al. (1984). "A Language and Axioms for Explicit Mathematics." In: *Journal of Symbolic Logic* 49.1, pp. 308–311.

Feynman, Richard P. (1965). *The Character of Physical Law*. Cambridge, Mass.: MIT Press.

Flori, Cecilia (2013a). *A first course in topos quantum theory*. Vol. 868. Springer.

— (2013b). "Review of the topos approach to quantum theory 1." In: *Canadian Journal of Physics* 91.6, pp. 471–473.

Floridi, Luciano (Jan. 2013). *The Philosophy of Information*. Reprint edition. OUP Oxford.

Fodor, Jerry A. (2008). *Lot 2: The Language of Thought Revisited*. Oxford University Press.

Folina, Janet (1992). *Poincaré and the Philosophy of Mathematics*. Martins Press.

Frege, Gottlob (1980). *The foundations of arithmetic: A logico mathematical enquiry into the concept of number.* Northwestern University Press.

Freyd, Peter (Dec. 1999). *Real coalgebra.* URL: https://www.mta.ca/~cat-dist/catlist/1999/realcoalg.

Fuller, R. Buckminster, Arthur L. Loeb, and E. J. Applewhite (Apr. 1982). *Synergetics: Explorations in the Geometry of Thinking.* New edition. New York: Macmillan.

Galilei, Galileo (Aug. 1988). *The Discoveries and Opinions of Galileo.* Ed. by Stillman Drake. Anchor Books.

Girard, Jean-Yves (1995). "Linear logic: its syntax and semantics." In: *London Mathematical Society Lecture Note Series*, pp. 1–42.

Goldblatt, Robert (May 2006). *Topoi: The Categorial Analysis of Logic.* Revised edition. Dover Publications Inc.

Goldman, Alvin I. (1986). *Epistemology and Cognition.* Harvard University Press.

Goodman, Nelson (Sept. 1951). *The Structure of Appearance.* 1st edition. Harvard University Press.

Gray, John W (1966). "Fibred and cofibred categories." In: *Proceedings of the Conference on Categorical Algebra.* Springer, pp. 21–83.

Greeno, James G. (1970). "Evaluation of Statistical Hypotheses Using Information Transmitted." In: *Philosophy of Science* 37.2, pp. 279–294.

Gregory, Richard L. (Feb. 1991). "Putting Illusions in their Place." en. In: *Perception* 20.1, pp. 1–4.

— (Aug. 1997). "Knowledge in perception and illusion." en. In: *Philosophical Transactions of the Royal Society of London B: Biological Sciences* 352.1358, pp. 1121–1127.

Grothendieck, Alexander (1959). "Technique de descente et théorèmes d'existence en géométrie algébrique. I. Généralités. Descente par morphismes fidèlement plats." In: *Séminaire Bourbaki* 5, pp. 299–327.

— (1983). *Pursuing stacks.* unpublished manuscript.

— (1985). "Récoltes et semailles: Réflexions et témoignage sur un passé de mathématicien."

Hadamard, Jacques (June 1945). *The Psychology of Invention in the Mathematical Field.* 1st edition. Dover Publications.

Haken, Hermann (1984). *The Science of Structure: Synergetics.* Van Nostrand Reinhold.

Hardy, G. H. (Jan. 1940). *A Mathematician's Apology*. 1st edition. Cambridge University Press.

Harnad, Stevan (1990). "The Symbol Grounding Problem." In: *Philosophical Explorations* 42, pp. 335–346.

Hartimo, Mirja (2010). *Phenomenology and Mathematics*. Springer.

Hartimo, Mirja and Mitsuhiro Okada (2016). "Syntactic Reduction in Husserl - Early Phenomenology of Arithmetic." In: *Synthese* 193.3, pp. 937–969.

Hawley, Katherine (2001). *How things persist*. English. Oxford: Clarendon Press.

— (2009). "Identity and Indiscernibility." English. In: *mind Mind* 118.469, pp. 101–119.

Heisenberg, Werner (Mar. 1971). *Physics and Philosophy: The Revolution in Modern Science*. New edition. Allen & Unwin.

Hempel, Carl G. (1942). "The Function of General Laws in History." In: *Journal of Philosophy* 39.2, pp. 35–48.

— (1966). "Aspects of Scientific Explanation and Other Essays in the Philosophy of Science." In: *Synthese* 16.1, pp. 110–122.

Hempel, Carl G. and Paul Oppenheim (1948). "Studies in the Logic of Explanation." In: *Philosophy of Science* 15.2, pp. 135–175.

Hintikka, Jaakko (1962). *Knowledge and Belief*. Ithaca: Cornell University Press.

— (1969). "The modes of modality." In: *Models for Modalities*. Springer, pp. 71–86.

— (1996). *Ludwig Wittgenstein : half truths and one-and-a-half-truths*. Kluwer Academic Press.

— (1997). "No scope for scope?" In: *Linguistics and Philosophy* 20.5, pp. 515–544.

Hodges, Wilfrid (1997). *A shorter model theory*. Cambridge University Press.

— (2008). "Tarski's Theory of Definition." In: *New Essays on Tarski and Philosophy*. Ed. by Douglas Patterson. Oxford University Press, p. 94.

— (2009). "Set Theory, Model Theory, and Computability Theory." In: *The Development of Modern Logic*. Ed. by Leila Haaparanta. Oxford University Press, p. 471.

Hofstadter, Douglas R. (1980). *Godel, Escher, Bach: An Eternal Golden Braid*. Vintage.

Hopcroft, John E. and Jeffrey D. Ullman (Apr. 1979). *Introduction to Automata Theory, Languages and Computation*. 1st edition. Addison-Wesley Publishing Company.

Hughes, Jesse, Steve Awodey, and Dana Scott (May 2001). "A Study of Categories of Algebras and Coalgebras." PhD thesis. Carnegie Mellon University - Department of Philosophy.

Huxley, Aldous (1929). *Do what you will: Essays*. Garden City: Doubleday, Doran.

— (1931). *Music at night, and other essays*. London: Chatto & Windus.

Hyland, Martin and John Power (2007). "The category theoretic understanding of universal algebra: Lawvere theories and monads." In: *Electronic Notes in Theoretical Computer Science* 172, pp. 437–458.

Isham, C. J. (1997). "Topos Theory and Consistent Histories: The Internal Logic of the Set of all Consistent Sets." In: *International Journal of Theoretical Physics* 36.4. arXiv: 9607069 [gr-qc], pp. 785–814.

Israel, David J. and John Perry (1990). "What is Information?" In: *Information, Language and Cognition*. Ed. by Philip P. Hanson. University of British Columbia Press.

Jacobs, Bart (2001). *Categorical Logic and Type Theory*. Butterworth-Heinemann.

— (Dec. 2016). *Introduction to Coalgebra: Towards Mathematics of States and Observation*. 1st edition. Cambridge: Cambridge University Press.

Jeffrey, Richard C. (1969). "Statistical Explanation Vs. Statistical Inference." In: *Essays in Honor of Carl G. Hempel*. Ed. by Nicholas Rescher. Reidel, pp. 104–113.

Johnstone, Peter T (2002a). *Sketches of an elephant: a topos theory compendium*. English. Vol. 1. Oxford: Clarendon Press.

— (2002b). *Sketches of an elephant: a topos theory compendium*. English. Vol. 2. Oxford: Clarendon Press.

— (1977). *Topos theory*. Vol. 8. Academic press London.

— (1986). *Stone spaces*. Vol. 3. Cambridge University Press.

Johnstone, Peter T and Gavin C Wraith (1978). "Algebraic theories in toposes." In: *Indexed categories and their applications*. Springer, pp. 141–242.

Joyal, André and Ieke Moerdijk (Sept. 1995). *Algebraic Set Theory*. Cambridge University Press.

Kock, Anders (2006). *Synthetic Differential Geometry*. Cambridge University Press.

Kock, Anders and Gonzalo E Reyes (1977). "Doctrines in categorical logic." In: *Studies in Logic and the Foundations of Mathematics*. Vol. 90. Elsevier, pp. 283–313.

Kolmogorov, Andrei N. (1965). "Three approaches to the quantitative definition of information." In: *Problems of information transmission* 1.1, pp. 1–7.
Kripke, Saul A. (1981). *Naming and Necessity*. Wiley-Blackwell.
— (2013). *Reference and Existence: The John Locke Lectures*. Oxford University Press USA.
Kuhn, Thomas (1993). "The resolution of revolutions." In: *A postmodern reader*, pp. 376–388.
Kuhn, Thomas S. (1962). *The Structure of Scientific Revolutions*. University Of Chicago Press.
Kunen, Kenneth (2009). *Set Theory*. Studies in Logic: Mathematical Logic and Foundations. College Publications.
Kursunoglu, Behram N. and Eugene Paul Wigner, eds. (1990). *Paul Adrien Maurice Dirac: Reminiscences about a Great Physicist*. Cambridge University Press.
Lawvere, F. William (1963). "Functorial semantics of algebraic theories." In: *Proceedings of the National Academy of Sciences of the United States of America* 50.5, p. 869.
— (1964). "An elementary theory of the category of sets." In: *Proceedings of the national academy of sciences* 52.6, pp. 1506–1511.
— (1966). "The category of categories as a foundation for mathematics." In: *Proceedings of the conference on Categorical Algebra*. Springer, pp. 1–20.
— (1969a). "Adjointness in foundations." In: *Dialectica* 23.3, pp. 281–296.
— (1969b). "Diagonal arguments and cartesian closed categories." In: *Category theory, homology theory and their applications*. Vol. 2. Lecture Notes in Mathematics. Springer, pp. 134–145.
— (1969c). "Ordinal sums and equational doctrines." In: *Seminar on Triples and Categorical Homology Theory*. Springer, pp. 141–155.
— (1970). "Equality in hyperdoctrines and comprehension schema as an adjoint functor." In: *Applications of Categorical Algebra* 17, pp. 1–14.
— ed. (1972). *Toposes, Algebraic Geometry and Logic: Dalhousie University, Halifax, January 16-19, 1971*. Springer-Verlag.
— (1979). "Categorical dynamics." In: *Topos theoretic methods in geometry* 30, pp. 1–28.
— (1986). "Taking categories seriously." In: *Revista Colombiana de Matemáticas, XX*, pp. 147–178.

Bibliography

— (1992). "Categories of space and of quantity." In: *The Space of Mathematics*, pp. 14–30.
— (1994). "Cohesive Toposes and Cantor's 'lauter Einsen'." In: *Philosophia Mathematica* 2.1, pp. 5–15.
— (1996a). "Adjoints in and among bicategories." In: *Lecture notes in pure and applied mathematics*, pp. 181–190.
— (1996b). "Grassmann's dialectics and category theory." In: *Hermann Günther Graßmann (1809–1877): Visionary Mathematician, Scientist and Neohumanist Scholar*. Springer, pp. 255–264.
— (2005). "An elementary theory of the category of sets (long version) with commentary." In: *Reprints in Theory and Applications of Categories* 11, pp. 1–35.
— (2007). "Axiomatic cohesion." In: *Theory and Applications of Categories* 19.3, pp. 41–49.
— (2011). "Euler's continuum functorially vindicated." In: *Logic, Mathematics, Philosophy, Vintage Enthusiasms*. Springer, pp. 249–254.
Lawvere, F. William and Robert Rosebrugh (2003). *Sets for Mathematics*. Cambridge University Press.
Lawvere, F. William and S. H Schanuel (2009). *Conceptual mathematics: a first introduction to categories*. English. Cambridge: Cambridge University Press.
Leary, L.G. (1955). *The Unity of Knowledge*. Doubleday.
Leinster, Tom (July 2001a). "A Survey of Definitions of n-Category." In: *arXiv: 0107188 [math]*.
— (Sept. 2001b). "Structures in higher-dimensional category theory." In: *arXiv: 0109021 [math]*.
— (Nov. 2004a). "A general theory of self-similarity I." In: *arXiv: 0411344 [math]*.
— (Nov. 2004b). "A general theory of self-similarity II: recognition." In: *arXiv: 0411345 [math]*.
— (Nov. 2004c). "General self-similarity: an overview." In: *arXiv: 0411343 [math]*.
— (July 2004d). *Higher Operads, Higher Categories*. Cambridge University Press.
Lewis, David (1973). "Counterfactuals and Comparative Possibility." In: *Journal of Philosophical Logic* 2.4, pp. 418–446.
— (1986). *On the Plurality of Worlds*. Wiley-Blackwell.
Lo Vetere, Davide (1996a). "Appunti di filosofia della matematica." Unpublished Paper.

Lo Vetere, Davide (1996b). "Appunti sulla mente matematica." Unpublished Paper.
— (1996c). "Osservabilità di concetti matematici." Unpublished Paper.
— (2008a). "Modes of Thought." Unpublished Paper.
— (2008b). "The Yarn-Ball Experiment." Unpublished Paper.
— (2009a). "On The Structure of Mathematical Thinking." Unpublished Paper.
— (2009b). "The Metaphysical Structure of Intentionality." Unpublished Paper.
— (2011). "Identity and Perturbation of Definitional Structures." Unpublished Paper.
— (2013a). "A mathematical allegory." Unpublished Paper.
— (2013b). "Observational Structures." Unpublished Paper.
— (2013c). "Redefining Computability." Unpublished Paper.
— (2014). "The 4 Lenses: Towards a Phenomenology of Mathematics." Unpublished Paper.
— (2015a). "Consciousness and Spatialization." Unpublished Paper.
— (2015b). "The Structure of Conceptualization." Unpublished Paper.
— (2016a). "A basic type of metaknowledge." Unpublished Paper.
— (2016b). "A principle of epistemic continuity." Unpublished Paper.
— (2016c). "Categorial (E)pistemology." Unpublished Paper.
— (2016d). "How knowledge becomes reality." Unpublished Paper.
— (2017). "Metaphysical Realism." Unpublished Paper.
Locher, J. L and M. C Escher (1978). *Il mondo di Escher*. Milano: Garzanti.
Lucas, J. R. (2000). *Conceptual Roots of Mathematics*. Routledge.
Lurie, Jacob (Oct. 2009a). "Derived Algebraic Geometry VI: E_k Algebras." In: *arXiv:0911.0018 [math]*.
— (July 2009b). *Higher Topos Theory*. Princeton University Press.
MacLane, Saunders (1986). *Mathematics Form and Function*. Springer.
— (1998). *Categories for the working mathematician*. English. 2nd edition. New York: Springer-Verlag.
MacLane, Saunders and Ieke Moerdijk (1994). *Sheaves in Geometry and Logic: A First Introduction to Topos Theory*. Springer.
Maddy, Penelope (Oct. 1992). *Realism in Mathematics*. Clarendon Press.

Margolis, Eric and Stephen Laurence, eds. (July 1999). *Concepts: Core Readings*. Cambridge, Mass: A Bradford Book.

Maturana, H. R. and F. J. Varela (1980). *Autopoiesis and Cognition: The Realization of the Living*. D. Reidel Publishing Company.

McCabe, Gordon (2007). *The Structure and Interpretation of the Standard Model, Volume 2*. Elsevier Science.

McGurk, Harry and John MacDonald (1976). "Hearing lips and seeing voices." In: *Nature* 264.5588, pp. 746–748.

McLarty, Colin (2003). "The Rising Sea: Grothendieck on simplicity and generality."

— (2011). "The large structures of Grothendieck founded on finite order arithmetic." In: *arXiv: 1102.1773 [math]*.

McLuhan, Marshall and Lewis H. Lapham (1964). *Understanding Media: The Extensions of Man*. MIT Press.

Merleau-Ponty, Maurice (1958). *Phenomenology of Perception*. Routledge.

Ming, Li and Paul Vitányi (1997). *An introduction to Kolmogorov complexity and its applications*. Springer Heidelberg.

Minsky, Marvin (Mar. 1988). *The Society of Mind*. Pages Bent edition. Simon & Schuster.

— (Nov. 2006). *The Emotion Machine: Commonsense Thinking, Artificial Intelligence, and the Future of the Human Mind*. Simon & Schuster.

Moore, Gregory and Nathan Seiberg (1989). "Classical and quantum conformal field theory." In: *Communications in Mathematical Physics* 123.2, pp. 177–254.

Nagel, Thomas (1986). *The View From Nowhere*. Oxford University Press.

Peirce, Charles S. (1976). *The New Elements Of Mathematics*. English. Ed. by Carolyn Eisele. Vol. 3. Hague : Atlantic Highlands, N.J: Mouton Publishers / Humanities Press.

Peirce, Charles Sanders (1992). *Reasoning and the logic of things: The Cambridge conferences lectures of 1898*. Harvard University Press.

Penrose, Roger (1989). *The emperor's new mind: concerning computers, minds, and the laws of physics*. English. Oxford: Oxford University Press.

— (Sept. 1995). *Shadows Of The Mind: A Search for the Missing Science of Consciousness*. New edition. Vintage.

Poincaré, Henri (1913). *The Foundations of Science: Science and Hypothesis; The Value of Science; Science and Method*, trans. by GB Halsted. The Science Press.

Popper, Karl R. (1959). *The Logic of Scientific Discovery*. New York: Routledge.

Putnam, Hilary, ed. (Apr. 1975). *Mind, Language and Reality*. Vol. 2. Philosophical papers. Cambridge University Press.

Quine, Willard Van Orman (1960). *Word and object*. English. Cambridge, Mass.: MIT Press.

Rescher, Nicholas (Feb. 1996). *Process Metaphysics: An Introduction to Process Philosophy*. State University of New York Press.

— (Dec. 2000). *Process Philosophy: A Survey of Basic Issues*. 1st edition. University of Pittsburgh Press.

Resnik, Michael D. (2000). *Mathematics As a Science of Patterns*. Oxford University Press.

Rodin, Andrei (2005). "Identity and Categorification." In: *arXiv: 0509596 [math]*.

— (2013). *Axiomatic Method and Category Theory*. Vol. 364. Synthese Library. Springer International Publishing.

Rosch, Eleanor and Barbara L. Lloyd, eds. (Sept. 1978). *Cognition and Categorization*. 1st edition. Hillsdale, N.J., New York: Lawrence Erlbaum.

Russell, Bertrand (1908). "Mathematical logic as based on the theory of types." In: *American journal of mathematics* 30.3, pp. 222–262.

— (1927). *Analysis of Matter*. Routledge & Kegan Paul.

— (1940). *An Inquiry into Meaning and Truth*. English. Pp. 352; London: Allen & Unwin.

Ryle, Gilbert (Jan. 1949). *Concept of Mind*. University paperbacks. Barnes & Noble.

Sagan, Hans (1994). *Space-filling curves*. New York: Springer.

Salmon, Nathan U. (2005). *Metaphysics, Mathematics, and Meaning: Philosophical Papers: v. 1*. Oxford University Press.

— (2007). *Content, Cognition, and Communication: Philosophical Papers II*. Oxford University Press.

Salmon, Wesley C. (Sept. 1971). *Statistical Explanation and Statistical Relevance*. 1 edition. University of Pittsburgh Press.

— (Dec. 1984). *Scientific Explanation and the Causal Structure of the World*. Princeton University Press.

— (Jan. 1998). *Causality and Explanation*. 1st edition. Oxford University Press.

— (June 2006). *Four Decades of Scientific Explanation.* 1st edition. University of Pittsburgh Press.
Searle, John R. (1968). "Austin on Locutionary and Illocutionary Acts." In: *Philosophical Review* 77.4, pp. 405–424.
— (May 1983). *Intentionality: An Essay in the Philosophy of Mind.* Cambridge University Press.
— (Jan. 1990). *The Mystery of Consciousness.* 1st edition. The New York Review of Books.
— (July 1992). *The Rediscovery of the Mind.* Cambridge, Mass.: MIT Press.
— (Jan. 2001). *Rationality in Action.* Cambridge, Mass.: MIT Press.
— (2002). *Consciousness and Language.* Cambridge University Press.
— (2004). *Mind: A Brief Introduction.* Oxford University Press.
— (2007). "2 Illocutionary Acts and the Concept of Truth." In: *Truth and Speech Acts: Studies in the Philosophy of Language.* Ed. by Geo Siegwart and Dirk Griemann. Routledge, pp. 5–31.
— (May 2009). *Foundations of Illocutionary Logic.* 1st edition. Cambridge University Press.
Seely, Robert AG (1983). "Hyperdoctrines, natural deduction and the Beck condition." In: *Mathematical Logic Quarterly* 29.10, pp. 505–542.
Shannon, Claude (1948). "A Mathematical Theory of Communication." In: *Bell System Technical Journal* 27, pp. 379–423.
Shannon, Claude E. and Warren Weaver (1949). *The Mathematical Theory of Communication.* University of Illinois Press.
Sherburne, Donald W. (1981). *A Key to Whitehead's Process and Reality.* University Of Chicago Press.
Shields, Paul (2012). *Charles S. Peirce on the Logic of Number.* Docent Press.
Shulman, Michael A. (2008). "Set theory for category theory." In: *arXiv: 0810.1279 [math].*
Stapp, Henry P. (Feb. 2009). *Mind, Matter and Quantum Mechanics.* 3rd edition. Springer Berlin Heidelberg.
— (June 2011). *Mindful Universe: Quantum Mechanics and the Participating Observer.* 2nd edition. Springer.
Steiner, Mark (1978a). "Mathematical Explanation." In: *Philosophical Studies* 34.2, pp. 135–151.
— (1978b). "Mathematics, Explanation, and Scientific Knowledge." In: *Noûs* 12.1, pp. 17–28.

Strawson, Galen (Oct. 1994). *Mental Reality*. Cambridge, Mass.: MIT Press.

Streicher, Thomas (1999). "Fibred categories à la Bénabou." In: *Munich Summer School*.

Tarski, Alfred (1941). *Introduction to Logic and to the Methodology of the Deductive Sciences*. Oxford University Press, USA.

Theodore Sider (2001). *Four-Dimensionalism: An Ontology of Persistence and Time*. English. Oxford University Press.

Thom, René (1971). "'Modern' Mathematics: An Educational and Philosophic Error? A distinguished French mathematician takes issue with recent curricular innovations in his field." In: *American Scientist* 59.6, pp. 695–699.

— (1972). *Stabilité Structurelle et Morphogénèse: Essai D'une Théorie Générale des Modèles*. French. Reading, Mass: Benjamin-Cummings.

Troelstra, Anne S., D. Van Dalen, and Ranis (Dec. 1999). *Constructivism in Mathematics Vol.1: An Introduction*. Butterworth-Heinemann.

Univalent Foundations Program, The (2013). *Homotopy Type Theory: Univalent Foundations of Mathematics*. Institute for Advanced Study: https://homotopytypetheory.org/book.

Varela, Francisco J., Evan Thompson, and Eleanor Rosch (1993). *The Embodied Mind: Cognitive Science and Human Experience*. MIT Press.

Watzlawick, Paul, Janet Beavin Bavelas, and Don D Jackson (1967). *Pragmatics of human communication, a study of interactional patterns, pathologies and paradoxes*. London: Faber and Faber.

Weil, Andre (1950). "The Future of Mathematics." In: *The American Mathematical Monthly* 57.5, pp. 295–306.

Weil, S. and S. Miles (1986). *Simone Weil, an Anthology*. Weidenfeld & Nicolson.

Whitehead, Alfred North (1898). *A Treatise on Universal Algebra: With Applications*. Cambridge University Press.

— (1967). *Adventures of Ideas*. The Free Press.

— (1968). *Modes of Thought*. The Free Press.

— (1979). *Process and Reality*. The Free Press.

Whitehead, Alfred North and Bertrand Russell (1925). *Principia mathematica*. 2nd edition. Vol. 1. Cambridge University Press.

Whorf, Benjamin Lee (1956). *Language, thought, and reality: selected writings*. English. Cambridge: MIT Press.

Wiener, Norbert (1949). "Cybernetics. Or Control and Communication in the Animal and the Machine." In: *Journal of Philosophy* 46.22, pp. 736–737.
— (1952). "The Human Use of Human Beings. Cybernetics and Society." In: *Philosophy* 27.102, pp. 249–251.
— (1961). *Cybernetics*. New York: MIT Press.
Wigner, Eugene (1960). "The Unreasonable Effectiveness of Mathematics in the Natural Sciences." In: *Communications in Pure and Applied Mathematics, New York: John Wiley & Sons, Inc.* 13.1.
Wittgenstein, Ludwig (1958). *Philosophical investigations. R. Rhees and G.E.M. Anscombe (eds.), trans. by G.E.M. Anscombe.* English. 2nd edition. New York: Macmillan.
Woodward, James (1989). "The Causal Mechanical Model of Explanation." In: *Minnesota Studies in the Philosophy of Science* 13, pp. 359–83.
— (2003). *Making Things Happen: A Theory of Causal Explanation*. Oxford University Press.
Worrall, John (1989). "Structural realism: The best of both worlds?" In: *Dialectica* 43.1-2, pp. 99–124.
Yanofsky, Noson S. (1999). "The Syntax of Coherence." In: *arXiv: 9910006 [math]*.
— (2003). "A universal approach to self-referential paradoxes, incompleteness and fixed points." In: *Bulletin of Symbolic Logic* 9.3, pp. 362–386.
Zukav, Gary (1979). *Dancing Wu Li Masters: An Overview of the New Physics*. Bantam Books.

Index

∞-diagram, 19, 242, 254, 299, 303, 311, 319, 320, 323, 324, 552
1-sorted representation, 307

Abstract Mind
–analytic interpretation of, 526
–sound analogy, 254
–synthetic interpretation of, 527
Actuation Mechanism, 316, 331, 355, 371, 415, 465
Analytic Structure of Intentionality, 157
Actuation Mechanism, 229
abstract dictionary, 138
acquiring knowledge, 272
agent, 60
Algebraic Set Theory, 78
algebras, 468
alphabet, 17
anthropic worldview, 53
–beginning-and-end, 54
–space, 54
–time, 54
Aristotle, 22
autopoietic system, 556
awareness, 315
axiom, 18

Banach-Tarski paradox, 115, 152
bounded identity, 134
Brahman, 8
buzzer circuit, 57

Cognitive Architecture, 239, 265, 311, 355, 386, 431, 483, 540
concrete concept, 226
Cantor's theorem, 168
categorial logic, 78
categorification, 85, 94
–horizontal, 94
–vertical, 94
category, 92
Category Theory, 78
causal relation, 62
causality, 264
centipede effect, 221
chaos, 8
classifier, 289
cognitive state, 250
cognitive superstructure, 272
comprehension
–characterization of, 278
–superstructure of, 279
computational philosophy, 534
concept
–1-sorted representation of, 307
–Actuation Mechanism of, 229
–Definitional Model of, 230
–Organizing Principle of, 228
–**abstract**, 222
–**concrete**, 120, 221
–definitional interference of, 129
–image, 318
–lens, 317
–metaphysical structure of, 227
–no apriori necessity of, 524

Index

–observ. mode of existence of, 123
–observational identity of, 126
–stipulative mode of existence of, 122
–substrate, 220
conception of knowledge, 241
conception of Mathematics, 354
coordinate system
Cartesian, 147
correspondence theory of truth, 262, 304
cosmos, 9

Definitional Model, 266, 316, 350, 372, 417, 469
Definitional Model, 230
decategorification, 85
definition of knowledge, 240
definitional interference, xxiv, 128, 129, 132, 182, 227, 367, 563
definitional model, 250
Dependent Type Theory, 79
despatialization, 95, 252, 333, 378, 423, 475
Dharma, 8
duality
–Isbell, 96
–as spatialization, 98

Elementary Theory of the Category of Sets, 78
endoexplanations, 315
endomorphism, 385
epistemic continuity, 301, 346, 405, 451, 503, 543
epistemic relativity, 298
ETCS, 78
Eugene Wigner, 23
exoexplanation, 315, 522

Fundamental Correspondence, 185

false causation, 51
finitism, 454
form computer, 504
formal system, 16, 215
–alphabet, 17
–axioms, 18
–completeness, 167
–consistency, 167
–inference rules, 18
–syntax, 17
foundation of mathematics, 75
foundational crisis of mathematics, 82, 535
Fregean senses, 128
fully explainable reality paradox, 313
functor factorization system, 279

Gödel's incompleteness theorems, 168
Galileo Galilei, 22
Galois Theory, 393
generative knowledge, 231, 373, 418, 469
geometry without points, 90

horizontal categorification, 94
HoTT, 79

Interaction Model, 165
–cognitive, 250
–pre-cognitive, 249
–pre-conceptual, 249
Identity-based mode of thought, 204
identification, 140
identity, 170, 225
–through change, 18
–to observe OSP, 204
identity of indiscernibles, 367
identity principle for buses, 130
identity through change, 200

Index

ignorance, 312
illocutionary act, 223
image, 318
incidence geometry, 90
indiscernibility of identicals, 367
inference rules, 18
information, 283
intelligibility, 179, 187, 218, 222
intelligible universe hypothesis, 13, 307
Intensional Type Theory, 79
internal logic, 78, 96, 392
Isbell duality, 96

Knowledge-Reality Interface, 318
Kan extension, 279, 423, 539
knowledge
–acquisition of, 272
–conception of, 241
–definition of, 240
–generative, 569
–procedural, 263
–propositional, 263
–without comprehension, 275
koan, 368

Local Identity, 171, 226, 382, 427, 479
law of Nature, 302
Leibniz's Laws, 367
lens, 317
level of consciousness, 184
Lo Vetere's
–propeller diagram, 94
local realism, 50
locality, 50
logic of knowledge, 313
Loop Quantum Gravity, 311
Lotka-Volterra, 473, 495

Minimal Interaction Model, 235

–as object of thought, 256, 385, 429, 482
mathematical gadget, 194
–classification, 192
–observational structure, 196
Mathematical Platonism, 565
Mathematical Realism, 565
mathematics, 14
–as a cognitive technology, 15
–foundational crisis of mathematics of, 535
–phone analogy, 21
–use of, 19
meaning, 19
–dual nature of, 282
–phenomenology of, 281
–superstructure of, 288
–without comprehension, 276
measure, 290
mentalist theories of meaning, 282
metaillusion, 115, 117
metamathematics, 536
model of truth, 62
monoidal, 540
morphism, 92

natural number object, 83
non-commutative geometry, 90
non-mentalist theories of meaning, 282
noos, 241
Nousaurus, xxix

Object Classifier, 157
Organizing Principle, 228, 370, 415, 464
Observational Structure of Intentionality, 60
OSP Pattern, 242, 321, 346, 354, 542
Organizing Principle, 228

Index

objects of perception, 139
—formal systems interpretation of, 201
—identity through change, 199
—mediated presentation, 143
—resolution, 197
observational pattern, 60
oidification, 94

Principle of Epistemic Relativity, 298, 521, 563, 568
panpsychism, 369
partial order, 291
participatory, 333
participatory mode of thought, 254
participatory worldview, 56
—part, 57
—relation, 58
—similarity, 58
Paul Dirac, 23
Per Martin-Löf, 79
periodic table of higher categories, 539
perlocutionary act, 222
Perturbation Theory, 116
Philosophy of Linguistics, 216
Philosophy of Process, 257
Physicalism, 79, 369, 565, 571
Plato, 22
possibilist realism, 556
pre-cognitive Interaction Model, 249
pre-conceptual Interaction Model, 249
principium individuationis, 562
principle
—of epistemic relativity, 298
procedural knowledge, 263
process computer, 405
propositional knowledge, 263
protosemantic transformation, 286

proximity computer, 452
pursuit of knowledge, 14
Pythagoras, 22

quiddity, 8

reconstruction of the success condition, 180, 231, 245, 251, 266, 272, 320, 344, 380, 425, 477
reality, 392, 438, 489
respatialization, 95
Richard Feynman, 23
Richard's paradox, 168
Russell's paradox, 82

Synthetic Structure of Intentionality, 160
substrate concept, 220, 224
self-referential paradox, 167
semantic mirage, 344, 400, 446, 497
semantogenesis, 286, 330, 334, 341, 388, 400, 433, 446, 485, 497, 520, 547
—definition of, 287
—examples of, 289
Ship of Theseus, 367
solvitur ambulando, 221
space
—consistency, 151
—one-dimensional concept of, 150
spatial interpretation of identity, 54
spatialization, xx, 94, 95, 229, 366, 526
speed of light, 148
strong math-reality conection, 24
strong math-reality connection, 307, 322
subject-object dichotomy, 57

subobject, 422
superadditivity, 272, 308
superstructure of meaning, 288
supertasks, 455
syntax, 17
syntax-semantics interface, 216

The 4 Lenses, xxiii, 66, 71, 539
The Sieve, xxi, 284
Tarski's undefinability theorem, 168
The Assayer, 22
The superadditive knowledge paradox, 274
theory, 33
theory of everything, 397
thought
–metaphysical structure of, 212
topos, 78
Topos Quantum Theory, 293
Turing's halting problem, 168

unity of consciousness, 254
univalent foundation, 79
Universal Turing Machine, 555
universality of computation, 24

Validation System, 62, 161
vertical categorification, 94

weak math-reality connection, 24, 308
Whitehead's actual entity, 254
worldview, 13
–process-centric, 50

yarn-number, 146

ZFC, 78

Culture Hacking

Every civilization is defined by its interpretation of the human experience. It is shaped by the fundamental worldviews that guide the intellectual, cultural, spiritual, scientific and technological development of our species at that moment in the history of humanity, in what is essentially an evolutionary process that involves ideas rather than organisms. And like any evolutionary process, it has turning points: pockets of creative intuition that outgrow the worldviews that nurtured human progress up to that point and inspired our vision of human purpose. There are times when some of the worldviews that steered human progress no longer serve the purpose for which they were originally adopted, and become a hindrance to the evolutionary process they ignited. The last big change of this kind in Western culture culminated with the Age of Enlightenment at the end of the 17th century, which marked a radical shift started about 2500 years ago in the way we see ourselves and the world, and that ratified the supremacy of reason over everything else. But that shift came at a price: it separated us from the unity of knowledge and reality. The modern conception of rationality, is based on a view of the world according to which everything that exists is a process. A process is anything that can be described as specific changes occurring in specific objects that result in a specific outcome. This is how modern Physics sees the Universe, and this is how we explain reality. With this worldview in mind, we say that a claim is rationally sound, if it describes how a certain object becomes what it is through change. This is the basic structure of the modern rational knowledge of the world. And this is the reason why we think that knowledge must be predictive: because, since everything that exists is a process, knowing something must necessarily mean knowing how it becomes what it is. The formulation of a unified theory of knowledge I have given in this book, allows me to generalize and reprogram the conceptual architecture of modern thought so as to integrate seamlessly uncharted types of intellectuality that are usually regarded as intuitive, mystical, revelatory or otherwise non-rational. There is a sense in which what I do bears some resemblance to computer hacking, because I reprogram a cultural system to expand it and make it work differently, and to allow it to incorporate, internalize and manifest new ideas and new goals. I am, in this sense, a "culture hacker".

ALSO BY DAVIDE LO VETERE

The Journey of the Self: Seven laws to reclaim your nature and create spiritual wealth.
Seven timeless principles that distill the wisdom to create a better version of you, and to live consciously.
TRUTH BEAUTY WONDER PRESS (2019)

Archetypes of the Modern World.
The psychological patterns that define the era we live in, and that inspire our future and the evolution of humanity.
TRUTH BEAUTY WONDER PRESS (2019)

The Mystery of Mathematics.
A treasure trove of cutting-edge knowledge to understand the subtle structures underlying rational thought, intuition and invention.
TRUTH BEAUTY WONDER PRESS (2019)

TRUTH BEAUTY WONDER PRESS

Truth Beauty Wonder Press is an independent press dedicated to publishing writings that push the boundaries of modern thought and catalyze the evolution of culture and society.

https://truthbeautywonder.com

www.ingramcontent.com/pod-product-compliance
Lightning Source LLC
Chambersburg PA
CBHW021049080526
44587CB00010B/188